TECHNOLOGIES FOR SMALL WATER AND WASTEWATER SYSTEMS

TECHNOLOGIES FOR SMALL WATER AND WASTEWATER SYSTEMS

Edward J. Martin
Edward T. Martin

ENVIRONMENTAL ENGINEERING SERIES

Library of Congress Catalog Card Number 90-21743
ISBN 0-442-23829-0

Manufactured in the United States of America

Published by Van Nostrand Reinhold
115 Fifth Avenue
New York, New York 10003

Chapman and Hall
2-6 Boundary Row
London, SE1 8HN

Thomas Nelson Australia
102 Dodds Street
South Melbourne 3205
Victoria, Australia

Nelson Canada
1120 Birchmount Road
Scarborough, Ontario M1K 5G4, Canada

16 15 14 13 12 11 10 9 8 7 6 5 4 3 2

Library of Congress Cataloging-in-Publication Data

Martin, E. J. (Edward J.)
Technologies for small water and wastewater systems / Edward J. Martin, Edward T. Martin.
p. cm.
Includes bibliographical references and index.
ISBN 0-442-23829-0
1. Water-supply engineering. 2. Sewage disposal. 3. Appropriate technology. I. Martin, Edward T. II. Title.
TD353.M37 1991
628—dc20 90-21743
CIP

Dedication

To our friend Ricardo Nitroso who is now where all theories have answers.

VNR ENVIRONMENTAL ENGINEERING SERIES
Nelson L. Nemerow, Consulting Editor

Water Pollution Control

CONTROL AND TREATMENT OF COMBINED-SEWER OVERFLOWS, edited by Peter Moffa
TECHNOLOGIES FOR SMALL WATER AND WASTEWATER SYSTEMS by Edward J. Martin and Edward T. Martin
INDUSTRIAL AND HAZARDOUS WASTE TREATMENT by Nelson Nemerow and Avijit Dasgupta

Water Resources Development

CURRENT TRENDS IN WATER-SUPPLY PLANNING by David W. Prasifka
HANDBOOK OF PUBLIC WATER SYSTEMS by Culp/Wesner/Culp
DESIGN AND CONSTRUCTION OF WATER WELLS by the National Water Well Association
CORROSION MANAGEMENT IN WATER SUPPLY SYSTEMS by W. Harry Smith
DISINFECTION ALTERNATIVES FOR SAFE DRINKING WATER by Ted Bryandt and George Fulton

Solid Waste Management

SMALL-SCALE MUNICIPLE SOLID WASTE ENERGY RECOVERY SYSTEMS by Gershman, Brickner and Bratton, Inc.

Hazardous Waste Treatment

GROUNDWATER TREATMENT TECHNOLOGY by Evan Nyer
HAZARDOUS WASTE SITE REMEDIATION by O'Brien & Gere, Inc.
SUBSURFACE MIGRATION OF HAZARDOUS WASTES by Joseph S. Devinny, Lorne G. Everett, James C.S. Lu and Robert L. Stollar

General Environmental

COMPUTER MODELS IN ENVIRONMENTAL PLANNING by Stephen I. Gordon
ENVIRONMENTAL PERMITS by Donna C. Rona
BEYOND GLOBAL WARMING by A. John Appleby
AIR POLLUTION MODELING by Paolo Zannetti
CONTINGENCY PLANNING FOR INDUSTRIAL EMERGENCIES by Piero Armenante

CONTENTS

PREFACE

The book contains 75 technologies, descriptions, capital and operation and maintenance costs (current to 1990), and design criteria. The presentation of cost data and design criteria will be especially useful to the engineer and practitioner who is developing and designing water supply and wastewater systems. The reader may use the book as a reference because the presentation is largely encyclopedic in nature. On the other hand, the process descriptions and presentation of design criteria make the book useful as an advanced text, particularly in a graduate course where the students have some background knowledge about the field and requirements of some applications.

A series of case studies is presented which may be used as the basis for preliminary designs on the part of practitioners including design engineers contemplating the use of the given technologies in applications in the field. The case studies will also be useful as upperclass or graduate student exercises for design analysis. In most cases, performance data are presented for influent and effluent quality characteristics.

The presentation of technologies is suitable both for analysis of water supply or wastewater treatment applications.

The technology sections were developed based on a complete review and analysis of several hundred technology possibilities in the technical literature, from field visits and analyses in the U.S. and abroad, and personal experience with the design and design analysis of these and similar technologies over the last 30 years.

Thousands of small treatment plants for wastewater will be built through the end of the century and into the next. The discharges of these plants have several times the pollution potential for the nation's surface and groundwaters than the relatively few large plants serving major cities. Furthermore, large cities tend to be near the sea rather than discharging to inland rivers, lakes, and streams. Thus the potentially adverse environmental impact from small plants is much greater. It is vital that the appropriate technology, i.e., technology that will work in small town situations be selected rather than "copy catting" larger facilities where maintenance and operation personnel are likely to be more sophisticated and consistently available.

The technologies are completely presented with critical design and analysis information in several categories: description including design criteria and schematic diagrams, costs; operation and maintenance considerations, availability, special factors, and recommendations.

In summary, the text is an encyclopedia of technologies which may be used for water supply and wastewater treatment plants. The selection criteria for each are presented in detail and cost and usability information are given.

Edward J. Martin

ACKNOWLEDGMENTS

There were several persons who provided data and information on projects outside the U.S. I want to especially thank Mr. Juan Alfaro at the Interamerican Development Bank. Many thanks to Ing. Jose M. de Azevedo Netto of São Paulo, Brazil, Ing. Julio Burbano of Bogota, Colombia, Ing. Adilio Luiz Monteiro de Barros of Rio de Janeiro, Ing. Humberto Olivero of Guatemala City, and Dr. Eugene J. Kazmierczak of Washington, D.C.

PART I
INTRODUCTION

APPROPRIATE, INNOVATIVE, ALTERNATIVE

The word "appropriate" may mean many things especially in the context of technology applications, as in this case for development and management of water and wastewater—quantity and quality. The technology is appropriate if it is suitable to the functional application for which it is intended from the viewpoints of cost, operability, and simplicity of design.

Functional success is probably the most important consideration. If all engineered systems work properly then enough water of suitable quality is available at all times for all water uses, and the quality of the ground and surface waters receiving treated wastewater is preserved. Rather than insisting on lowest cost technology, the goal should be to acquire lowest cost technology that works. Unfortunately, there are many more examples of the former than the latter.

Other words have crept into use by practitioners of water technology applications and are thus important. Two are especially important—alternative and innovative. The U.S. Environmental Protection Agency (EPA) funds special technology projects for wastewater management; the effort is called the innovative and alternative technology (I/A) program. Some of the I/A technologies are presented herein.

The word innovative applied to water management suggests newness. The innovative applications presented herein are more likely to be new applications of old technologies, rather than new technologies per se. The word alternative suggests an option to what is currently being done. Technologies tend to fall in and out of fashion. Imhoff tanks are not popular and currently are not often designed for application to wastewater treatment. Many older cities in the world have Imhoff tanks, vintage the 1930s, and some are still functioning. In any case, whatever the terms being applied, there is room for consideration of "other" technologies, especially if the same objectives of water supply and/or quality can be achieved at lower cost.

There is a danger to applying "new" technologies if they have not been proven in practice. There is sometimes an urge on the part of the designer to "do something different." The authors include herein technologies only if there have been applications.

NEED FOR PILOT-PLANT ANALYSIS

Design criteria are presented for most of the technologies. The design criteria represent the experience of designers of systems requiring the same or similar applications to those given. To the extent that other experience is desired to serve as the basis for new designs, the encyclopedic nature of the information given should be helpful.

There is a temptation to apply design criteria developed for specific applications directly to designs elsewhere. Often pilot plant studies are not conducted because of the cost involved. If pilot plant studies are designed properly, much of the physical facility may become a part of the final design. Pilot studies allow local conditions of influent quality and effluent requirements to be taken into account. The time and resources expended for pilot plant analysis are never wasted.

APPLICATIONS AND SUITABILITY OF TECHNOLOGY

Some technologies were included which have strong potential for application, even though the applications up to this time have been limited, i.e., there is a high probability for success. Some technologies, such as steep slope sewers, are virtually necessary applications in the future because of the high potential cost savings and changes in construction materials over the past several years allow such applications.

The book is in two parts. Part I contains technologies which are more certain of applicability because there are more data and information available. Part II contains case studies or technologies which appear to be very promising.

This is not a manual to be used for detailed design. Thus, details such as inlet and outlet structures, etc., are not included. The reader is expected to be familiar with the basic kinetics and theory of the technologies. The information herein should be used for preliminary design, planning, review of submitted designs, and comparison of alternatives. Thus, the book should be especially useful to practitioners and students.

Process technologies are not presented separately for water and wastewater as is conventional practice for books of this kind. The distinction between technologies suitable for potable water production and those suitable for wastewater treatment is disappearing, but unfortunately not fast enough. The technologies presented are suitable for application to any situation requiring improved water quality. Thus, any unit process (a technology performing one function in the series referred to as a treatment plant) will perform equally well on either water or wastewater, provided the influent characteristics to the unit process and the expected effluent quality are within the limits of its performance capability. In other words "a filter is a filter" whether treating water or wastewater. Furthermore, it is expected to perform adequately, as long as the suspended solids loading to the filter is controlled, and the system is properly operated. Today, there are many existing examples of filtration applied to wastewater treatment.

WATER VERSUS WASTEWATER TREATMENT

A distinct separation between water management for water supply purposes and approaches to wastewater management is a luxury reserved for water rich geographic areas. Others must consider water needs and services as a continuum—ranging on the one hand from treating a groundwater source perhaps, to on the other hand treating water used once for waste carriage, especially in preparation for direct or indirect reuse.

In developing countries wastewater treatment is often considered to be "very expensive," or "not a high priority." Wastewater treatment may be considered pretreatment for water reuse. Once-through water usage (treatment followed by disposal) is very costly over the long term. Many areas are experiencing potentially serious water shortages, or at least shortages of water of sufficient quality for required uses. One example is falling groundwater tables; another is increasing salinity in available supplies. Even countries with relatively high rainfall, but characterized by high intensity—short duration storms, are becoming water short as populations grow and water use expands. It is difficult to maintain consistent water supplies in such situations.

ORGANIZATION OF THE SECTIONS

To help the engineer and system designer in the evaluation and selection process, the discussion of each technology is presented in several subsections which are considered as selection criteria to be used in configuring treatment systems. These are the following:

Description; written material which describes the function and operation of the technology. Design criteria are usually presented in this section. Where design criteria are presented elsewhere; e.g., in the operation and maintenance (O&M) section, the criteria have special importance to O&M.

Limitations/advantages and disadvantages; processes have differing application potential to specific locations depending on water quality, effluent requirements, etc. Limitations should be evaluated before application from the perspective of the actual performance expected given the water quality and service requirements.

Costs; costs are often considered to be the most important consideration in the selection process. Function should be considered primarily and then if equal functional performance is expected from two different processes, cost is evaluated, and the lower cost option may be chosen. Often a process is applied because it has a lower first cost. Operating and maintenance costs should also be considered.

Availability; this criterion addresses the availability of materials, equipment and supplies, a special consideration for remote areas. Many applications of technology have been made without consideration of follow-up services, replacement parts, and regular maintenance programs. These judgments must usually be made from the perspective of the local area where the technology will be applied.

Operation and maintenance (O&M); O&M factors are addressed in this subsection. Processes may be easier to operate or more complex, thus more or less operator training may be required for successful function. Preventive maintenance is often very important to continued function. These evaluations should be made before a process is chosen.

Reliability; this is an especially important selection factor in remote areas and developing countries. Processes with more mechanical equipment may be expected to require more attention—the more complex the equipment, the more attention. Nevertheless, selection should be made on the basis of functional performance.

Special factors; considerations which are not included elsewhere are mentioned here. Special chemical requirements or maintenance features.

Recommendations; a summary statement about the selection criteria and the technology is sometimes given. Recommendations are only general in nature. It is important to consider local water quality, flow variability, and other factors.

The range of information available for the technologies varies for each. Thus all of the subsections are not always presented. A subsection on "control" is sometimes presented in those cases where process control is a special consideration in the process selection. Other subsections may be found in special cases.

COST ESTIMATES AND ADJUSTMENTS

Cost data have been adjusted to January 1990, using the Engineering News Record (ENR), Construction Cost Index (CCI). Costs should be adjusted by the reader to current values by using an appropriate index. The adjustment may be made to cost data provided by applying the following relationship:

$$C_Y = C_X (I_Y/I_X)$$

where

C_Y = Cost in Year Y
C_X = Cost in Year X
I_Y = Index Value for the Year Y
I_X = Index Value for the Year X

The index values are for any appropriate index of cost as long as both I_X

and I_Y are from the same index. Since the costs presented herein were adjusted from original data using the ENR CCI, some aberrations may be introduced if other CCIs are used for adjustment.

The costs presented should be considered unit process installed construction cost (capital costs) and unit process operation and maintenance (O&M) costs in the case of O&M costs. Thus, to arrive at an estimate for treatment plant capital cost, other cost considerations must be included. Estimates for these additional costs are best made for the country and area where application is being considered and have therefore not been included.

The approach which may be used to arrive at total plant construction costs is as follows:

Unit Process Construction Cost			
Unit Process #1		$________	
Unit Process #2		$________	
Unit Process #3		$________	
etc.		$________	
Miscellaneous Structures (office, lab, garage, etc.)		$________	
SUBTOTAL 1			$________

Non Unit Process Costs	*Range*	
Piping	8–15% of SUBTOTAL 1	$________
Electrical	5–125% of SUBTOTAL 1	$________
Instrumentation	3–15% of SUBTOTAL 1	$________
Site Preparation	1–10% of SUBTOTAL 1	$________
SUBTOTAL 2		$________

Non-Construction Costs		
Engineering and Construction Supervision	@ 15% of SUBTOTAL 2	$________
Contingencies	@ 15% of SUBTOTAL 2	$________
TOTAL CAPITAL COST		$________

Costs may be adjusted using available computer programs (11) or may easily be calculated for large or repetitious applications using conventional spreadsheet software.

How should the costs be applied to states in the U.S. and other countries? This question is especially important for applications in developing countries where technologies such as those presented herein are suitable. The authors struggled with this question for a long time and on several assignments for the Interamerican Development Bank for Latin America and elsewhere. One approach is to use the Area Cost Factor Indices (ACFs) developed by the U.S. Army Corps of Engineers (COE) for applications in many parts of the world (78). The ACFs were developed to be applied to unit cost values in U.S. dollars, to arrive at a site cost estimate. A review of the ACFs is instructive for making adjustments to cost estimates which were developed for the U.S. and applying them elsewhere.

The range of ACFs for the states in the U.S. is 0.84 to 1.32. Thus, the estimated adjusted values of cost estimates using the ACFs within the continental U.S. would vary from approximately −20 to +30%. The range of ACF values worldwide (for the countries and locations included in the COE analysis) is 0.80 to 1.68, or −20 to +70%. The worldwide range excludes remote locations, such as Iceland and Diego Garcia and the U.S. range excludes Alaska. Remote locations should be considered special in any case. Therefore the range of the cost estimate variations are about the same as the variability expected for engineering preliminary cost estimates in general, −30 to +50%. The difference between the U.S. and the world range is on the high side of the estimate, and this should be taken into account in transposing values.

In another analysis the authors have examined in detail the costs from bidding documents in Latin America. The cost estimates from the foreign studies have been compared to the U.S. cost curves for many of the components and technologies presented herein and variations are within the range of preliminary engineering cost estimates (11).

It is clearly shown that using bid estimates from projects in the location of interest will improve the accuracy of estimating. When the Latin American estimates were added to a U.S. data base of costs and included in a computerized data base to be used by engineers working in Central and South America (11), the time-adjusted estimates from the combined data base compared favorably with location specific engineering estimates (one estimate was within +20%).

But comparable bid data from locations of interest are not generally available, especially where projects are being developed for the first time. For example, water treatment facilities have been built in many countries and throughout the U.S. but not to the extent of wastewater treatment facilities. Indeed, wastewater treatment facilities are rare in Latin America and in most of the developing world.

The cost data for past projects are difficult and very costly to obtain. Even in those cases where comparable facilities have been built in a region, records of past projects are usually poorly kept or have been lost.

For projects abroad, the problem of currency is a major one. Records are sometimes available for projects which include estimates in local currency. Cost data in local currency for developing countries and countries without market based economies (including those with high inflation rates) are virtually useless for estimating purposes for new projects because of the difficulty in assessing the value-for-price relationship.

One of the major objections raised when using U.S. cost data abroad is that in developing countries especially, project construction cost is dominated by low-cost hand labor versus use of construction equipment in the U.S. and developed countries. If this were true, project construction costs abroad would be lower, especially in those countries with high currency exchange rates and lower labor rates. There is no evidence to indicate that project costs for comparable facilities are lower abroad—the opposite is true. There are many reasons for project costs abroad being higher than in the U.S. and not all of them technical or economic.

In summary, there is often no practical alternative to using U.S. cost data for estimating project costs abroad at the preliminary engineering level of design. The range of cost variability in U.S. dollars, around the world is similar to that within the U.S. except on the high side of an estimate and in very remote locations. Thus, there are probably no special estimating considerations applicable to projects abroad that are not applicable to projects within the U.S. Engineers familiar with projects in the U.S. may analyze and apply the same factors for projects anywhere: do-ability in remote locations, availability and rates for local labor, transport costs for equipment, utility costs, service lifelines, and other factors as necessary.

Additionally, differences among developed country market economies are becoming easier as time passes because of international competition on a price basis. For example, the impact of equipment costs on water and wastewater projects is significant. But the cost estimate variations among various bidders regardless of country of origin and currency cannot be significantly different whenever international tenders are required (for most projects involving international lending and granting agencies). Thus, estimates for projects in original currencies that have reasonably stable exchange rates with the U.S. dollar may be used as well.

TECHNOLOGY EVALUATION AND SELECTION

This section is provided to further aid the process of evaluation and selection of technologies for application to water and wastewater treatment. There are many factors which may be used. The factors may be general in nature or specific to locations and problems.

The format of this handbook is designed to help in selection. The material is presented in two parts. The first part presents data and information which is applicable to the technology for general application. The second part is in a case study format illustrating specific applications. The fact that a technology has been used to solve specific problems lends credibility to its application elsewhere.

Each technology section is presented using evaluation and selection criteria. The criteria are: description (usually containing design criteria), limits (advantages and disadvantages when enough information is available), operation and maintenance (process control considerations), availability (ease of implementation), special factors, and recommendations. In some cases these criteria would not fit the specific technology and other factors were used. Sometimes not all of the factors were applicable to a technology.

The user of this book may compare one technology to another or prepare evaluations of groups of technologies to meet a certain need or application. Cost is often over-emphasized in the evaluation process. A technically based evaluation should always be prepared. Process performance is the most important consideration. In other words, the technology must function in the application for which it is intended. Costs should only be used to compare processes when the performance of one or more technologies will meet the requirements of a given application. In fact, if this evaluation procedure is followed strictly, cost will often not be a consideration, because conditions from location to location are very different.

Selection of technologies can only be made once, and the utility management and plant operators must provide consistent service using the facility during the life of the plant. It is well worth the time and effort to perform a complete evaluation when a facility is being planned and when upgrading of an existing facility is being considered. Shorthand approaches to the selection process should be avoided by engineers, and leaving the technology selection to the consulting engineering company should be avoided. Selection should be a joint process.

TECHNOLOGY SELECTION GUIDELINE MATRIX

Recognizing that shorthand selection is to be avoided, it is nevertheless useful to consider evaluation and selection criteria in a summary fashion. Most technologies which are presented in Part I are included in Table I.1, along with evaluation criteria. A summary assessment of each technology is given on Table I.1.

Evaluations may be qualitative or quantitative. The evaluation provided on Table I.1 is qualitative, using high (H), medium (M), and low (L); or yes (Y)/no (N) for Table entries. The evaluations performed for specific applications may be in more detail and quantitative. It is not possible in a book of limited scope to address specific applications. The evaluations given are to be considered general. The factors in Table I.1 will be applicable to specific situations and may be used for detailed analyses, but the entries in Table I.1 may not be applicable.

Most of the factors are self evident. The following comments may be useful as elaboration.

Life Cycle Cost—The total cost of acquisition and ownership over the full useful life of a facility. The total cost to acquire, operate, maintain, and salvage. The evaluation of life cycle cost is a technique which allows the consideration of total cost and local factors in the selection. Selections based on capital cost alone are unsatisfactory, as are those which do not consider major equipment replacements, for example during a plants lifetime. The costs presented in the book are capital and operating and maintenance (O&M), from which the user may construct life cycle cost estimates taking into account specific site and project factors.

Cost Effectiveness—The cost per unit of treatment and/or per unit performance. For example, $/lb of solids removed and processed. Note that it is not sufficient to consider removal alone but processing to the point where the pound of solids has been itself treated and permanently disposed. Cost effectiveness estimates may be developed by computing total annual costs

Table I.1. Technology Selection Guidelines

TECHNOLOGIES	LIFE CYCLE COST	COST EFFECTIVENESS	RELIABILITY	SIMPLICITY OF OPERATION	EASE OF MAINTENANCE	PERFORMANCE	ABILITY TO MEET WATER QUALITY OBJECTIVES	ADAPTABILITY TO CHANGE IN INFLUENT QUALITY	PERFORMANCE DEPENDENT ON PRETREATMENT	ADAPTABILITY TO VARYING FLOW RATE	EASE OF CONSTRUCTION	ADAPTABILITY TO UPGRADING	AVAILABILITY OF MAJOR EQUIPMENT	EQUIPMENT/SUPPLIES AVAILABLE LOCALLY	POST-INSTALLATION SERVICE/CHEMICAL DELIVERY	PERSONNEL SKILL LEVEL	ENERGY UTILIZATION	RESIDUE PRODUCTION	COST OF RESIDUE DISPOSAL	POTENTIAL FOR EFFLUENT USE/REUSE	IMPORTANCE OF AIR EMISSIONS
Slow Sand Filter	L	M	H	H	L	H	H	H	Y	H	M	L	H	H	M	M	L	H	H	H	L
Rapid Sand Filter	H	M	M	L	L	M	H	M	Y	M	L	H	H	H	M	H	H	H	H	H	L
Chemical Precip. & Filtration	M	H	H	L	L	H	H	H	N	M	L	H	H	H	H	H	M	H	H	H	L
Dual Media Filter	H	M	H	L	L	H	H	H	Y	M	L	M	H	L	L	H	H	H	H	H	L
Sludge Vacuum Filter	M	M	H	L	L	M	H	H	Y	H	L	M	M	M	L	H	H	H	M	L	H
Sedimentation—Rect. Primary	M	M	H	M	M	M	L	H	N	H	L	H	H	L	L	M	L	H	M	L	M
Upflow Filter	H	H	M	L	M	M	H	H	Y	M	L	M	H	L	L	H	M	H	H	M	L
Flocculation	M	H	H	L	M	H	H	H	N	M	L	M	H	H	M	H	L	H	H	H	M
Gravity Sewers	H	L	H	H	M	M	L	H	N	H	L	L	H	H	H	L	L	L	L	L	L
Pressure Sewers	L	M	M	M	L	H	H	M	Y	M	H	L	M	M	L	H	H	M	L	L	H

Table I.1. *(continued)*

	LIFE CYCLE COST	COST EFFECTIVENESS	RELIABILITY	SIMPLICITY OF OPERATION	EASE OF MAINTENANCE	PERFORMANCE	ABILITY TO MEET WATER QUALITY OBJECTIVES	ADAPTABILITY TO CHANGE IN INFLUENT QUALITY	PERFORMANCE DEPENDENT ON PRETREATMENT	ADAPTABILITY TO VARYING FLOW RATE	EASE OF CONSTRUCTION	ADAPTABILITY TO UPGRADING	AVAILABILITY OF MAJOR EQUIPMENT	EQUIPMENT/SUPPLIES AVAILABLE LOCALLY	POST-INSTALLATION SERVICE/CHEMICAL DELIVERY	PERSONNEL SKILL LEVEL	ENERGY UTILIZATION	RESIDUE PRODUCTION	COST OF RESIDUE DISPOSAL	POTENTIAL FOR EFFLUENT USE/REUSE	IMPORTANCE OF AIR EMISSIONS
Lagoons—Facultative	L	M	M	M	M	L	L	H	N	M	M	H	H	H	H	M	L	M	L	L	M
Aquaculture	L	H	L	M	L	L	L	H	N	H	H	L	H	H	H	L	L	L	L	L	M
Preliminary Treatment	L	H	L	H	L	H	L	H	N	H	M	H	H	H	H	H	L	H	L	L	L
Rotary Screen	H	H	M	M	L	M	H	H	Y	M	L	M	M	M	L	H	H	H	M	H	L
Wedgewire Screen	H	H	M	M	L	M	H	H	Y	M	L	M	M	M	L	H	H	H	M	H	L
Trickling Filter—Plastic	M	H	M	M	H	M	L	H	N	M	M	M	M	M	M	M	M	M	L	L	M
Trickling Filter—Rock	M	H	L	M	M	M	L	M	N	M	M	H	H	H	H	M	M	M	L	L	M
Trickling Filter—Low Rate	M	H	L	M	M	M	L	H	N	L	M	H	H	H	H	M	L	M	L	M	M
Aerated Lagoons	M	M	L	L	L	M	M	H	N	M	M	M	H	H	H	M	H	M	L	L	H

Sludge Drying Bed	L	H	H	H	H	H	H	H	N	L	H	H	H	H	H	L	L	H	M	M	H
Land Application Sludge	L	H	H	M	H	M	H	M	Y	L	H	M	M	M	M	M	L	L	L	H	H
Chlorination	H	H	H	L	L	H	H	H	Y	L	L	L	M	L	L	H	M	L	L	H	L
UV Disinfection	H	M	M	L	L	M	M	M	Y	L	L	L	L	L	H	H	H	L	L	H	L
Plate Settler	M	H	H	M	L	H	M	H	N	L	L	L	H	H	M	H	L	H	M	M	L
Land Application																					
Waste Water:																					
Irrigation	M	H	M	M	H	M	M	M	N	M	M	M	H	H	H	M	M	L	L	H	H
Overland Flow	L	M	L	M	L	L	L	H	Y	L	H	H	H	H	H	M	L	M	L	L	H
Percolation	H	H	M	M	H	M	M	M	N	M	M	L	H	H	H	M	M	L	L	H	H
Lagoon—Anaerobic	L	M	L	M	L	M	M	M	N	M	M	H	H	H	H	M	L	L	L	L	M
Lagoon Aerobic	M	M	M	L	L	M	M	H	N	M	M	L	H	M	M	M	H	L	L	M	M
Water Collector	M	M	M	L	L	M	•	M	Y/N	M	L	•	L	L	M	M	H	L	L	H	L
GAC Adsorption	H	VH	H	L	L	M	H	L	Y	L	L	L	L	L	L	H	H	M	H	M	M
Flotation	M	M	H	M	L	M	M	L	N	L	L	L	L	L	L	H	H	M	M	M	H
Imhoff Tank	L	H	H	M	L	M	M	M	N	H	M	H	H	H	M	M	L	M	M	L	M
Roughing Filter	L	L	L	M	M	M	M	M	N	H	M	H	H	H	H	M	L	M	L	L	M
RBC	M	M	M	L	L	M	M	M	Y	M	L	L	L	L	L	H	M	M	M	M	H
Activated Sludge	M	M	M	M	L	M	H	H	Y	M	L	H	M	M	M	H	M	M	M	M	H
Steep Slope Sewers	L	H	M	M	M	M	H	M	Y/N	M	H	L	L	L	L	L	L	L	L	L	L
SBR	M	M	M	L	L	M	M	•	Y	M	L	H	M	L	L	H	M	M	M	M	H
Intermittent Sand Filter	M	L	M	L	M	M	M	M	Y	H	L	•	H	H	L	H	M	H	H	H	L
Pulsed Bed Filtration	M	M	M	L	L	H	M	H	Y	H	L	•	H	H	L	H	M	H	H	H	L
Biological Nutrient Removal	M	L	L	L	L	M	M	M	Y	L	L	•	H	H	L	H	M	M	M	H	H
Hydrograph Release Lagoons	M	M	M	L	L	H	M	M	N	H	M	•	H	H	H	M	M	L	L	L	M
Vacuum Assisted Dewatering	H	H	M	L	L	H	H	M	N	H	L	•	H	L	L	H	H	H	H	L	H
Vacuum Sewers	L	M	L	L	L	H	L	L	Y	M	VL	•	L	L	L	H	H	M	L	L	H
Sludge Composting	L	M	M	M	M	H	M	M	N	H	H	•	H	H	H	M	L	VH	M	•	H
Methane Recovery	M	H	H	L	M	H	L	M	N	H	L	•	H	M	H	H	L	M	L	•	H
Dual Digestion	M	H	H	H	M	H	L	M	N	H	L	•	H	H	H	H	L	M	M	•	H

Table I.2. Technology Removal Capability

TECHNOLOGIES	PESTICIDES	ARSENIC	ASBESTOS	CADMIUM	CHROMIUM	COPPER	CYANIDES	LEAD	ZINC	BIS-CHLORO METHYL ETHER	PTHTHALATES	AMINES	TOTAL PHENOLS	BENZENE	CHLOROFORM	CARBON TETRA CHLORIDE	XYLENES	BENZO(A) PYRENE	PCBs	METHYLENE CHLORIDE	THMS	THM PRE-CURSORS	HARDNESS	TASTE & ODOR	COLOR	TURBIDITY
Coagulation and Flocculation	M	M/H	•	H	M/H	•	M	L/H	L/H	•	L/H	•	L	•	•	L	•	•	•	M	•	H	•	•	•	H
PAC Adsorption with Activated Sludge	•	•	•	•	L/H	M/H	M	•	M/H	•	•	•	H	•	•	•	•	•	•	•	H	H	•	•	•	•
GAC Adsorption	•	L/H	•	M/H	L/H	L/M	L/H	L/M	L/H	•	M/H	•	M/H	M	M/H	•	•	H	•	M/H	M	H	•	•	•	•
Chemical Prcipitation with Sedimentation	•	•	•	•	•	•	•	•	•	•	•	H	•	•	•	•	•	•	•	•	•	H	•	•	•	M/H
Lime	L	M/H	H	L/H	M/H	L/H	M/H	L/H	M/H	•	M/H	•	L/H	H	H	•	•	H	•	L	•	•	M/H	•	•	•
Ferric Chloride	•	H	•	•	•	•	•	•	•	•	L	•	L	•	•	•	•	•	•	H	•	•	H	•	•	•
Polymer	•	•	•	L/H	M/H	L/H	L/H	M/H	L/H	•	•	•	•	L	•	H	•	•	•	H	•	•	H	•	•	•
Sodium Carbonate	•	H	•	M/H	H	M	•	H	H	•	•	•	•	•	L	•	•	•	•	•	•	•	•	•	•	•

Barium Chloride	•	L	M	•	M	M	•	•	M	•	•	•	•	•	•	•	•	•	•	•	•	•	•	•	•	•
Sodium Hydroxide	•	•	•	L/H	M/H	L/H	•	H	H	•	M/H	•	H	H	H	•	•	•	•	H	•	•	M/H	•	•	•
Combined Precipitants	•	L/H	•	L/H	M/H	M/H	•	L/H	L/H	•	H	•	M/H	H	H	H	•	•	•	L	•	•	•	•	•	•
Alum	L	L	•	M	L/H	L/H	M	L/H	M	•	L/H	•	L/H	H	M/H	H	•	•	•	H	•	•	M/H	•	•	•
Chemical Precipitation with Filtration	•	L	•	L/H	M/H	L/H	M	H	H	•	M/H	•	H	•	•	•	•	•	•	•	•	•	•	•	•	H
Filtration	L/H	L/H	H	L/H	L/H	L/H	L/H	L/H	L/H	•	M/H	M	L/M	•	H	H	M	•	L	L/H	•	•	•	•	•	M/H
Sedimentation	L/H	L/H	H	L/H	L/H	L/H	L/H	L/H	L/H	•	M/H	•	L/H	M/H	M	H	•	M/H	•	•	•	•	•	•	•	L/H
Activated Sludge	M	L/H	•	L/H	L/H	L/H	L/H	L/H	L/H	M	L/H	M/H	L/H	M/H	L/H	H	•	•	•	•	•	•	•	•	•	•
Lagoons																										
Aerated	•	H	•	H	L/H	M/H	M	H	M/H	•	L/H	•	L/H	M	H	•	•	•	•	•	•	•	•	•	•	•
Non-Aerated	•	•	•	•	H	•	•	H	H	•	•	•	L	•	•	•	•	•	•	•	•	•	•	•	•	•
Chemical Oxidation																				•	•	•	•	H	H	•
Chlorine	•	•	•	•	•	L	H	•	•	•	•	•	•	•	•	•	•	•	•	•	•	•	•	•	H	•
Ozone	•	M	•	•	•	L	L/H	L	H	•	M/H	M	•	M	•	•	•	H	•	•	H	H	•	•	H	•
Trickling Filter	•	•	•	•	•	•	M	•	•	•	M	•	H	•	•	•	•	•	•	•	•	•	•	•	•	•
Roughing Filter	•	•	•	•	•	•	M	•	•	•	M	•	H	•	•	•	•	•	•	•	•	•	•	•	•	•
RBC	•	•	•	•	•	M	•	•	•	•	•	•	•	•	•	•	•	•	•	•	•	•	•	•	•	•
Chemical Reduction	•	•	•	•	•	L/H	•	M	M	•	•	•	•	•	•	•	•	•	•	•	•	•	•	•	•	•

(O&M plus amortized capital cost, for the removal treatment technology and the solids processing technology) divided by the total amount processed (in the case of the solids example, total solids removed, dewatered, landfilled, etc.). Cost effectiveness is a term often used and almost never correctly applied to process selection and evaluation. The cost effectiveness determination must consider local factors.

Performance-and-Ability to Meet Water Quality Treatment Objectives —These two factors may be considered together. Design should be based on the expectation for the production of water with required effluent quality characteristics. Expected performance is the key to the evaluation. Quantitative performance information is given later in this submission.

1. Adaptability to Changes in Influent Quality—Performance,
2. Dependability on Pretreatment,
3. Adaptability to Changes in Flow Rate.

These three factors may be considered together. They relate to how the technology may be expected to function separately. Typically, process performance is dependent on the performance of upstream processes. How the technology relates to the operation of other—including downstream—processes is also important.

1. Availability of Major Equipment,
2. Equipment/Supplies Available Locally,
3. Post Installation Service.

These three relate heavily to local considerations. The qualitative summary evaluations in Table I.1 may have the least accuracy in special applications.

PERFORMANCE AND REMOVAL CAPABILITY

A qualitative summary of technology performance capability is given in Table I.2 (67). The information herein should not be used as a substitute for testing. Jar tests, pilot scale evaluations, and even full scale tests at existing plants should always be used whenever capital expenditures are planned. It is likely that the tests will pay for themselves in saved capital and O&M costs.

Both water and wastewater treatment are considered in Table I.2. For general purposes, the following may be considered typical:

L = 0 to 30%
M = 20 to 70%
H = 60 to 99%
L/H = widely various performance and probably not applicable

Table I.3 (3) contains largely qualitative information also, mostly pertinent to water treatment.

Table I.3. Process Selection Guidelines

WATER QUALITY PARAMETERS	PROCESS AND COMPONENTS	APPLICABILITY	COMMENTS
Turbidity	In-Line filtration Coagulation filtration	Low turbidity, low color	Greater operator attention required; shorter filter run than with direct filtration and conventional treatment; additional sludge-handling facilities may be required; pilot plant studies may be required; lower capital and O&M costs.
	Direct filtration Coagulation Flocculation Filtration	Low to moderate turbidity, low to moderate color	Greater operator attention required; greater sludge-handling facilities may be required; pilot plant studies may be required; lower capital and O&M cost; better filter run time than in-line filtration but shorter than convention treatment.
	Conventional coagulation flocculation sedimentation filtration	Moderate to high turbidity, moderate color	Detention time in sedimentation basins allows for adequate contact time for T&O and color removal chemicals; more operational flexibility and less operation attention required
	Microscreening	Removal of gross particulate matter (e.g., algae)	Process relies on straining mechanism; process could not meet water quality objective if used alone
Bacteria/virus	Chemical Disinfection Chlorine Chloramine Chlorine dioxide Ozone Other chemicals Bromine Iodine Potassium permangate	Disinfection of surface and ground waters	THM potential with chlorine needs to be assessed, chloramine treatment not as potent as chlorine but does eliminate THM formation; cost of treatment; Cl_2 < Chloramine < CLO_2 < ozone; other chemicals such as bromine, iodine, etc. limited to small application

Table I.3. *(continued)*

WATER QUALITY PARAMETERS	PROCESS AND COMPONENTS	APPLICABILITY	COMMENTS
	Nonchemical disinfection Ultraviolet Ultrasonic	Disinfection of surface and ground waters	Advantage of UV is that no residual is left, which is applicable to fish aquariums and hatchery disinfection; ultrasonic is expensive, some success when used with ozone in tertiary disinfection
Color	Coagulation	High color levels	Use of high coagulant dose and low pH (5–6) is cost-effective when high color levels exist
	Adsorption GAC PAC Synthetic resins	Moderate to low color levels	GAC bed life is in the order of 1–6 weeks; synthetic resins are expensive (capital and regeneration costs); PAC is used for handling short-term color problems (however, it is costly as a routine color control method)
	Oxidation Chlorine Ozone Potassium permanganate Chlorine dioxide	Low, consistent color levels	Effectiveness: Ozone > Cl_2 > ClO_2 > $KMnO_4$; Cl_2 and $KMnO_4$ typically used for other purpose (disinfection and T&O control) but are effective for color control
Taste and Order	Source control Copper sulfate Reservoir destratification	Used to prevent any T&O problems in plant	Most satisfactory way of controlling is to control the problem at the source; copper sulfate may require chelation of certain pH values
	Oxidation Chlorine Ozone	Low T&O levels	Chlorine may result in increased odor problems where odors are of industrial or algal origin; $KMnO_4$ widely used and very effective for odor control (however, overdosing may result in slightly pink color)

	Potassium permanganate Chlorine dioxide		
	Adsorption GAC PAC	Low to moderate levels, GAC used for industrial odor sources	PAC, in slurry form, is usually added at coagulation process for moderate T&O levels and ahead of filters for low T&O levels; GAC commonly used for odors caused by industrial sources—bed life usually very long
Hardness	Lime-soda softening, ion exchange	Extremely hard waters. Removes not only hardness but also selective constituents, THMs	Most common method of removal of hardness. Very expensive, especially in large-scale facilities
Organics (THM)	Alternative disinfectants Chloramine Chlorine dioxide Ozone Removal of precursors Chlorine dioxide Ozone GAC PAC Coagulation Removal of THM Ozone GAC PAC	THMs	Chloramines are not as powerful a disinfectant as free chlorine; ozone offers no appreciable residual protection in distribution system; modification of chlorine addition points may reduce THM formation; PAC has been found to yield only partial removals at very high doses

Table I.4. Removing Pollutants by Treatment Processes[1]

PARAMETER	PRIMARY TREATMENT	TRICKLING FILTER	ACTIVATED SLUDGE	LAGOON	ROTATING BIOLOGICAL CONTRACTOR	LAND TREATMENT INFILTRATION PERCOLATION
BOD	—	82	91	90	96	95
COD	8	74	83	75	93	93
Oil and grease	38	74	78	71	86	90
Total susp. solids	48	88	90	92	93	98
Ammonia nitrogen	—	4	25	53	37	87
Benzene	—	92	81	75	96	96
Bis(2-ethylberyl) phthalate	44	60	50	—	94	91
Butyl benzyl phthalate	44	55	—	—	72	72
Cadmium	35	93	80	—	96	96
Chloroform	44	78	74	56	97	94

Chromium	35	85	87	72	98	98
Copper	9	72	86	86	99	93
Cyanide	—	62	72	79	82	94
Diethyl phthalate	—	—	—	—	—	—
Ethylbenzene	10	87	78	72	95	94
Lead	—	58	57	—	56	56
Mercury	11	40	26	76	71	71
Methylene chloride	—	96	—	99	99	99
Nickel	23	66	36	72	92	92
Phenols, total	—	75	76	60	86	97
Solids, total	—	25	26	62	8	30
Solids, volatile, total	4	55	53	76	—	55
Tetrachloroethylene	11	92	80	95	98	96
TOC	5	70	73	65	84	94
Toluene	—	98	94	92	100	99
Trichloroethylene	16	96	81	92	99	98
Zinc	—	90	78	87	98	98

[1] Removal is calculated on the basis of average influent to the POTW and effluent of each particular process.

Table I.5. Median Removal Efficiencies (Percent)

	BOD	COD	TKN	TOC	TSS
Sedimentation	23	93	—	32	97
Polymer	50	71	—	82	99
Lime	52	32	—	18	71
Lime Polymer	—	99	—	22	99
Alum, Coag., Aid	37	59	—	47	66
Alum	61	10	—	63	84
Alum, Lime	41	86	—	80	93
Filtration	24	24	—	13	67
Activated Sludge	93	67	—	69	44
Aerated Lagoons	86	62	77	45	45
Granular Carbon	52	50	—	55	38
Powdered Carbon with Activated Sludge	—	—	—	—	—
Ozonation	—	50	—	9	15

COMPARING GROUPS OF PROCESSES

An example comparing a group of technologies is given in Table I.4 (67), in this case for biological treatment processes and land treatment. The values are considered average removals. Care should be exercised in using such average numerical values, because of variations in systems, influent quality, sampling and analytical measurements, and a variety of other factors. Example conclusions for comparison in Table I.4 are: a) RBC and land treatment are better for ethylbenzene (and similar organics) removal than the other processes—same for nickel, and b) primary treatment is not effective for removal of most of the contaminants shown.

The user may make similar comparisons for other groups of technologies using actual local data. Another example is given for filtration in Table I.3.

Median values for removal of common contaminant parameters and some technologies are given in Table I.5.

1. SLOW SAND FILTERS

1.1 DESCRIPTION

Slow (0.05 to 0.13 gpm/ft^2—0.12 to 0.32 m/h) (48) sand filters have a high degree of efficiency for solids and turbidity removal in the case of raw water with low turbidity and color (turbidity up to 50 NTU and color up to 30 units). Taste and odor are removed also. If raw water quality is poor, roughing filters may be used preceding the slow sand filters. The slower filtration rate also means a greater efficiency in the removal of bacteria (as compared to rapid filters). Bacterial removal is considered the strong point of slow sand filtration. Chemicals are typically not used. The flow rates for slow sand filters are many times slower than rapid and roughing filters, and the operating filter bed is not stratified.

The effective size of the sand used is about 0.2 mm and may range between 0.15 and 0.35 mm. The sand should also have a uniformity coefficient between 1.5 and 3.0. In contrast, the range of effective size for rapid sand filters is 0.35 to 1.0 mm, with a uniformity coefficient of 1.2 to 1.7. The media size used for roughing filters is much larger.

A distinguishing feature of slow sand filters is the presence of a thin layer at the surface of the bed called "schmutzdecke." This layer forms on the surface of the sand bed and is composed of a large variety of biologically active micro-organisms. The breakdown products (organic matter) fill the interstices of the sand, so that solid matter is retained more effectively than rapid and roughing filters. The cleaning of the filter bed is carried out by manually scraping (removing) the top layer of the filter bed when it becomes clogged with impurities.

In general, during filtration through a porous substance, the water quality is improved by mechanical straining of suspended and colloidal matter, by pretreatment to reduce the number of bacteria and other microorganisms, and sometimes by chemical treatment. The porous filter substance may be any chemically-stable material, but beds of sand (silica and garnet) are used for water supply and wastewater treatment in most cases. Sand is cheap, inert, durable, and widely available. It has been extensively tested and found to give excellent results. The design bed thickness varies from 1.2 to 1.4 m, but after successive cleaning, the resultant thickness may be 0.6 to 1.2 m. (Figure 1.1).

Essentially, a slow sand filter consists of a watertight box provided with an underdrain system which also serves the purposes of supporting the filtering

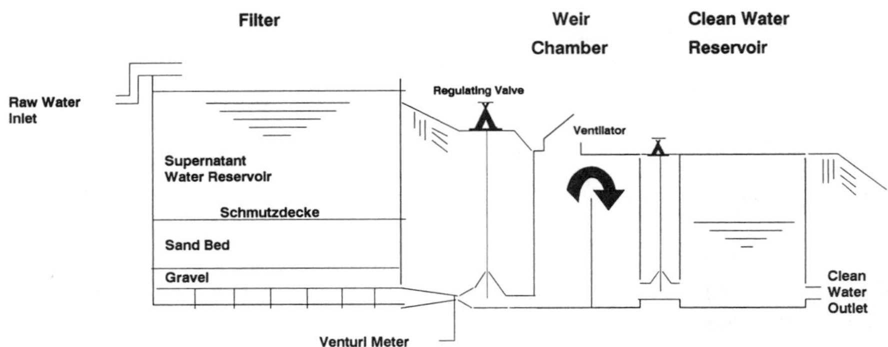

Figure 1.1. Slow sand filter.

material, and distributing the flow evenly through the filter. Many different media have been used for the underdrain system. Bricks, stone, and even bamboo have all been used for this purpose. Bamboo however, requires frequent renewal because it is organic and unstable over extended periods of use.

The successful performance of a slow sand filter is dependent mainly on the schmutzdecke layer. In a mature bed, the layer, generally consisting of algae, plankton, and bacteria forms on the surface of the sand. Inorganic suspended matter is retained by the straining action of the sand as well as by the schmutzdecke layer. The schmutzdecke organisms also may accomplish a certain amount of organic material breakdown.

The walls of the filter can be concrete or stone. Sloping walls dug into the earth, supported or protected by chicken wire reinforcement and sand or sand-bitumen could be a cost effective alternative to concrete. Inlets and outlets should be provided with controllers to keep the raw water level and the filtration rate constant.

Filtration rates usually employed are between 2.5 and 6.0 m^3/m^2/day. Higher rates may be used after a series of tests are performed which yield good effluent quality results.

The system should be designed for ease of operation and flexibility. The design should consist of a number of separate units to accommodate down-time for cleaning and repairs. A filter must be taken out of service for cleaning. The suggested number of units for given populations is as follows (9):

POPULATION	UNITS
2000	2
10000	3
60000	4
200000	6

Table 1.1. Lateral and Manifold Pipe Sizes and Flows for Bottom Drains

FILTRATION RATE (m³/m²/DAY)		FILTER AREA DRAINED IN SQUARE METERS						
		2.80	3.75	4.70	5.60	7.50	9.35	14.00
Lateral	2″	7.4	6.5	6.0	5.5	4.9	4.5	3.7
	3″	16.8	14.9	13.7	12.8	11.4	10.6	8.7
	4″	30.1	26.8	24.6	22.8	20.3	18.6	15.6
	5″	48.2	42.8	39.1	36.3	32.0	29.4	24.8
	6″	69.7	62.3	56.8	53.0	46.5	42.8	36.2
	8″	112.0	112.0	102.0	94.0	84.0	76.0	64.0
Manifold	10″	320	280	250	230	205	185	160
	12″	455	400	360	335	300	270	220
	15″	720	640	575	540	475	430	360
	18″	1 040	930	840	770	690	620	520
	21″	1 420	1 260	1 145	1 060	930	850	710
	24″	1 860	1 650	1 500	1 390	1 230	1 120	930
	27″	2 360	2 080	1 890	1 750	1 540	1 105	1 120
	30″	2 930	2 580	2 355	1 180	1 925	1 750	1 460

Source: Reference 9.

Bottom drains consist of a system of manifold and lateral pipes sized as shown in Table 1.1 (9).

1.2 ADVANTAGES AND DISADVANTAGES

In order to get good results from slow filtration, the raw water must not be too contaminated with suspended solids (influent quality generally below 50 mg/L total suspended solids, TSS). Such levels of TSS include raw water from lakes and reservoirs but generally only pretreated water from flowing streams.

Advantages for small systems and developing countries are: low construction cost using manual labor, simplicity of design and operation, unskilled maintenance labor, no chemicals required, sand can usually be found locally, power is not required, large quantities of water for backwashing not required, backwash water treatment is not required. The disadvantages are: operation is suggested only with low influent contamination levels, pretreatment is probably required in many applications. Close operational control of head loss is required to prevent air binding which is a potential problem in all types of filters.

Finally, provisions must be made for storing used sand permanently or temporarily until it is washed. If sand is to be washed, a separate facility and a water supply are required; treated water may be used for washing.

1.3 COSTS

See Table 1.2 (1) and Table 1.3 for operating and maintenance costs (57).

Table 1.2. Per Capita Costs and Operation and Maintenance Manpower Requirements. Process: Slow Sand Filter

POPULATION	TYPE OF COST	$ PER CAPITA* COST RANGE	NO. UNSKILLED WORKERS REQUIRED
500–2,499	Construction	17.08–27.00	
	Operation and Maintenance	1.80–6.75	1
2,500–14,999	Construction	12.19–19.28	
	Operation and Maintenance	0.81–3.04	2
15,000–49,999	Construction	8.55–13.51	
	Operation and Maintenance	0.45–1.69	5
50,000–100,000	Construction	5.33–8.44	
	Operation and Maintenance	0.27–1.01	8

* at 120 gallons per capita per day (gpcd)
Source: Reference 1.

1.4 AVAILABILITY

Slow sand filtration is perhaps the most common of the filtration technologies in developing countries. It has been proven both mechanically and economically many times. Some slow sand systems are being replaced with rapid sand filters. A separate section for rapid sand filtration is included.

1.5 OPERATION AND MAINTENANCE

The initial resistance (loss of head) of the clean filter bed is about 6 cm. During filtration, impurities deposit in and on the surface layer of the sand bed, and the loss of head begins to increase. When the loss of head has reached a pre-set limit (the head loss is usually not allowed to exceed the depth of water over the sand, about 1 to 1.5 m), the filter is put out of service and cleaned. The period between cleaning is typically 20 to 60 days, shorter for higher influent TSS concentrations. Filters may be cleaned by either scraping off the surface layer of sand and washing and storing cleaned sand for periodic resanding of the bed, or washing the surface sand in place with a washer traveling over the sand bed. If replacement sand is readily available, the former method is favored. Manual cleaning is with flat wide shovels for scraping and removal of from 1 to 2 cm of top material. Cleaning by hand is usually completed in one or two days. After washing the sand is either stored or replaced on the bed. Replacement is made when as a result of successive

Table 1.3. Estimated Operation and Maintenance Costs for Slow Sand Filters*

LOCATION	AVERAGE OPERATIONAL FLOW (MGD)	LABOR FOR SCRAPING[1] (MAN-HOURS/YEAR)	LABOR FOR RESANDING[1] (MAN-HOURS/YEAR)	LABOR FOR DAY-TO-DAY ACTIVITIES (MAN-HOURS/YEAR)	TOTAL LABOR COSTS ($/YEAR)	TOTAL OPERATION AND MAINTENANCE UNIT/COST (C//1000 GAL)
Auburn	6.0	1007	618	365	11,400	0.5
Geneva	2.5	374	218	365	8,000	1.2
Hamilton	0.3	224	NA	365	6,400	5.7
Ilion	1.5	905	563	365	20,000	3.6
Newark	2.0	143	226	365	8,200	1.2
Ogdensburg	3.6	8736	—	365	25,700	2.2
Waverly	1.2	582	420	365	14,800	4.0

* All cost figures are based on a $10/hr wage rate except at Auburn, where $3/hr was used.
[1] Scraping and resanding may be done simultaneously.
Source: Reference 57.

Table 1.4. Summary of Filter Scraping Data

LOCATION	PLANT SIZE MGD	AVERAGE FILTER RUN WATER PRODUCTION (GAL/FT^2)	AVERAGE FREQUENCY OF FILTER SCRAPING OPERATIONS (NUMBER PER YEAR)	AMOUNT OF SAND REMOVED IN SCRAPING OPERATION (IN)	METHOD(S) USED IN REMOVING SAND FROM FILTER SURFACE	TIME REQUIRED TO SCRAPE FILTERS ($MAN\text{-}HOURS/100\ FT^2$)
Auburn	6.0	6,844	4.3*	0.5	Shovels, hydraulic	4
Geneva	2.5	15,718	2.0	1.0	Shovels, motorized buggy	4–5
Hamilton	0.3	4,302	2.0	1.0	Shovels, 50 gal drums, backhoe	8–9
Ilion	2.0	15,487	1.8	3–4	Shovels, hydraulic	23–42
Newark	3.6	10,122	3.3	1.0	Shovels, motorized buggy	2
Ogdensburg	N/A	2,978	12.0	1.0	Shovels, hydraulics	4–5
Waverly	1.2	3,200[1]	9.7[1]	1.0	Shovels, wheelbarrows	5

* Two scraping operations per year are actually occasions when the filters are raked and no sand is removed.
[1] Water production and scraping frequency estimated by the Waverly personnel for the future using data from a 9-month operations study.
Source: Reference 57.

cleaning, the thickness of the sand bed has been reduced to about 50 to 80 centimeters. About 0.2 to 0.6% of the water filtered is generally required for washing purposes. A process of "throwing over" is carried out during re-sanding. During this process, a layer of old sand is added on top of the cleaned sand. The purpose of this process is to retain some of the active material in the top surface layer and enable the resanded filter to become operational with a minimal amount of "reripening."

After being cleaned, the beds are slowly refilled with filtered water from below until the sand is completely covered. This prevents entrainment of air in the sand. Cleaning experience is given in Table 1.4 (57).

1.6 CONTROL

Process control is based on effluent quality, usually some established level of suspended solids for given water use or in the case of wastewater, some level acceptable for discharge or for subsequent treatment. The desired effluent quality is typically related to head loss in the filter. Thus quality may be indirectly measured by process performance. If an indirect technique is used and the product water quality is important; e.g., for water supply, the correlation between effluent quality and operational head loss should be checked regularly and often.

1.7 SPECIAL FACTORS

Whenever raw water influent turbidity values of higher than 50 JTU are encountered, slow sand filtration should be preceded with pretreatment; such as sedimentation, rapid filtration, or roughing filters.

1.8 RECOMMENDATIONS

Slow sand filters have a significant water cleaning capacity. Product water is usually bacteriologically safe and free from solid impurities. Disinfection of filtered water should always be practiced for water supply applications. Some typical removal rates achieved in small systems are as follows: turbidity, 97 to 99%; color, 20 to 30%; iron (removed incidentally), 50 to 60%; and bacteria, 95 to 97%.

2. RAPID SAND FILTERS

2.1 DESCRIPTION

When large amounts of water or very turbid water must be treated, slow sand filters are at a disadvantage because solids may be stored only in the relatively thin layer at the top of the bed. More rapid filtration and filtration of more turbid waters is possible by making available more of the bed depth during filtration. Accordingly, using coarser and in particular more uniform sand grains is preferred for more demanding applications. Rapid sand filter media may range in size from 0.35 to 1.0 mm. A typical size might be 0.5 mm, with an effective size of 1.3 to 1.7 mm (1, 51). This range of media size has a demonstrated capability for turbidities in the range of 5 to 10 NTU at rates up to 2 gpm/sq ft (4.88 m/h) (51). Common filter rates for rapid filters may be as high as 100 to 300 m^3/m^2/day (m/day), that is about 50 times the rates used with a slow rate filter.

Rapid sand, dual media, or multi media filters operating in the rapid mode require backwashing and often downtime for repairs and maintenance is significant. Extra filters are usually required to maintain system capability and system filtration capacity. The minimum number of filters in a system is two. The surface area of a unit is normally less than 150 m^2. The ratio of length to width is 1.25 to 1.35. The number of filters used for a specific plant treatment capacity is as follows (1):

PLANT CAPACITY (L/SEC)	NO. OF FILTERS
50	3
250	4
500	6
1000	8
1500	10

In general, coagulation and sedimentation (settling) may require pretreatment for rapid sand filtration since filter influent suspended solids concentrations must be maintained at no more than about 30 to 59 mg/L. Therefore, rapid sand filtration plants include chemical pretreatment, followed by rapid sand filtration, and disinfection (see Figure 2.1). The gravity rapid sand filter is commonly used to remove nonsettleable floc and other impurities remaining after coagulation and settling. The removal action is a combination of mechanical straining, flocculation and sedimentation, all of which participate within the filter to remove particles (1).

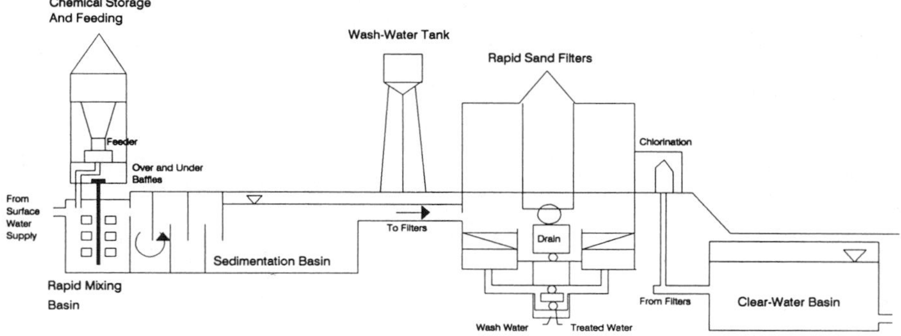

Figure 2.1. A rapid sand filtration plant.

A rapid sand filter consists of an open basin containing a layer of sand 60 to 80 cm thick, supported on layers of gravel. The gravel in turn, is supported by an underdrain system (Figure 2.2). In contrast to a slow sand filter, the sand is graded in a rapid rate filter media configuration. The sand is regraded each time the filter is backwashed—the lowest specific gravity material (usually the finest media) at the top of the bed. The underdrain system, in addition to supporting the overlying media, serves to uniformly distribute the washwater to the bed. The underdrain system may be of various types; perforated pipes, pipe and strainer, vitrified tile block with orifices, porous plates, etc. There are as many types as there are manufacturers. A clear well is

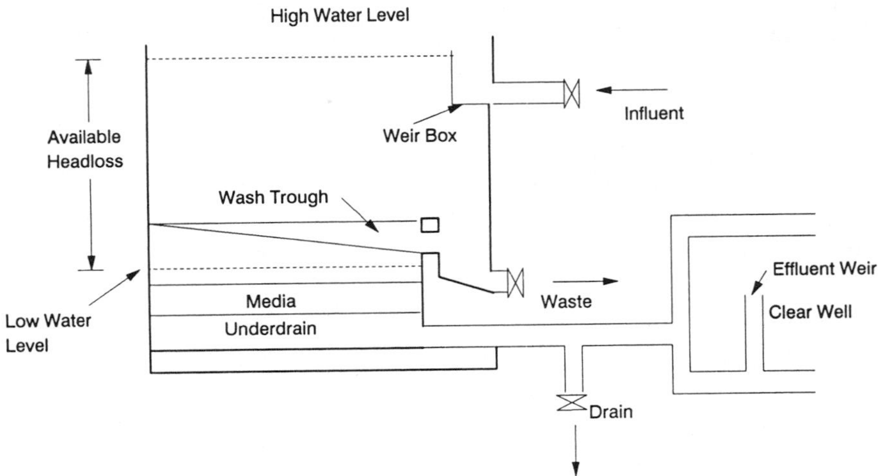

Figure 2.2. Typical rapid filter and clearwell.

Table 2.1. Filter Piping Design Flows and Velocities

DESCRIPTION	VELOCITY FT/SEC (M/S)	MAXIMUM FLOW, GPM/SQ FT (M/HR) OF FILTER AREA
Influent	1–4 (0.305–1.22)	8–12 (19.5–29.3)
Effluent	3–6 (0.92–1.83)	8–12 (19.5–29.3)
Washwater supply	5–10 (1.52–3.05)	15–25 (36.6–6–61)
Backwash waste	3–8 (0.92–2.44)	15–25 (36.6–61)
Filter to waste	6–12 (1.83–3.66)	4–8 (9.8–19.5)

Source: Reference 48.

typically located beneath the filters (or in a separate structure), to provide consistent output quantity.

Filter design flows and velocities are given in Table 2.1.

2.2 ADVANTAGES AND DISADVANTAGES

Rapid sand filtration plants are simple in function but complicated to operate because of the valving and controls to allow backwashing and restarting. Operator training is required for consistent output quality and quantity.

Filters require frequent backwashing to maintain satisfactory operating heads in the system (filter runs may vary from only a few hours, to as many as 24 to 72 hours, depending on the suspended solids in the influent. During filtration, the depth of water above the bed is 1.0 to 1.5 m. The total head available for filtration is represented by the difference in water levels between the water surface above the filter and the level in the treated water clear well, typically 3 to 4 m. Backwashing rates are usually 0.6 $m^3/min/m^2$ and higher, for a period of several minutes, to as much as 15 to 20 minutes. In addition, the initial production following backwashing may be wasted for several minutes if quality is unsatisfactory. Thus, the water usage for backwashing can be significant—ranging from a few percent to as much as 10 or 15 percent of the total plant output.

Rapid sand filter plants (including chemical treatment) can effectively treat higher solids loadings and produce higher outputs than slow sand filters. The land area requirements are significantly lower.

2.3 COSTS (See Table 2.2.)

2.4 AVAILABILITY

Conventional rapid sand filtration plants are widely available and widely used all over the world. Data and information related to design, construction, and operation can be found at many operating utilities.

Table 2.2. Per Capita Cost and Operation and Maintenance Manpower Requirements for Rapid Sand Filtration

POPULATION	TYPE OF COST	$ PER CAPITA* COST	RANGE	NO. UNSKILLED WORKERS REQUIRED
500–2499	Construction	12.84 –	15.12	
	Operation and Maintenance	2.43 –	5.40	1
2500–14,999	Construction	10.08	11.88	
	Operation and Maintenance	1.22	2.70	1
15,000–49,999	Construction	5.73	6.75	
	Operation and Maintenance	5.72	2.36	8
50,000–100,000	Construction	3.04	3.58	
	Operation and Maintenance	0.91	2.03	10

* at 120 gallons per capita per day (gpcd).
Source: Reference 1.

2.5 OPERATION, MAINTENANCE, AND CONTROL

There are a number of problems which can interfere with the consistent operation of rapid rate filters. These problems often result from poor design. Often design problems can be overcome by consistent operation (48, 56).

Surface clogging and cracking—caused by overloading of solids at the relatively thin surface filter layer. Often rapid increases in head loss are evident also. The problem is alleviated by installing dual or multiple media, which allow deeper penetration of solids into the bed, and generally longer run times. The same problem may result from use of filter aids, e.g., polyelectrolytes. Lower dosage may help in the latter case providing effluent quality is maintained.

Short runs due to floc breakthrough—this problem is avoided by use of mixed media. Mixed media or "coarse-to-fine particle" filters are more suitable where breakthroughs occur on a regular basis. There is typically a much greater media surface area in mixed media systems than in sand or even dual media. Also, there is a greater number of fine particles, and thus smaller pore openings at the bottom of mixed media beds, than with dual or sand filters. Floc storage depths in filter beds are greater (thus allowing more of the available bed depth to be used) in mixed media beds, perhaps as much as twice that of dual media.

Wide variations in effluent quality with changes in flow rate or input quality—a dual or mixed media system will improve the operational consistency. Modification of the backwash frequency and duration may also help.

Gravel displacement or mounding—mounding within the media bed can be alleviated by placing a 3 in. (76 mm) layer of coarse garnet (any of several silicate materials which are generally crystallized; usually red or brown in color) between the gravel supporting the media and the fine bed material. Reducing the total flow and head available for backwashing will also help. Sand leakage problems can be alleviated by the addition of a garnet layer.

Mudball formation—increasing the backwash flow rate (e.g., up to 20 gpm/ft^2, 48.8 m/hr), and providing for auxiliary water or air scour or surface wash capability will help. Very fine size sand particles are found to a higher degree in mudballs.

Growth of filter grains, bed shrinkage, and media pulling away from sidewalls—these are related problems which can be alleviated again, by providing adequate backwashing capability. Calcium carbonate adherence to filter grains may be controlled by adding chemical filtration aids to the filter influent.

Loss of media—coal grains are often lost from dual media filters, and loss may be difficult to control. Increasing the distance between the top of the expanded bed during backwashing and the wash water troughs may help. If used, auxiliary scour should be cut off a few minutes before the end of the backwash cycle.

Negative head and air binding—the more depth between the top of the expanded bed and the wash water troughs, the better, e.g., 5 ft (1.5 m). Filtration cycles should not be operated to terminal headlosses which are greater than the depth of water over the filter media. When the input water contains high concentrations of dissolved oxygen, and the pressure is reduced by siphon action, the potential for air binding increases (there is a discussion of siphon based filtration in the case study section). Accumulation of bubbles in the bed increases significantly the resistance to flow. Maintaining high water depths in the filters and frequent backwashing may help.

2.6 SPECIAL FACTORS

There are a number of considerations to be taken into account for good filter design. An approach is presented in Table 2.3.

2.7 RECOMMENDATIONS

Rapid sand filters are more complex than slow sand filters to operate and maintain, but they are widely used in developing countries in areas of high turbidity and where land requirements are an important consideration.

Table 2.3. Filter Evaluation Checklist

1. Filter media sizing and selection should be based on pilot tests. If this is not possible, data should be obtained from similar applications to determine the suitability of the media design.
2. In dual- and mixed-media filter systems, provisions should be made for the addition of polyelectrolytes directly to the filter influent.
3. The turbidity of each filter unit should be monitored continuously and recorded.
4. The flow and headloss through each filter should be monitored continuously and recorded.
5. Provisions should be made for the optional addition of disinfectant directly to the filter influent.
6. Provisions should be made for complete filter backwash cycle. The filter controls and pipe galleries should be housed.
7. The backwash rate selected must be based upon the specific filter media used and the wastewater temperature variations expected.
8. Filter backwash water supply storage should have a volume at least adequate to complete two filter backwashes.
9. Adequate surface wash or air scour facilities must be provided.
10. There should be adequate backwash and surface wash pump capacity available with the largest pump(s) out of service.
11. Backwash supply lines must be equipped with air release valves.
12. A means should be provided to indicate the backwash flow rate continuously and to enable positive manual control of the filter backwash rate. A means should also be provided to limit the filter backwash rate to a preset maximum value.

Source: Reference 48.

3. CHEMICAL TREATMENT FOR USE WITH FILTRATION

3.1 DESCRIPTION

Chemical treatment (coagulation and settling) is usually required as pretreatment for filtration, especially rapid rate filtration. Rapid rate filtration operates on higher turbidity influent water than slow rate filters. Actually, the rapid rate systems can treat higher input suspended solids concentrations at higher rates than slow rate systems, largely due to the use of chemical pretreatment ahead of the filters. The gravity filtration step may be viewed as "polishing" to remove impurities remaining after coagulation and settling. Chemical treatment is often used for treatment of domestic and industrial wastes. The design approach is similar for all applications.

Chemical treatment takes place in three stages: 1) rapid or flash mixing, 2) coagulation (usually taking place partly in the mixing stage and partly in the flocculation stage), 3) followed by flocculation or slow mixing. During rapid mixing a coagulant is rapidly and uniformly dispersed through the mass of water. In the subsequent flocculation process, a readily settleable floc is built up (floc growth). Since wastewater is often treated biologically before application of chemicals, flocculation is generally not required. Chemical addition and mixing followed by settling is usually sufficient for nutrient removal or pretreatment for filtration.

The flocculation stage involves slow and gentle stirring with sufficient time allowed to build up the floc. Detention times range from 20 to 60 min, and velocity gradients range from 5 to 100 L/sec with optimum values between 30 and 60 L/sec. A velocity gradient that is too high will shear floc particles, and a gradient that is too low will fail to provide sufficient agitation to allow floc formation. Baffled flocculation basins of horizontal (around-the-end) or vertical (over-and-under) types are the most suitable for small plants in rural areas. Little short-circuiting occurs in baffled flocculators and no mechanical flocculating equipment, e.g., rotating paddles are required. Basin depth is about 3 to 5 m. Spacing between the baffles is about 60 cm to facilitate cleaning.

Design for sedimentation which follows flocculation depends on the settling characteristics of the floc formed in the coagulation process. A general range of detention time is 2 to 4 hr. The overflow rates (surface loading) used for floc settling vary from 20 to 40 $m^3/m^2/day$ (m/day), and the horizontal velocity is commonly below 30 cm/min to minimize the disturbances caused by such things as density currents and eddy currents in the basin. The depth of the basin is about 2 to 5 m—3 m being preferred. The ratio of length to width is commonly between 3 : 1 and 5 : 1. Control of the outflow is generally by a weir(s) attached to one or both sides of a single or multiple outlet trough.

3.2 ADVANTAGES AND DISADVANTAGES

Raw waters contain colloidal particles (characterized by electrical charges which inhibit the agglomeration and subsequent removability by settling). Accordingly, colloids are relatively stable and non-settleable. These colloid systems may be destabilized (neutralization of the electrical forces) by adding chemical coagulants and supplying energy through mixing. Chemicals typically used include aluminum and iron salts and polyelectrolytes. Coagulation is the charge neutralization stage while flocculation is the floc-building or agglomeration stage of the chemical treatment unit process. Common coagulants are listed in Table 3.1 (55). Most polyelectrolytes are classified as nonionic, cationic, or anionic depending on the primary operative neutralization mechanism for the given molecule. The cationic types typically have molecular weights below 100,000 and are available as aqueous solutions. The others have weights above 1,000,000, and are generally available as powders and/or emulsions (55).

3.3 COSTS

The design criteria used for the development of cost estimates are given on Table 3.2. The costs are given on Figure 3.1.

Table 3.1. Properties of Common Coagulants

COMMON NAME	FORMULA	EQUIV. WEIGHT[1]	pH AT 1% SOLUTION	AVAILABILITY
Alum	$Al_2(SO_4)_3 \cdot 14H_2O$	100	3–4	Lump 17% Al_2O_3 Liquid 8.5% Al_2O_3
Lime	$Ca(OH)_2$	40	12	Lump asCaO Powder 93–95% Slurry 15–20%
Ferric chloride	$FeCl_3 \cdot 6H_2O$	91	3–4	Lump 20% Fe Liquid 20% Fe
Ferric sulfate	$Fe_2(SO_4)_3 \cdot 3H_2O$	51.5	3–4	Granular 18.5% Fe
Copperas (Ferrous Sulfate)	$FeSO_4 \cdot 7H_2O$	139	3–4	Granular 20% Fe
Sodium aluminate	$Na_2Al_2O_4$	100	11–12	Flake 46% Al_2O_3

[1] Relative weights based on Alum.
Source: Reference 55.

Table 3.2. Conventional Alum Coagulation Treatment

PROCESS	DESIGN CRITERIA
Raw water pumping	100 ft (30.5 m) TDH
Alum feed-liquid	50 mg/L design; 30 mg/L operating
Polymer feed	1 mg/L design; 0.2 mg/L operating
Rapid mix	60 sec detention; G = 900
Flocculation	30 min detention; G = 80
Clarifers-rectangular	900 gpd/sq ft (1.525 m/h)
Gravity filtration-mixed media	5 gpm/sq ft (12.2 m/h)
Chlorine feed-gas chlorine	5 mg/L design; 2 mg/L operating
Product water pumping	300 ft (91.44 m) TDH

Source: Reference 48.

3.4 AVAILABILITY

Most of the chemicals required are widely available with the exception of some polyelectrolytes. Availability should be examined in detail for planned chemicals before a plant is designed.

3.5 OPERATION AND MAINTENANCE

Tank cleaning can be carried out either mechanically during treatment (i.e., a sludge removal device) or manually. Where manual labor is readily available, manual clean out should be considered. In manually cleaned basins, the time lapse between cleanings varies from a few weeks to a year or more. The accumulated sludge may be sluiced by a fire hose after the tank has been taken out of service and dewatered. Sludge removal/collection mechanisms must be regularly serviced and cleaned.

3.6 CONTROL

For optimum alum floc formation the pH should be within the range of 5.0 to 7.0. Sufficient alkalinity must be present for reaction with the coagulant. If sufficient alkalinity is not present in the water, lime is generally added. While polyelectrolytes are not equally effective in all waters, when used in conjunction with the common metal coagulants they yield large dense floc which settles rapidly.

3.7 SPECIAL FACTORS

As untreated water flows through the various units, color, turbidity, tastes, odors, and bacteria are removed from the influent water. Additional pretreatment in cases where fish, stones, or other large debris are possible, may

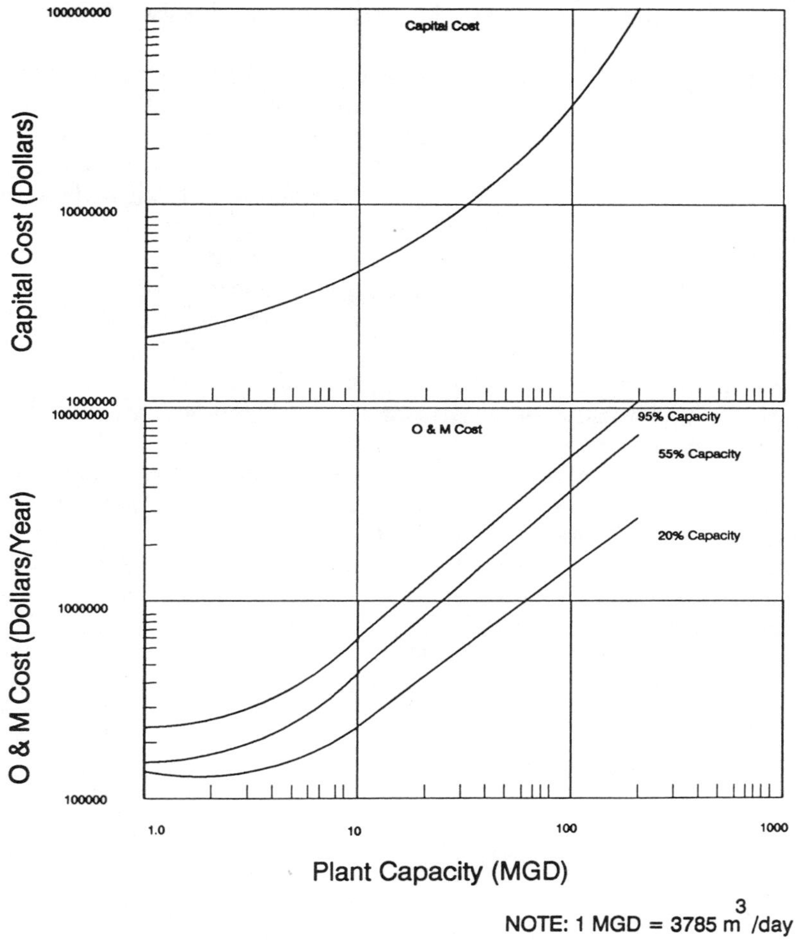

Figure 3.1. Conventional alum coagulation treatment costs (*source:* reference 48).

include bar racks and course screens. Aeration has been shown to be economical and beneficial for treatment of tastes and odors; plain sedimentation pretreatment if the water is highly turbid; and softening if the water is high in hardness. Often the need for such auxiliary treatment becomes apparent only after plants are designed and built and thus must be added to existing treatment plant capability.

Iron salts are effective over a wider pH range than alum and are generally more effective in removing color from water but are usually more costly. Coagulants and dosages should be chosen on the basis of jar tests. Careful testing will generally eliminate the need for later process additions and complicated operation.

When floc particles have grown in size during flocculation, they become weaker and more subject to being torn apart. Therefore, baffled flocculation is frequently used. The around-the-end type of basin is commonly applied to plants with capacities below 76,000 m^3/day, and the over-and-under type, with the advantage of more continuous turbulence, is applied where sufficient head is available and land is limited.

If a circular settling tank is used, the diameter may be as large as 70 m but is generally held to 30 m or less to reduce wind effects.

Settled coagulation sludge and backwash water have been disposed of without treatment, but these residues may promote buildup of deposits of sludge banks in the backwaters of slowly moving portions of streams and/or interfere with downstream use. Discharge of residues is not only aesthetically objectionable but may also contain polluting concentrations of the chemicals used in treatment and removed solids. Possible disposal/management procedures include: direct discharge to a receiving stream with adequate flow, lagoons or sludge drying beds, hauling away for surface land applications, discharge into a municipal sewerage system with treatment, dewatering of sludge, and reclamation of alum or other useful constituents. Sludge production depends on coagulant dose, quantity of solids removed, and character of the water being treated (pH, salts, etc.). Sludge production quantities and characteristics which may be used for design of disposal options are determined by testing.

4. DUAL MEDIA FILTRATION

4.1 DESCRIPTION

Gravity dual media (coal-sand) filtration is an effective form of granular media rapid flow filtration. Granular media filtration involves the passage of water through a bed of heterogeneous filter media with resulting removal of solids largely by straining, as with other filtration processes.

Dual media filtration involves the use of both sand and coal (anthracite) in two layers as filter media, with the anthracite being placed on top of the sand (see Figure 4.1) (2). Garnet is often placed between the sand layer and the underdrain to help support the media. Typical media sizes are shown on Table 4.1. Dual media filtration will remove residual biological floc in settled effluents from secondary wastewater treatment, and residual chemical-biological floc after alum, iron, or lime precipitation in water treatment plants. It is also used for tertiary or independent physical-chemical wastewater treatment in the U.S. and other countries.

Gravity or pumped-flow filters may include a flow splitter box, after which the wastewater flows by gravity to the filter cells. Pressure filters utilize pumping to increase the available head. The filter unit consists of a containing vessel, the filter media, structures to support the media, distribution and collection devices for influent, effluent and backwash water flows, supplemental cleaning devices, and necessary controls for maintaining flows, water levels, and backwash sequencing.

Design criteria are as follows: filtration rate is usually 2 to 8 gpm/ft^2 (5 to 20 m^3/hr/m^2); bed depth, 24 to 48 in. (61 to 122 cm); depth ratios of 1 : 1 to 4 : 1 sand to anthracite; backwash rates, 15 to 25 gpm/ft^2 (37 to 62 m^3/hr/m^2); filter run length, 8 to 48 hr; and terminal head loss 6 to 15 ft.

4.2 LIMITATIONS

Consistent effluent quality is dependent on consistent pretreatment quality and eliminating wide flow variations. Increasing suspended solids loading will reduce run lengths, and large flow variations will adversely effect effluent quality. Chemical pretreatment is often required.

Rapid rate filter systems can be easily converted to mixed media, using the same equipment and filter galleries. Since mixed media will successfully remove and store solids from high turbidity waters, it is often not necessary to add settling basin capacity when plants are being expanded. There is no

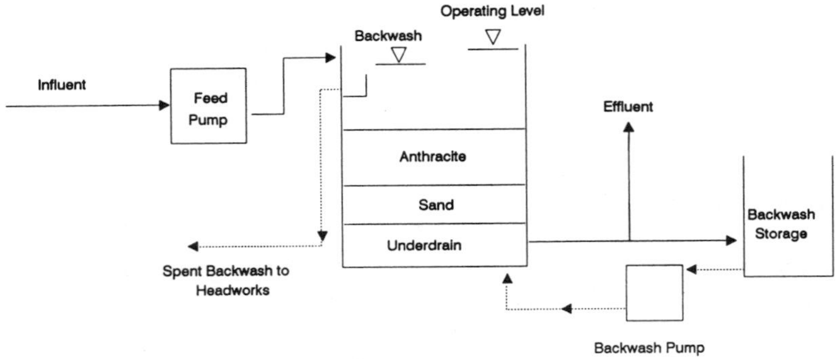

Figure 4.1. Flow diagram of dual media filtration.

fixed distribution of grain sizes for mixed media operation. Table 4.2 presents sizes and applications for mixed media.

4.3 COSTS

Construction cost includes facilities for backwash storage, all feed and backwash pumps, piping, and structure (see Figure 4.2) (2, 11). Operation and maintenance costs are also shown in Figure 4.2.

Table 4.1. Typical Coal and Sand Distribution by Sieve Size in Dual Media Beds

COAL DISTRIBUTION BY SIEVE SIZE	
U.S. SIEVE NO.	PERCENT PASSING SIEVE
4	99–100
6	95–100
14	60–100
16	30–100
18	0–50
20	0–5
SAND DISTRIBUTION BY SIEVE SIZE	
U.S. SIEVE NO.	PERCENT PASSING SIEVE
20	96–100
30	70–90
40	0–10
50	0–5

Source: Reference 48.

Table 4.2. Illustrations of Various Media Designs for Removal of Different Types of Floc

	GARNET		SILICA SAND		COAL	
TYPE OF APPLICATION	U.S. SIEVE SIZE	DEPTH, INCHES (mm)	U.S. SIEVE SIZE	DEPTH, INCHES (mm)	U.S. SIEVE SIZE	DEPTH, INCHES (mm)
Very heavy loading of fragile floc	−40 + 80	8(203)	−20 + 40	12(305)	−10 + 20	22(559)
Moderate loading of very strong floc	−20 + 40	3(76)	−10 + 20	12(305)	−10 + 16	15(381)
Moderate loading of fragile floc	−20 + 80	3(76)	−20 + 40	9(229)	−10 + 20	8(205)

Source: Reference 48.

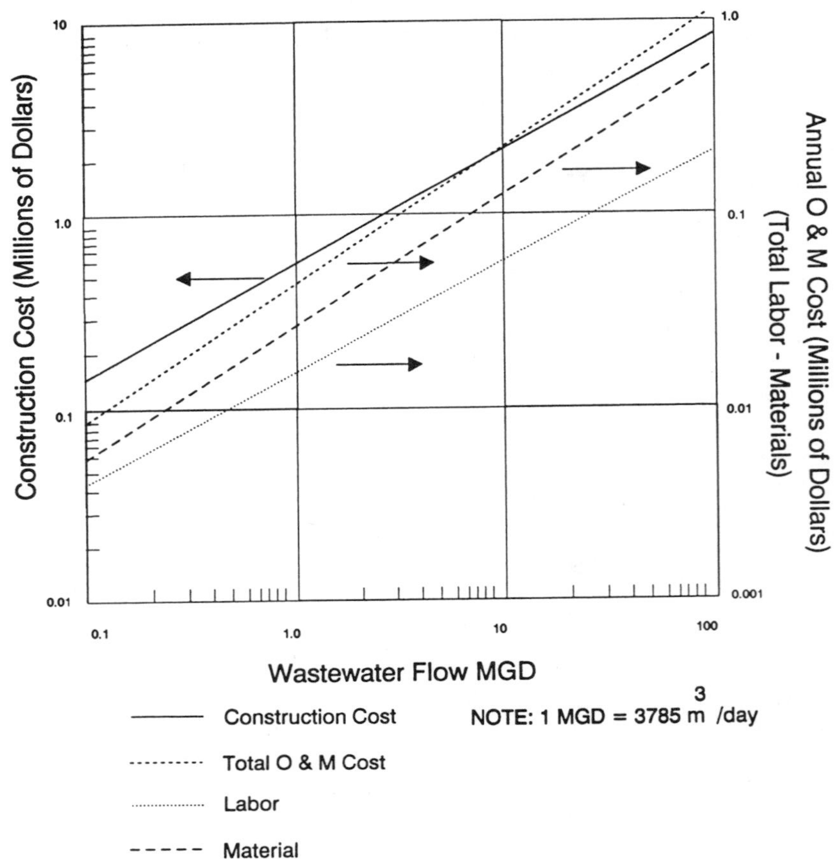

Figure 4.2. Construction, operation, and maintenance costs for dual media filtration.

4.4 AVAILABILITY

Application depends on a consistent supply of anthracite at reasonable costs. Trained operators are required to maintain system efficiency and high effluent quality. Backwashing uses a significant fraction of the treated effluent and close control of filter run lengths is required for cost effective operation.

4.5 OPERATION AND MAINTENANCE

In dual-media and mixed-media beds, floc is stored deeper within the media bed depth than rapid sand filters—to within a few inches of the bottom of the fine media. Thus, effective backwashing is necessary to maintain long term bed effectiveness. Auxiliary air or water spray surface wash devices should be included in designs of dual media filters. These should be used immediately before and/or during the backwash cycle.

Typical backwash flow rates are 15 to 20 gpm/sq ft (36.6 to 48.8 m/h). A 20 to 50% expansion of the filter bed is usually adequate to suspend the bottom grains. Backwash water requirements vary somewhat with temperature. The time required for a complete backwash cycle varies from 3 to 15 min.

Following the washing process, water should be filtered to waste for a short time until the effluent turbidity drops to an acceptable value. The filter-to-waste portion of the cycle may require from 2 to 20 min, depending on pretreatment and type of filter. This practice was discontinued for many years, but recording turbidimeters have shown that wasting is useful to preserve consistent production of a high-quality water. Operating the washed filter at a slow rate at the start of a filter run may accomplish the same purpose. A recording turbidimeter for continuous monitoring of the effluent from each individual filter unit is of value in controlling filtration during the filter run and for detecting solids breakthrough at the end of a run.

Backwash water flow rates should be gradually increased as backwashing begins to prevent "boiling" from uneven media lifting. The supporting gravel can be overturned and mixed with the fine media, which requires removal and replacement of the bed. Time is required for the media to equilibrate at expanded spacing in the upward flow of washwater. The time from start to full backwash flow should be at least 30 seconds and perhaps longer. Devices designed into the control system are desirable. Control is frequently accomplished by means of an automatically regulated master wash valve, controlled hydraulically or electrically and designed so that it cannot open too fast. Alternatively, a speed controller could be installed on the operator of each washwater valve.

Air binding of the media can significantly reduce the process throughput and must be avoided. Air can be unintentionally introduced to the bottom of the filter in a number of ways. If a vertical pump is used for the backwash supply, air may collect in the vertical pump column between backwashings. The air can be eliminated by starting the pump against a closed discharge valve and bleeding the air out from behind the valve through a pressure air release valve.

Washwater may be supplied by gravity flow from a storage tank located above the top of the filter boxes. Washwater supply tanks usually have a minimum capacity equal to the volume required for two complete backwash cycles. The bottom of the tank must be high enough above the filter wash troughs to supply water at the rate required for backwashing as determined by a hydraulic analysis of the washwater system. The height is usually at least 10 ft (3.05 m) but more often is 25 ft (7.6 m) or greater. Washwater tanks should be equipped with an overflow line, and a vent for release and admission of air above the high-water level.

4.6 SPECIAL FACTORS

Filters usually include multiple compartments. This allows for the filtration system to continue to operate while one compartment is being backwashed. No less than two multicompartment filter units should be designed into even a small system. In systems with two filters for example, the flow rate would be doubled in the operating unit when the other is being backwashed.

Filtration systems can be constructed of concrete or steel, with single or multiple compartments. Systems can be manually or automatically operated. If automatic systems are included in a design, complete capability for manual operation should be included to allow full system operation—should controllers fail.

4.7 RECOMMENDATIONS

Dual or mixed media filtration should be considered for applications where high turbidities are common. Also, these systems are preferred when operating problems with rapid sand or other single or mixed media are consistent and can not be solved in other ways.

5. SLUDGE VACUUM FILTRATION

5.1 DESCRIPTION

Vacuum filters are used to dewater sludges so as to produce a cake having the physical handling characteristics and moisture content required for subsequent processing, incineration, or disposal. A typical configuration is shown on Figure 5.1. Solids capture ranges from 85 to 99.5%, and sludge product cake moisture is usually 60 to 90% depending on feed type, solids concentration, chemical conditioning, and operation. Dewatered cake is suitable for landfill, heat drying, incineration or land spreading.

The vacuum filter unit includes a rotary vacuum filter which consists of a cylindrical drum covered with a cloth or other filter medium, rotating partially submerged in a vat or pan of pre-conditioned sludge. The drum is divided radially into a number of different sections which are connected through internal piping to ports in a valve body (plate) at the hub. This plate rotates in contact with a fixed valve plate with similar ports, which are connected to a vacuum supply, a compressed air supply, and an atmospheric vent. As the drum rotates each section is sequentially connected to the appropriate service. Various operating zones within the drum are encountered during a complete revolution. In the pickup or cake forming section, vacuum is applied to draw liquid through the filter covering (media) and form a cake or partially dewatered sludge. As the drum rotates the cake emerges from the liquid sludge, while suction is still maintained to promote further dewatering. A lower level of vacuum often exists in the cake drying zone (the top of the drum). If the cake tends to adhere to the media, a scraper blade may be provided to assist removal.

The three principal types of rotary vacuum filters are the drum type, coil type, and belt type. They differ primarily in the type of covering used and the cake discharge mechanism employed. Cloth media are used on drum and belt types while stainless steel springs are used on the coil type. Infrequently, metal media are used on belt types. The drum filter also differs from the other two in that the cloth covering does not leave the drum but when necessary is washed in place. Effective design of the drum filter provides considerable latitude in the amount of cycle time devoted to cake formation (usually by varying rotational speed), washing and dewatering capability; while inactive time is minimized.

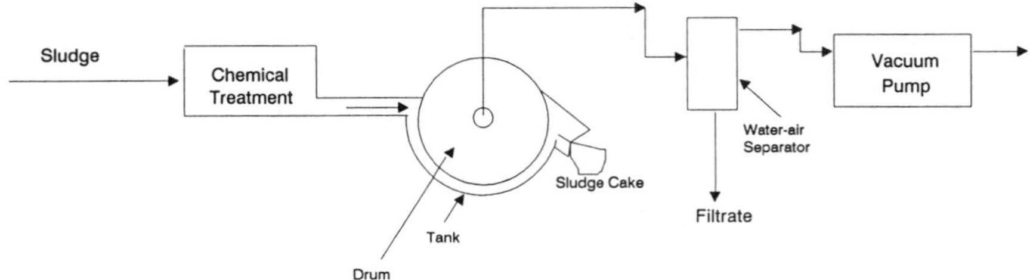

Figure 5.1. Generalized diagram of sludge vacuum filtration.

The coil type vacuum filter uses two layers of stainless steel coils arranged in corduroy fashion around the drum. After a dewatering cycle, the two layers of springs leave the drum and are separated from each other so that the cake is lifted off the lower layer of springs and discharged from the upper layer. The coils are then washed and reapplied to the drum.

Media on the belt type filter leave the drum surface at the end of the drying zone and pass over a small diameter discharge roll to facilitate cake discharge. Washing of the media then occurs before return to the drum and to the vat for another cycle.

Design criteria for vacuum filtration are as follows: typical loadings in pounds dry solids/hr/ft^2 are 7 to 15 (34 to 74 kg/m^2/hr) for raw primary sludges, and 3.5 to 5 (17 to 24.5 kg/m^2/hr) for mixed digested sludges. The solids loading is a function of feed solids concentration, subsequent processing requirements and chemical preconditioning.

5.2 LIMITATIONS

Vacuum filters are used in facilities when space for alternative sludge treatment (e.g., sludge drying beds) is limited or when high efficiency dewatering is required (e.g., for incineration) for maximum volume reduction. Chemical conditioning costs can sometimes be large if a sludge is hard to dewater.

5.3 COSTS

The typical construction costs of sludge vacuum filtration for lime and biological sludges are shown in Figure 5.2. (2, 11). Operation and maintenance of both options are shown in Figures 5.3 and 5.4 (2, 11).

5.4 AVAILABILITY

Vacuum filters possess many moving parts and require the continuous attention of trained operators. Available services for maintenance from factory representatives and availability of spare parts should be carefully

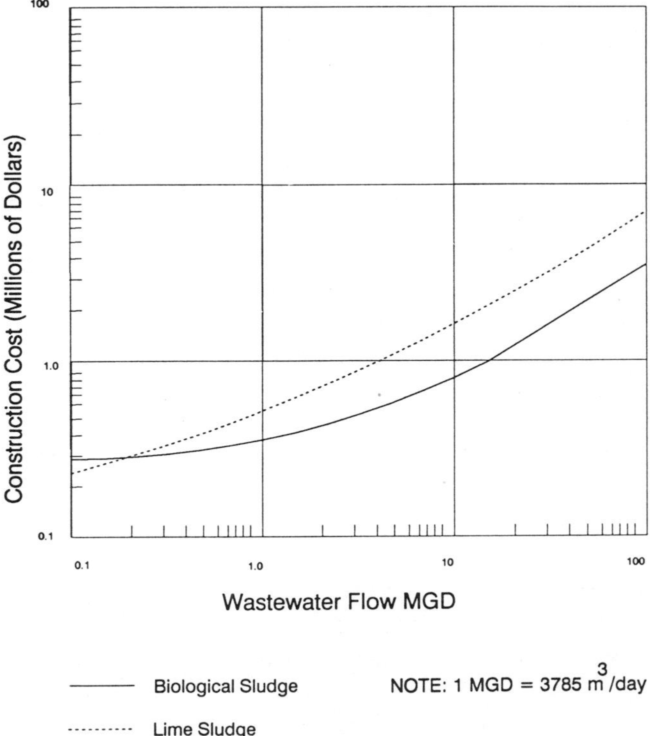

Figure 5.2. Construction cost of sludge vacuum filtration (lime and biological sludges).

checked in the area of interest before selecting this technology. Frequent replacement of filter media, cloth, wire, etc., is required, and these must be kept in adequate supply between shipments. If shipments are expected to be uncertain, it may be necessary to double the inventory of spare parts.

5.5 OPERATION AND MAINTENANCE

Operation is sensitive to type of sludge and chemical conditioning procedures. As raw sludge ages (3 or 4 hr) after thickening, vacuum filter performance decreases. Poor release of the filter cake from the belt is occasionally encountered.

When sludge is difficult to filter, chemical conditioning may be required. Information on the types of chemicals which may be used with specific media may be obtained from manufacturers. On the other hand, the best way to determine dosages of such chemicals is by testing, using the actual

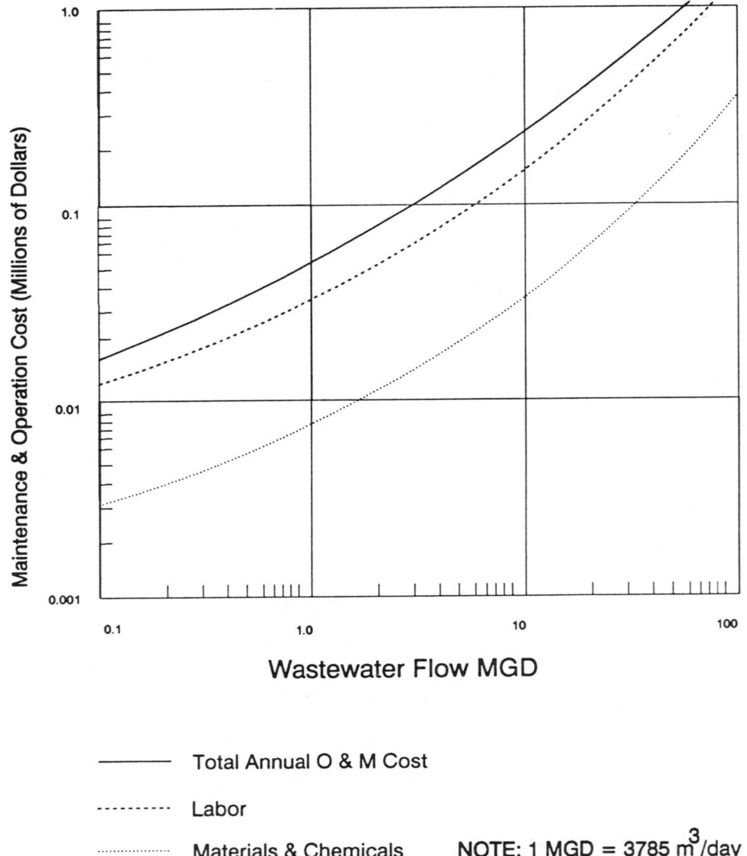

Figure 5.3. Annual operation and maintenance cost for sludge vacuum filtration (lime sludge).

sludge to be treated. Also, changes in conditioning practice may be required as sludge quality changes. Unsatisfactorily treated quantities produced during testing may be recirculated to the treatment plant.

Vacuum pumps, chain drives, media scraping mechanisms, and the media itself require frequent maintenance. A preventive maintenance program is required.

5.6 CONTROL

Large doses of lime, ferric chloride or other chemicals, and even polyelectrolytes in some cases required for good sludge yields from the filters. Frequent washing of drum filter media may also be required. Special measures, such as

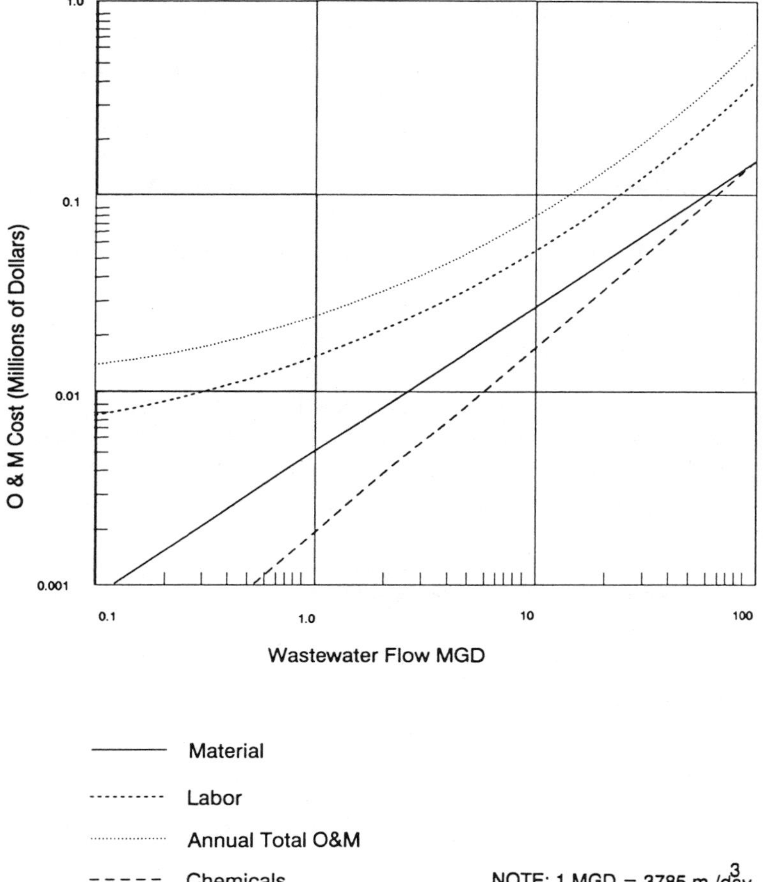

Figure 5.4. Annual operation and maintenance cost for sludge vacuum filtration (biological sludge).

addition or redesign of scraping mechanisms are frequently required to obtain "clean" cake release from belt filters. Filter media may be blinded under especially difficult operating conditions and require removal and cleaning. Operator training is required to maintain a high level of system reliability.

5.7 SPECIAL FACTORS

Many types of filter media are available for the felt and drum filters. Operating vacuums greater than 15 in. of Hg (381 mm Hg) are probably rarely justified, and vacuum increases should be based on yield and increased power costs. Increasing available filter area may be more cost effective. However, an increase in vacuum from 15 to 20 in. of Hg (381 to 510 mm of Hg), about 25%, is reported to provide about 10% greater yield in 3 full-scale installations.

Vacuum filters have limited applicability in most small systems because of the large energy requirement and high user skill requirements. Power requirements are in the range of 10,000 to 40,000 kWh/yr/MGD (230,000 to 910,000 kWh/yr/m^3/sec). On the other hand processed sludge is usually suitable for soil conditioning, especially if chemical additions are designed to enhance its usefulness.

5.8 RECOMMENDATIONS

This is the most common method of sludge dewatering in the United States. It should be applied to chemical and biological sludges in cases where recovery of chemicals for reuse or soil conditioning potential is desired. Its applicability to small systems is probably reserved to areas of high population where higher operating skills and technical expertise are available. Vacuum filtration should be used instead of drying beds or other less complex technology when high volumes of sludge are being produced.

6. SEDIMENTATION—CIRCULAR PRIMARY CLARIFIER

6.1 DESCRIPTION

Clarifiers described in this section are used for solids removal from wastewater, such as raw sewage or highly contaminated raw water supplies. The process provides for removal of settleable solids and floating material while reducing suspended solids concentrations to levels suitable for subsequent treatment. The process accepts high solids loadings and is generally employed as a preliminary step in the treatment sequence, or following treatment by softening or coagulation and flocculation. A circular clarifier cross section is shown in Figure 6.1.

A conical bottom (1 in./ft slope, 2.54 cm/meter; about 8%) is equipped with a rotating mechanical scraper (see Figure 6.1) that plows sludge to a central hopper. A feed system in the center of the clarifier distributes the influent radially near the top, and a peripheral weir overflow system carries the effluent. Floating scum is trapped inside a peripheral scum baffle and squeegeed into a scum discharge box. The unit contains a center motor-driven turntable supported by a bridge spanning the top of the tank, or it is supported by a vertical steel center pier. The turntable gear rotates a vertical cage or torque tube, which in turn rotates the truss arms. The truss arms carry multiple flights (plows) on the bottom chord which are set at a 30° angle and literally "plow" heavy fractions of sludges and grit along the bottom slope toward the center blowdown hopper. An inner diffusion chamber receives influent flow and distributes this flow inside of the large diameter feed well skirt. Approximately 3% of the clarifier surface area is used for the feed well. The depth of the feed wells are about one-half of the tank depth. The center sludge hopper should be less than 2 ft (0.61 m) deep and less than 4 ft (1.2 m) in cross section. Clariflocculators combine the two functions of coagulation in the center section near the hub and settling in the section more distant radially from the center.

Design criteria include: surface loading rates from 500 to 1200 gal/d/ft^2 (21 to 50 m^3/d/m^2) for untreated wastewater; 360 to 600 (15 to 25) for alum floc; 540 to 800 (22 to 33) for iron floc; 540 to 1200 (22 to 50) for lime floc. Detention times are usually between 1.5 to 3 hr. Weir loadings are from

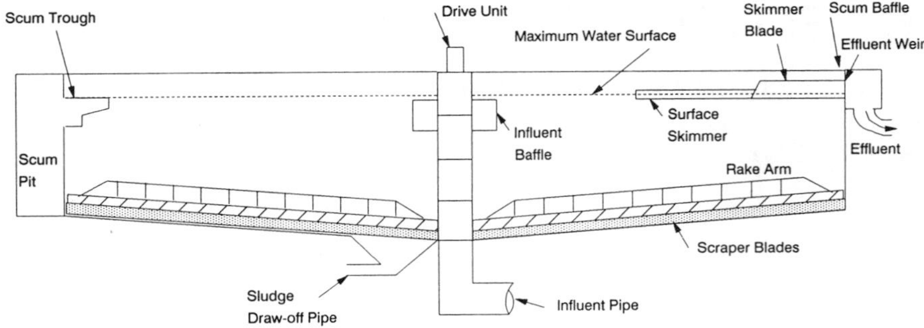

Figure 6.1. A primary circular clarifier.

10,000 to 30,000 gal/d/lineal ft of weir (120 to 360 m^3/d/m). Sludge collector tip speed is 10 to 15 ft/min (3 to 4.6 m/min). Heads of 2 to 3 ft (0.61 to 0.9 m) of water are required to overcome losses at the inlet and effluent controls and in connecting pipes. Forward or radial velocity should be less than 9 to 15 times the particle settling velocity to avoid scour. Scum handling equipment should be sized for 6 ft^3/Mgal (45 m^3/Mm^3). Sludge pumping rates range between 2500 and 20,000 gal/d/Mgal ($m^3/d/Mm^3$) depending on type of chemical addition and influent quality (2).

6.2 LIMITATIONS

The maximum diameter of a circular clarifier is 200 ft (61 m). Larger tanks are subject to unbalanced radial diffusion and wind action, both of which reduce efficiency. Horizontal velocities in the clarifier must be limited to prevent "scouring" of settled solids from the sludge bed and eventual escape of solids in the effluent. Circular clarifier installations have a larger land use requirement than rectangular clarifiers.

6.3 COSTS

Figures 6.2 and 6.3 show the construction, operation, and maintenance costs for circular clarifiers (2, 11).

6.4 AVAILABILITY

This technology is widely used in the United States and throughout the world. It is applied to treatment systems of any size in developing countries. There is a variety of mechanical parts required, but the same is true for rectangular clarifiers. Good, reliable performance rests on the availability of

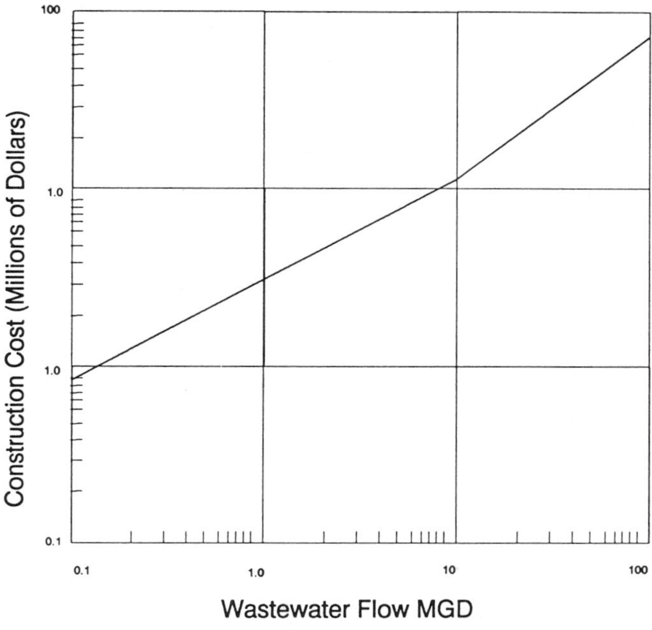

Figure 6.2. Construction cost for primary clarifier (circular with pump)

spare parts. The alternative is intermittent removal of sludge manually (most of the mechanism is for removal of accumulated solids). A spare parts inventory should be kept for all mechanical processes at any plant.

6.5 OPERATION AND MAINTENANCE

Primary clarification involves a period of quiescence (15 to 45 min) in a basin (depths of 10 to 15 ft; 3 to 4.6 m) where most of the settleable solids fall out of suspension by gravity; a chemical coagulant may be added to enhance settling. The solids are mechanically collected on the bottom and pumped as a sludge underflow.

The most important aspect of operation is regular, frequent removal of accumulated sludge solids. Typically, continuous removal mechanisms provide for better process performance than manual systems. Mechanical removal systems require regular maintenance for the chain drives, rakes, collectors, and pumps. Scum (floating solids) carried out of the sedimentation tanks in the discharge will significantly impact the effluent quality. Scum collecting mechanisms require very frequent attention (in cases of high

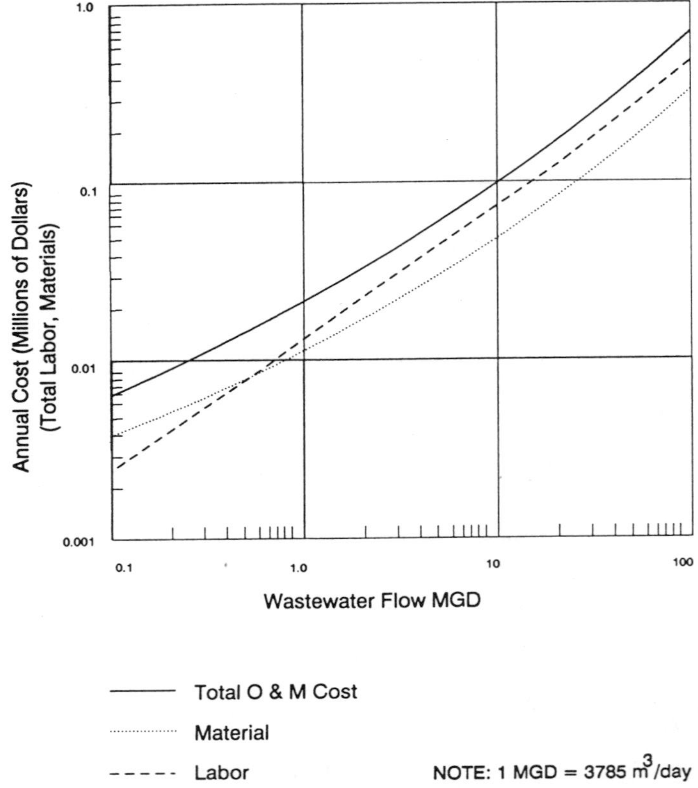

Figure 6.3. Operation and maintenance cost for primary clarifier (circular).

solids, perhaps hourly inspection), especially to remove accumulated deposits at overflows and other collecting points usually near mechanical parts. Accumulated floating solids may jam mechanisms and cause failure of parts and drives.

A consistent preventive maintenance program will ensure reliable operation over long periods of time.

6.6 RELIABILITY

Generally, reliability for circular primary clarifiers is high. However, clarification of solids into a packed central mass may cause collector arm stoppages. Attention to the design of the center area sludge hopper bottom slope, providing an adequate number of collector arms and proper alignment of the scraper blades is required to prevent such problems.

Sludge must be removed regularly. Often the failure of clarifiers to meet effluent expectations is due in turn to the failure of sludge removal equipment.

6.7 SPECIAL FACTORS

Two short auxiliary scraper arms are added perpendicular to the two long arms on large tanks. This makes practical use of deep spiral flights which aid in sludge plowing where ordinary shallow straight plows function poorly.

Peripheral influent feed systems are sometimes used instead of central feed. Also, central effluent weirs are used occasionally. Flocculating feed wells may also be provided if coagulants are added to assist sedimentation.

6.8 RECOMMENDATIONS

Coagulants such as alum, ferrous sulfate, and lime may be added to aid sedimentation. The dosage should be determined by jar tests, using the chemicals to be tested and the water to be treated.

Circular clarifiers require more land area but are potentially more reliable. Rectangular tanks may be constructed in less space, and may be designed in "common wall" fashion with other plant processes. They are especially effective for small plants which must be built in confined spaces, e.g., for industrial waste treatment. Equipment for circular tanks is more readily available since many manufacturers produce such hardware. Replacement parts and manufacturer service programs are also available. Package plants are often designed with the circular tank as the basis (together with chemical treatment).

7. RECTANGULAR PRIMARY CLARIFIER

7.1 DESCRIPTION

Rectangular clarifiers are used for the removal of settleable solids and floating material to reduce Total Suspended Solids (TSS) and Biochemical Oxygen Demand (BOD) and to treat raw water with high turbidity. Solids removal is generally applied as a preliminary step to further processing. Rectangular tanks lend themselves to "nesting" (common wall construction), often with preaeration tanks in water treatment plants and aeration tanks in activated sludge plants. A cross section of a rectangular clarifier is shown in Figure 7.1.

Efficiently designed and operated primary clarifiers should remove 50 to 65% of the TSS and 25 to 40% of the BOD while producing a sludge solids concentration of about 1 to 5%, generally at the lower end of this range. Skimmings volume rarely exceeds 1.0 ft^3/Mgal (7.5 m^3/Mm^3).

Settling involves a period of quiescence in a basin (depths of 10 to 15 ft; 3 to 4.6 m) where most of the settleable solids fall out of suspension by gravity; a chemical may also be added to the process influent to enhance settling and improve settling rates. The solids are mechanically collected on the bottom and pumped as a sludge underflow.

The maximum length of rectangular tanks is approximately 300 ft (about 90 m). Multiple bays with individual cleaning equipment may be employed, permitting tank widths of up to 80 ft (24 m) or more. Influent channels and effluent channels are typically located at opposite ends of the tank.

Sludge removal equipment usually consists of a pair of endless conveyor chains (see Figure 7.1). Attached to the chains at 10 foot intervals are cross pieces or flights, extending the full width of the tank or bay. Linear conveyor speeds of 2 to 4 ft/min (0.61 to 1.2 m/min) are common. The settled solids are scraped to sludge hoppers in small tanks and to transverse troughs in large tanks. The troughs in turn are equipped with cross-collectors, usually of the same type as the longitudinal collectors, which convey solids to one or more sludge hoppers. Screw conveyors have been used for the cross collectors.

Design criteria are as follows: loading rates equal to 600 to 1200 gal/d/ft^2 (25 to 49 $m^3/day/m^2$) for untreated wastewater; 360 to 600 (15 to 25) for alum floc; 540 to 800 (22 to 33) for iron floc; 540 to 1200 (22 to 49) for lime

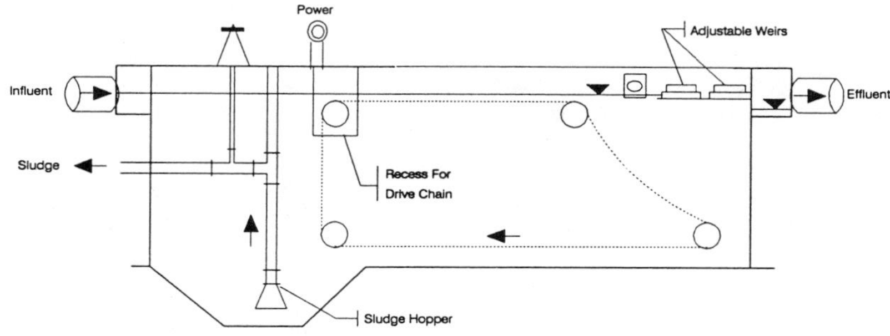

Figure 7.1. Typical rectangular primary clarifier.

floc. Detention times are usually between 1.5 to 3 hours. Weir loadings are 10,000 to 30,000 gal/d/lineal ft of weir (120 to 360 m^3/day/m). Individual bays of rectangular tanks should have a length to width ratio of at least 4. Forward velocities should be less than 9 to 15 times settling velocity to avoid scour of accumulated sludge solids in the bottom. Scum handling equipment should be sized for 6 ft^3/Mgal (45 m^3/Mm^3) of free decanted water. Sludge pumping rates range from between 2500 to 20,000 gal/d/Mgal ($m^3/day/m^3$) depending upon chemical addition and service. Other design information is

Table 7.1. Typical Design Information for Primary Sedimentation Tanks

ITEM	VALUE	
	RANGE	TYPICAL
Primary settling followed by secondary treatment:		
Detention time, hr	1.5–2.5	2.0
Overflow rate, $m^3/m^2 \cdot d$		
Average flow	32–48	
Peak flow	80–120	100
Weir loading, $m^3/m \cdot d$	125–500	250
Dimensions		
Primary settling with waste activated-sludge return:		
Detention time, hr	1.5–2.5	2.0
Overflow rate, $m^3/m^2/m \cdot d$		
Average flow	24–32	
Peak flow	48–70	60
Weir loading, $m^3/m \cdot d$	125–500	250
Dimensions		

Note: $m^3/m^2 \cdot d \times 24.5424 = gal/ft^2 \cdot d$
$m^3/m \cdot d \times 80.5196 = gal/ft \cdot d$
Source: Reference 4.

Table 7.2. Sedimentation Tank Detention Times

SURFACE LOADING RATE $M^3/M^2/DAY$	DETENTION TIME, HR			
	3.0 M DEPTH	3.5 M DEPTH	4.0 M DEPTH	5.0 M DEPTH
24	3.0	3.5	4.0	5.0
32	2.3	2.6	3.0	3.8
48	1.5	1.8	2.0	2.5
60	1.2	1.4	1.6	2.0
80	0.9	1.1	1.2	1.5
100	0.7	0.8	1.0	1.2
120	0.6	0.7	0.8	1.0

Note: $m^3/m^2/day \times 24.54 = gal/ft^2/day$
$m \times 3.281 = ft.$

given in Table 7.1. Detection times for clarification depend on surface loading rates as shown in Table 7.2.

7.2 LIMITATIONS

Rectangular clarification units will probably be cheaper to construct than circular ones, but the mechanical auxiliary equipment for circular units is more readily available and service is easier to obtain.

Horizontal velocities in the clarifier must be limited to prevent "scouring" of settled solids from the sludge bed and their eventual escape in the effluent.

7.3 COSTS

See Figures 7.2 and 7.3 for construction and operation and maintenance costs, respectively (2, 5, 11).

7.4 AVAILABILITY

This technology is widely used in the United States and throughout the world. It can be applied to treatment systems of virtually any size. Mechanical parts replacement on a fairly regular basis is necessary. It is necessary to keep a spare parts inventory to provide for repairs. Active preventive maintenance is necessary.

Rectangular tanks are preferred to circular designs where space is limited, especially in small industrial locations.

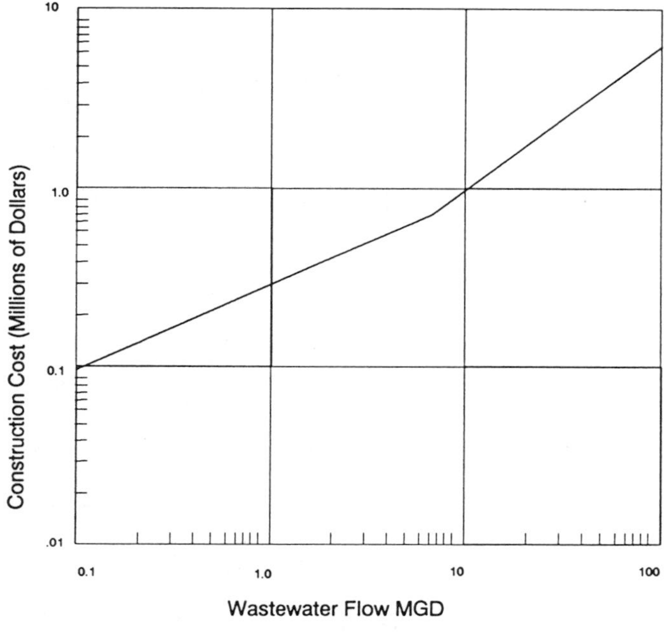

Figure 7.2. Construction cost of primary clarifier (rectangular).

7.5 OPERATION AND MAINTENANCE

Sedimentation systems, including rectangular clarifiers often fail in those cases where mechanical sludge collection systems are used. The sludge collection mechanisms require frequent cleaning and an active preventive maintenance program.

Scum is usually collected at the effluent end of rectangular tanks. The scum is moved by the flights to a point where it is trapped by baffles before removal, or it can also be moved along the surface by water sprays. The scum is then scraped manually from the tank end, up an inclined apron, or it can be removed hydraulically or mechanically. A number of means have been developed for removal including rotating slotted pipes, transverse rotating helical wipers, chain and flight collectors, and scum rakes. Tanks may also be cleaned by a bridge-type mechanism which travels up and down the tank on rails supported on the sidewalls. Scraper blades are suspended from the bridge and are lifted clear of the sludge on the return travel.

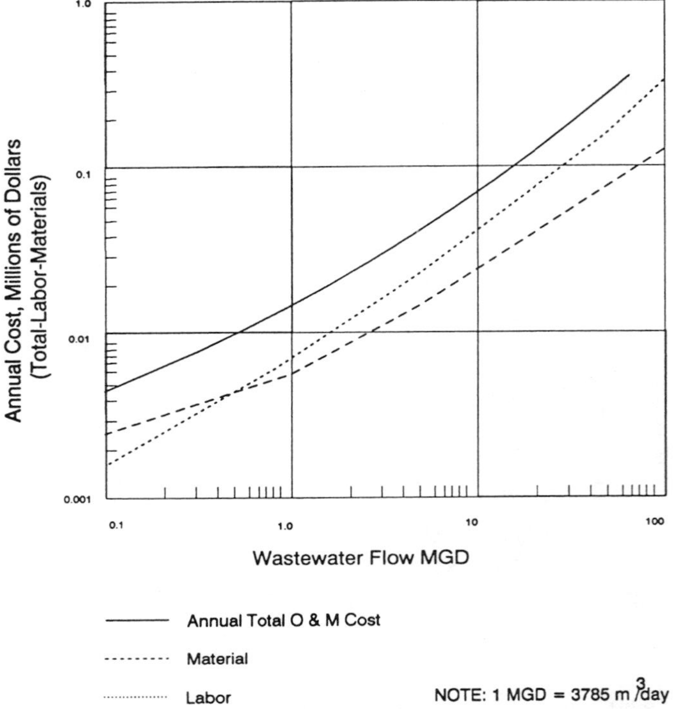

Figure 7.3. Operation and maintenance cost for primary clarifier (rectangular with pump).

7.6 CONTROL

Reliability for rectangular clarifiers is generally high. However, broken links in collector drive chains can cause system shutdowns. Plugging of sludge hoppers has also been a problem when cross collectors are not provided.

7.7 SPECIAL FACTORS

Coagulants such as alum, ferrous sulfate, and lime may be added to aid sedimentation. The dosage is determined by jar tests. Chemicals are almost always added in rapid mix systems. In the case of water treatment, flocculation is typically ahead of settling.

7.8 RECOMMENDATIONS

Multiple rectangular tanks require less area than multiple circular tanks and for this reason are used where ground area is at a premium. However, they require relatively large space for the level of treatment achieved. See recommendations for circular clarifiers (Chapter 6).

8. UPFLOW SOLIDS CONTACT CLARIFIER (FILTER)

8.1 DESCRIPTION

Upflow solids contact clarifiers combine mixing, coagulation, and flocculation, liquid-solids separation and sludge removal into a single unit process. These units may be classified as filters. They eliminate the need for separate flocculators and settling tanks. The process should be applied to raw water with low turbidity up to 50 JTU or has no more than 150 mg/L of suspended solids.

Many plants, especially in Brazil, use this type of process (see Figure 8.1) (9). These filters are designed for rates of filtration between 120 and 150 $m^3/day/m^2$.

8.2 LIMITATIONS

If applied hydraulic loading rates become too high, fine bed material will be lost with the treated effluent over the top of the upflow unit.

8.3 COSTS

See Figures 8.2 and 8.3 (5).

8.4 AVAILABILITY

Because of the simplicity and low cost of this technology, it should have widespread use throughout developing countries and for small applications where package plants are considered.

8.5 OPERATION AND MAINTENANCE

Upflow filters possess a major advantage in that the fine bed grains are at the top (with uniform density bed material), thus allowing the use of virtually the entire bed depth for bed solids storage. During filtration, the bed material is kept in place usually by a metal grid.

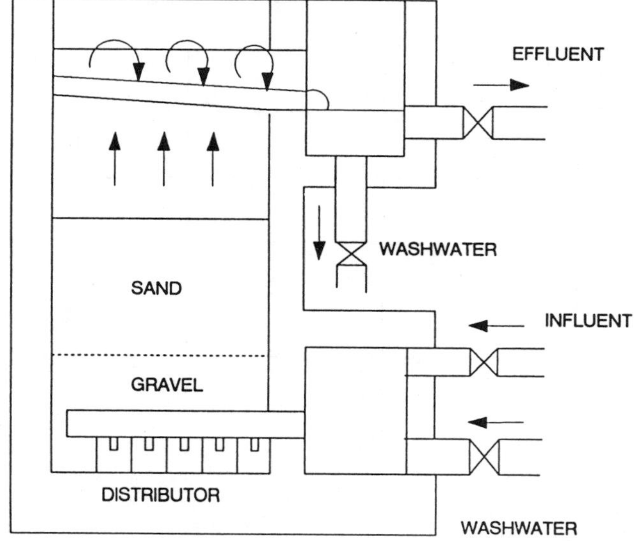

Figure 8.1. Generalized diagram of upflow solids contact clarifier.

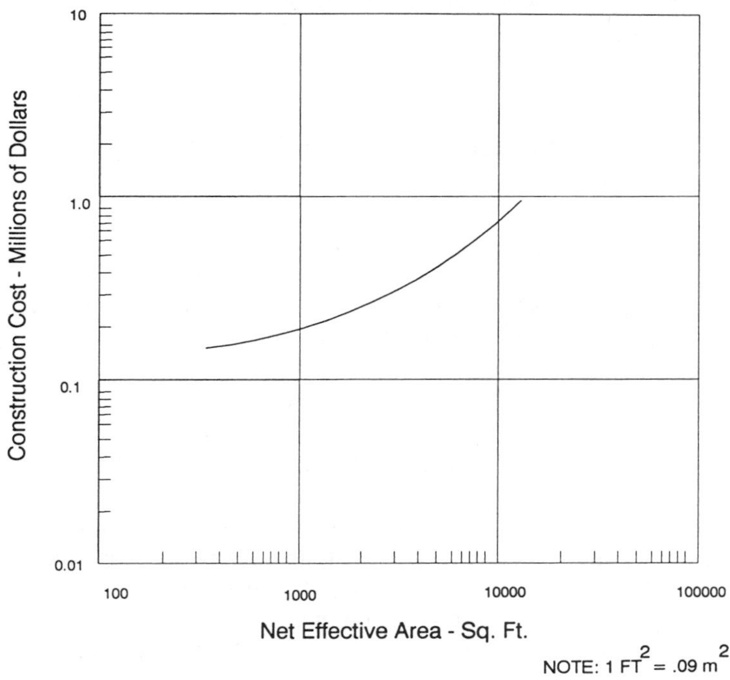

Figure 8.2. Construction cost for upflow solids contact clarifier.

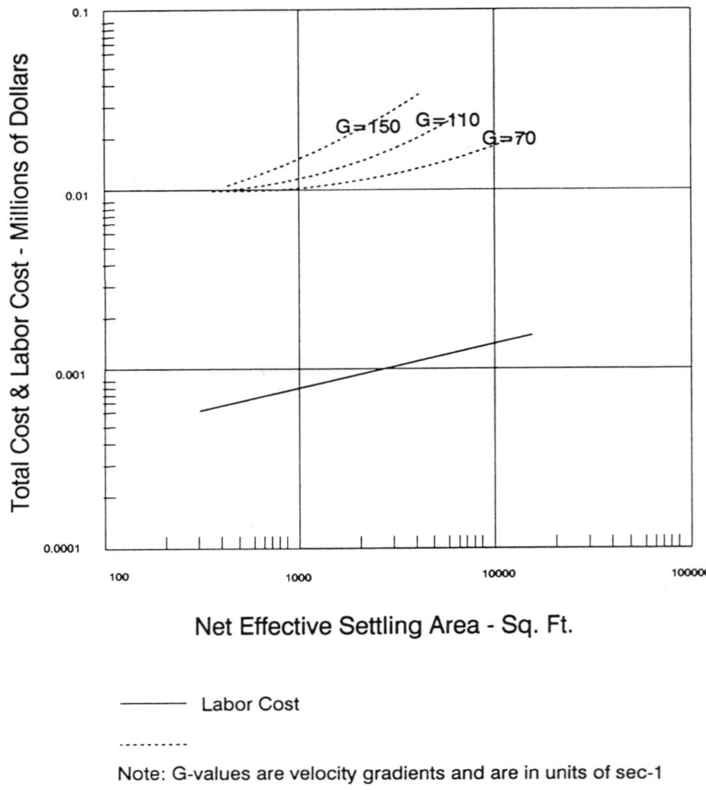

Figure 8.3. Costs for upflow solids contact clarifier.

The same water that is being filtered is used for backwashing, and this may be considered an advantage for this type of unit.

8.6 SPECIAL FACTORS

Coagulation and flocculation performed in a granular media (such as the layer of gravel under the sand bed) and in the presence of chemical compounds previously precipitated, improves filtration results. Thus, it may be economic in the use of chemicals. If more than one operation is expected in a single unit, the complication for the operator increases. More attention is required to monitor flow rates and system head loss for successful operation. Even with careful operator attention, the results may be mixed on certain waters.

8.7 RECOMMENDATIONS

Upflow solids and contact clarifiers are generally selected on the basis of a lower cost and the operational advantages of combining several processes into a single unit. There are advantages and disadvantages in operational complexity. Operator training is required for consistent process performance.

Upflow clarifiers with flocculation may be more appropriate for small systems since there are fewer moving parts. The sensitivity of successful operational performance especially if used with plate settlers beforehand, make it an attractive option for package plants.

9. FLOCCULATION—CHEMICAL TREATMENT

9.1 DESCRIPTION

The objective of flocculation is to provide for an increase in the number of contacts among coagulating particles suspended in water by gentle and prolonged agitation. Flocculation follows chemical addition. During agitation, particles collide, producing larger and more easily removed flocs.

There are many different types of flocculators currently in use. The four discussed here are gravel, baffled, and horizontal and vertical mechanical flocculators. Gravel and baffled types may be used for smaller plants and mechanical types for larger plants. There are no fixed size selection criteria.

In small and medium installations (up to 200 L/sec) the most common types of flocculation units in developing countries are the baffled flocculator shown in Figure 9.1 and the "Alabama" jet action flocculator (see Figure 9.2) (9,25). The Alabama flocculator was introduced into Brazil during the Second World War by North American engineers. For small plants, the gravel flocculator (a bed of gravel used to create tortuous flow paths in which flocculation is enhanced) is a good option.

The size and shape of a flocculation basin are generally determined by the type of flocculator selected and the type of sedimentation process employed. For example, if mechanical flocculators are paired with rectangular horizontal flow sedimentation basins, the width and depth of the flocculation basins should match the width and depth of the sedimentation basins. Common wall construction with the same dimensions lower construction costs.

In all cases, the size of the flocculation basin is determined by the required reaction or detention time, determined by testing. Although no theoretical relationship exists between basin area and water depth for optimal flocculation, the tank should be no deeper than 5 m. Basins with depths of greater than 5 m often display unstable flow patterns and poor flocculation.

Design Criteria (3,8,9,25)—for baffled units with vertical flow: detention time, 15 to 25 min (warm climate); flow velocity, 0.10 to 0.20 m/sec; velocity gradient *(G)*, 80 to 40 sec^{-1}. For Alabama flocculators: detention time, 15 to 25 min; G values 50 to 40 sec^{-1}; loss of head, 0.30 to 0.50 m; applied flow per chamber, 25 to 50 L/sec/m^2; velocity in curves or bends of baffle flocculators, 0.40 to 0.60 m/sec; useful depth 2.5 to 3.5 m (see Table 9.1 for additional design guidance for flocculation chambers). For the gravel flocculator: depth of stone bed, 3 m. For vertical shaft mechanical flocculators: mini-

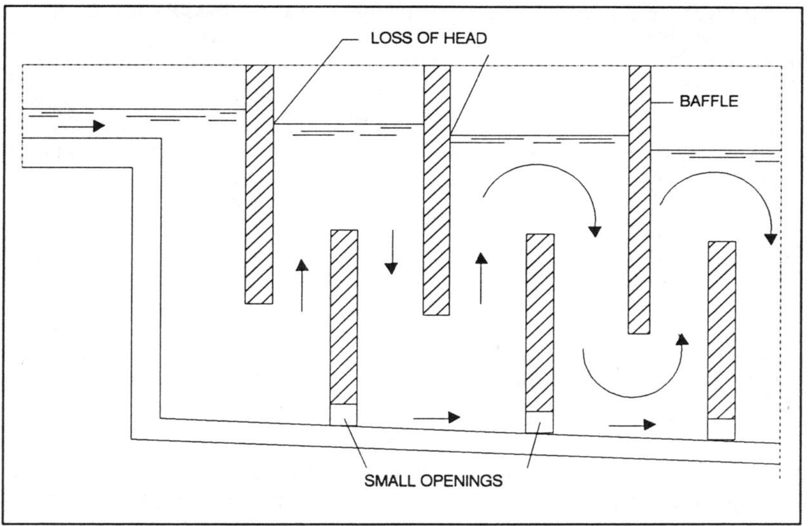

Figure 9.1. Typical baffled flocculation basin commonly used in developing countries.

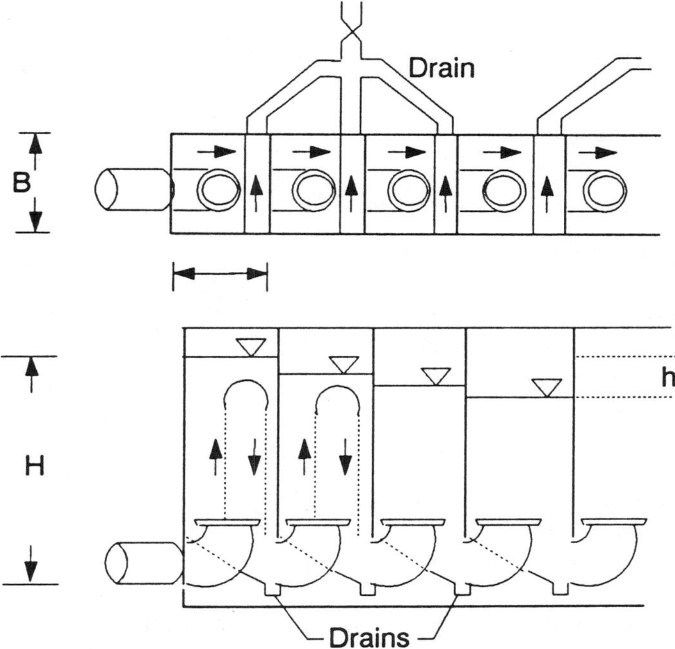

Figure 9.2. Generalized diagram of flocculation basin commonly used in Brazil.

Table 9.1. Typical Design Parameters for Flocculation Units

FLOW RATE Q L/sec	WIDTH B (m)	LENGTH L (m)	DIAMETER D (mm)	UNIT CHAMBER AREA (m^2)	UNIT CHAMBER VOLUME (m^3)
10	0.60	0.60	150	0.35	1.1
20	0.60	0.75	250	0.45	1.3
30	0.70	0.85	300	0.6	1.8
40	0.80	1.00	350	0.8	2.4
50	0.90	1.10	350	1.8	3.0
60	1.00	1.20	400	1.2	3.6
70	1.05	1.35	450	1.4	4.2
80	1.15	1.40	450	1.6	4.8
90	1.20	1.50	500	1.8	5.4
100	1.25	1.60	500	2.0	6.0

mum no. of successive chambers, 3; detention time: 20 to 30 min; G, 70 to 20 sec^{-1}; maximum tip speed 2 m/sec; approximately 5m × 5m to 10m × 10m basin surface area per unit. For horizontal shaft mechanical flocculators: G values up to 50 sec^{-1}; max. tip speed of 1 m/sec, paddle area should not exceed 20% of tank section area.

The velocity gradient *(G)* for flocculators is determined from the equations:

$$G = (Q \rho \mathrm{g} h/uV)^{1/2}$$

for hydraulic flocculation, and

$$G = [P/(uV)]^{1/2}$$

for mechanical flocculation, where:

G = velocity gradient (sec^{-1})
ρ = density of water (kg/m^3)
h = head loss (m)
u = dynamic viscosity (kg/m · s)
t = detention time, Q/V, (s)
Q = flow (m^3/s)
p = power, $Q \cdot \rho \cdot g \cdot h$ (watts, kg · m^2/s^3)
V = volume of unit (m^3)
g = gravitational constant (9.81 m/s^2)

In the design of flocculation systems, the total number of particle collisions, and thus the floc formation action, is indicated as a function of the product of the velocity gradient and the detention time, *Gt*. The range of velocity gradient (*G*) and *Gt* values given in Table 9.2 are the most effective for plants using mechanical flocculators. Nonetheless, to obtain appropriate values for particular designs and water characteristics to provide for the optimal formation of flocs, laboratory jar testing, or pilot-plant studies should be conducted on the water to be treated.

In baffled channel flocculation, mixing is accomplished by reversing the flow of water through channels formed by around-the-end or over-and-under baffles. Baffled channel flocculators are limited to relatively large treatment plants (greater than 10,000 m^3/day capacity) where sufficient head losses can be maintained in the channels for slow mixing without requiring that baffles be spaced too close together (which would make cleaning difficult). A distinct advantage of such flocculators is that they operate under plug-flow conditions which result in few short-circuiting problems.

Horizontal-flow flocculators with around-the-end baffles are sometimes preferred over vertical-flow flocculators with over-and-under baffles because they are easier to drain and clean; also, the head loss, which governs the degree of mixing, is changed more easily by installing additional baffles or removing portions of existing ones. However, vertical-flow units have been used successfully in Brazil and in the U.S. and are appropriate for specific applications, such as where a scarcity of land prohibits the use of larger horizontal-flow flocculators.

The water depth in the channels of vertical flow units can be as high as 3 m, and, therefore less surface area is required than with horizontal units. The major problem with such flocculators is the accumulation of settled material on the chamber floors and the difficulty in removing it. To mitigate this problem, the Brazilian designs have included small openings (weep holes) in the base of the lower baffles of a size equivalent to 5% of the flow area of each chamber. The purpose is to allow the major portion of the flow of water to follow the over-and-under path created by the baffles, whereas a smaller

Table 9.2. Recommended *G* and *GT* Values for Flocculators

TYPE	VELOCITY GRADIENT G (SEC^{-1})	*Gt*
Turbidity or color removal (without solids recirculation)	20 to 100	20,000 to 150,000
Turbidity or color removal (with solids recirculation)	75 to 175	125,000 to 200,000
Softeners (solids contact reactors)	130 to 200	200,000 to 250,000

Source: Reference 25.

portion flows through the holes, creating some additional turbulence and preventing the accumulation of material (22). Weep holes also facilitate manual cleaning of the over-and-under flocculator.

For design purposes, the head loss in the bend of a baffle flocculator is approximated by the following:

$$h = kv^2/2g$$

where:

h = head loss (m)
v = the fluid velocity (m/s)
g = the gravitational constant (9.81 m/s^2)
k = empirical constant (varies from 2.5 to 4)

The value of k cannot be determined precisely in advance; therefore it is better to design for a low k value, because boards can always be added to the baffles if additional head loss is needed.

The number of baffles needed to achieve a desired velocity gradient for horizontal flow units can be calculated from the equation:

$$n = \{[(2ut)/\rho(1.44 + f)]\ [(HLG)/Q]^2\}^{1/3}$$

and for vertical flow units

$$n = \{[(2ut)/\rho(1.44 + f)]\ [uLG/Q]^2\}^{1/3}$$

where:

n = number of baffles in the basin
H = depth of water in the basin (m)
L = length of the basin (m)
G = velocity gradients (s^{-1})
Q = flow rate (m^3/s)
t = time of flocculation (s)
u = dynamic viscosity (kg/m · s)
ρ = density of water (kg/m^3)
f = coefficient of friction of the baffles
W = width of the basin (m)

The water velocity in both horizontal-flow and vertical-flow units generally varies from 0.3 to 0.1 m/sec. Detention time varies from 15 to 30 min. In general, velocity gradients for both types of baffled channel flocculators should vary between 100 to 10 sec^{-1}. In addition to the preceding design criteria, the practical criteria given below should be followed:

Guidelines for the Design and Construction of Baffled Channel Flocculators (25)

A. Around-The-End (Horizontal Flow)

1. Distance between baffles should not be less than 45 cm to permit cleaning.
2. Clear distance between the end of each baffle and the wall is about 1 ½ times the distance between the baffles; should not be less than 60 cm.
3. Depth of water should not be less than 1.0 m.
4. Decay-resistant timber should be used for baffles; wood construction is preferred over metal parts.
5. Avoid using asbestos-cement baffles because they corrode at the pH of alum coagulation.

B. Over-and-Under (Vertical Flow)

1. Distance between baffles should not be less than 45 cm.
2. Depth should be two to three time the distance between baffles.
3. Clear space between the upper edge of a baffle and the water surface, or the lower edge of a baffle and the basin bottom, should be about 1½ times the distance between baffles.
4. Material for baffles is the same as in around-the-end units.
5. Weep holes should be provided for drainage.

The Alabama-type flocculator is illustrated in Figure 9.2. The jet action is provided in each chamber via a cast iron pipe with its outlet turned upward. For effective flocculation, the outlet should be placed at a depth of about 2.5 m below the water level. Common design criteria are listed below:

Rated capacity per unit chamber	25 to 50 $L/s \cdot m^2$
Velocity at turns	0.40 to 0.60 m/s
Length of unit chamber (L)	0.75 to 1.50 m
Width (W)	0.50 to 1.25 m
Depth (H)	2.50 to 3.0 m
Detention time (t)	15 to 25 min

The head loss in this type of flocculator is estimated at about 2 times the velocity per chamber, or about 0.35 to 0.50 m for the entire unit. Velocity gradients range from 40 to 50 $sec.^{-1}$. Arrangements should be made for draining each chamber, since material tends to collect at the bottom and must be removed.

The gravel bed flocculator provides a simple and inexpensive design for flocculation in small water treatment plants (less than 5000 m^3/day capacity). It has been tested experimentally and employed successfully in several upflow-downflow plants in India and in package plants in Parana, Brazil. The packed bed of gravel provides ideal conditions for the formation of compact settleable flocs because of continuous recontacts provided by the sinuous flow of water through the interstices formed by the gravel. The velocity gradients that are introduced into the bed are a function of: (1) the size of the gravel, (2) rate of flow, (3) cross-sectional area of the bed, and (4) the head loss across the bed. The direction of flow can be either upward or downward and is usually determined from the design and hydraulic requirements of other process units in the plant.

9.2 ADVANTAGES AND DISADVANTAGES

The major shortcomings of hydraulic flocculators are: (1) little flexibility to respond to changes in raw water quality, (2) hydraulic parameters are a function of flow and cannot be changed within the process, (3) head loss can be significant, (4) cleaning may be difficult unless design incorporates cleaning provisions. There is a paradox for the application of these various types. While simpler hydraulic types would be desirable for small plant applications where operator skill is likely to be less, high potential head losses argue for application at large plants.

Baffled flocculators: a lack of flexibility for mixing intensity; a high head loss if over-and-under baffles are used; some plant flowrates may vary in the range of 1 : 4 within a single day so achieving good mixing in the entire flow range may be difficult.

Vertical shaft mechanical flocculators: many units are required in a large plant; high capital cost for variable speed reducers and support slabs.

Horizontal shaft mechanical flocculators: precise installation and maintenance is necessary; difficult to increase energy input; problems with leakage and shaft alignment.

The main problem with gravel bed flocculators is fouling, either by intercepted floc or by biological growth on the media, or both.

9.3 COSTS

See Table 9.3 (5) for capital costs for selected types. These costs are for mechanical units only, at different G values. Operation and maintenance costs for horizontal paddle systems on Table 9.4 provide O&M requirements. Other capital and O&M costs are given in Figures 9.3 and 9.4.

Table 9.3. Flocculation Capital Cost U.S. Dollars

	TOTAL BASIN VOLUME (ft^3)					
	1,800	10,000	25,000	100,000	500,000	1,000,000
Horizontal Paddle System			Capital Cost $			
G = 20	49,000	163,000	310,000	385,000	1,260,000	2,500,000
G = 50	50,000	163,000	313,000	425,000	1,400,000	2,600,000
G = 80	49,000	172,000	228,000	515,000	1,900,000	—

	TOTAL BASIN VOLUME (ft^3)								
	1,800			10,000			25,000		
Vertical Turbine	G = 20	G = 50	G = 80	G = 20	G = 50	G = 80	G = 20	G = 50	G = 80
					Capital Cost $				
		43,000			. 144,000		209,000	209,000	218,000

Table 9.4. Operation and Maintenance Summary Flocculation—Horizontal Paddle System

Total Basin	ENERGY	(KW-HR/YR)	MAINTENANCE	Material	Labor	TOTAL COST ($/YR)*		
Volume (ft^3)	G = 20	G = 50	G = 80	($/yr)	(hr/yr)	G = 20	G = 50	G = 80
1,800	330	2,070	6,100	615	99	2,100	2,200	2,400
10,000	1,960	11,870	33,660	1,600	199	4,650	5,100	6,100
25,000	4,900	29,630	84,080	1,600	199	4,800	5,900	8,300
100,000	19,600	118,720	336,550	6,000	397	12,900	17,300	27,000
500,000	98,020	593,590	1,682,750	21,600	496	33,500	56,000	105,000
1,000,000	198,230	1,188,300	—	43,000	990	67,000	112,000	—

* Calculated using $0.03/kw-hr and $10.00/hr of labor.
Note: G values are velocity gradients and are in units of sec^{-1}.

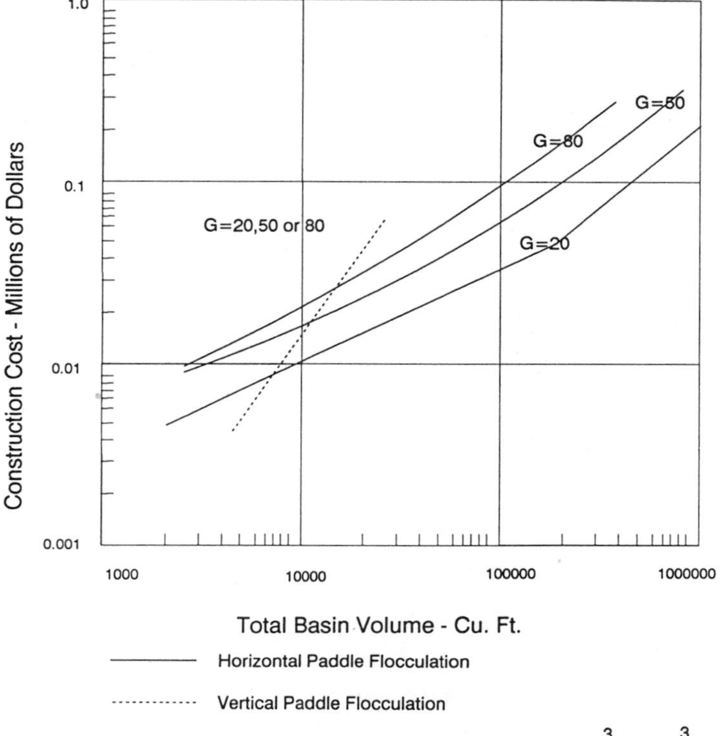

Figure 9.3. Construction cost for flocculation horizontal paddle system and vertical turbine flocculation.

9.4 AVAILABILITY

Many examples of baffle flocculators exist all over the world. All of these systems are used throughout developing countries as well as the U.S. Most materials (even the mechanical flocculator parts) can be purchased in many countries.

9.5 OPERATION AND MAINTENANCE

Baffled: this type provides plug-flow mixing conditions and is an effective flocculation system. Since tapered mixing is effective in forming large settleable flocs, the baffles should be properly arranged to reduce the mixing intensity and floc shearing force. It performs well if the plant flow rate is

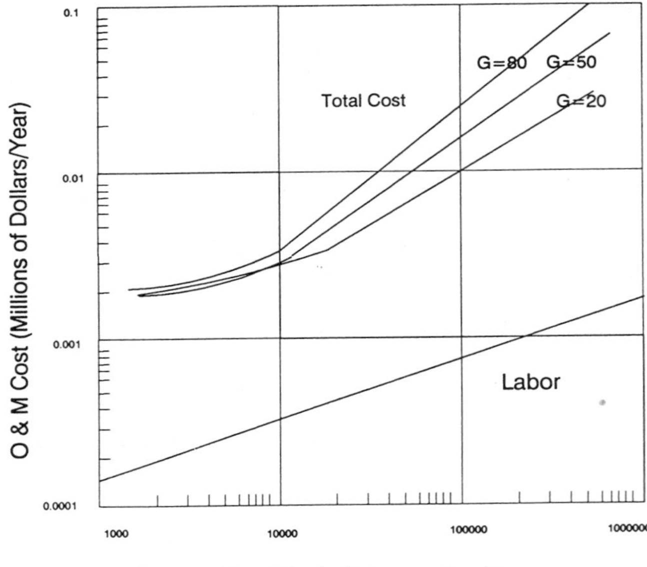

Figure 9.4. Operation and maintenance requirements for horizontal paddle flocculation.

reasonably constant. Removal of accumulations of material in chambers is required. Little maintenance is required because of few moving parts.

Gravel: this flocculator is composed of a concrete chamber full of small stones or pebbles where water is introduced by a distribution system at the bottom so that it flows upward through the bed of stones. With this process, little maintenance is required, but frequent cleaning is necessary.

Vertical mechanical: this system is suitable for high energy input, direct filtration, and conventional treatment. All mechanical systems require more maintenance than systems with few or no moving parts.

Horizontal mechanical: this type produces a large size floc, has a simple mixing unit and is suitable for conventional treatment. More data and experience is available than for vertical mixing types.

9.6 CONTROL

Mechanical systems require more intensive maintenance programs than nonmechanical systems. Thus, the gravel and baffled types are more appropriate for remote systems and for those where operator training is not available.

9.7 SPECIAL FACTORS

Baffled channel flocculators should be designed so that the cross section will provide the proper flow velocity. Adjustable baffles should be included to provide the desired degree of turbulence. Vertical shaft flocculators should be situated to cover a square or circular mixing zone for maximum efficiency.

Which of the two mechanical flocculator systems to use is mainly dependent on the type of filtration system used. Horizontal shaft flocculators are generally used if conventional rapid sand filtration is employed. Conventional rapid sand filtration requires a high degree of solids removal by the sedimentation basin before filtration. The horizontal shaft flocculators generally are suited to produce large and easily settleable floc with alum flocculation. However, they usually require more maintenance and expense largely because bearings and packings are typically submerged. High energy, vertical shaft flocculators are the unit of choice for application with high rate filtration systems. Since high rate filters allow floc penetration into the filter bed, the desired type of floc for these filters is small in size but physically strong to resist high shear forces in the filter bed.

9.8 RECOMMENDATIONS

All these systems can be effectively used in developing countries and for small system applications if the system selection is correctly matched to the size of the plant and expected operator skill. For instance, for small installations gravel could be used. For small and medium size installations baffled or Alabama flocculators could be used (provided available head allows). Alabama flocculators have been proven in Brazil; gravel bed units have been used in India for small plants. Finally, for larger installations the mechanical types are best. The vertical shaft flocculator is used with some frequency in South America.

10. GRAVITY SEWERS

10.1 DESCRIPTION

Gravity sewers are used for the transport of sanitary industrial wastewater, stormwater and any combination of wastewaters. Slope design is important and results in flow due to gravity. Access to a gravity sewer is by manholes spaced about every 300 to 500 feet (91 to 152 m) or at changes in slope or direction.

Design Criteria (2): size is dependent on flow, minimum 6 in. inside diameter for all laterals in collection systems. Slope: dependent on size and flow. Velocity: minimum of 2.0 ft/sec (0.6 m/sec) at full depth. Materials: must meet service application requirements. Additional Requirements: adequate ground cover, minimum scouring (self-cleaning) velocity; infiltration should not exceed 200 gal/d/in. diameter/mile (5.5×10^{-2} L/sec/cm diameter/km). Small diameter gravity sewers for transport of septic tank effluent have a typical diameter of 4 in. (10.2 cm) and are designed for one-half peak flow (corresponding to a gradient of 0.67% for a 4 in. sewer), although pipes as small as 1.5 in. (3.8 cm) are used. See Table 10.1 (2) for design criteria for the sizing of collector and interceptor sewers. Also see Case Study No. 4.

10.2 LIMITATIONS

High per capita cost in rural areas with low population density, in areas requiring removal of rock and where depths greater than 15 ft (4.6 m) are required. Possible explosive hazards can occur due to production of gas in sewers with shallow slopes. Blockage is also a possibility because of grease, sedimentation, tree root development and, in the case of combined sewers, debris. Excessive infiltration and inflow are the most common problems for both old and new systems.

10.3 COSTS

See Figures 10.1 and 10.2 (2,11).

10.4 AVAILABILITY

Gravity sewers are the oldest and most common wastewater transport system existing. Use of water for waste carriage results in contamination of the water and often of sources used for downstream supply or of the groundwater.

Table 10.1. Sizing of Collector and Interceptor Sewers

PIPE DIAMETER (in.)	MINIMUM SLOPE FOR PIPE VELOCITY OF 2 ft/sec	8 ft/sec	DESIGN WASTEWATER FLOW (MGD) WITH PIPE FLOWING FULL, VELOCITY, ft/sec: 2	3	4	8
6	0.0060	0.075	0.26	0.36	0.47	0.91
8	0.0038	0.45	0.48	0.69	0.91	1.76
10	0.0030	0.035	0.72	1.04	1.37	2.54
12	0.0022	0.026	1.04	1.46	1.94	3.51
15	0.0015	0.020	1.69	2.47	3.25	5.84
18	0.0012	0.016	2.41	3.45	4.42	9.43
21	0.0010	0.013	3.38	4.94	6.37	12.35
24	0.00078	0.011	4.10	6.24	8.13	15.28
27	0.00065	0.0095	5.20	7.80	10.08	18.85
30	0.00058	0.0080	6.50	9.75	13.00	24.05
36	0.00045	0.0060	9.75	14.63	18.20	37.05
42	0.0008	0.0050	13.00	19.50	25.36	48.10
48	0.00032	0.0045	16.25	24.70	31.85	59.80
54	0.00026	0.0039	20.80	31.85	39.65	84.50

Note: ft/sec × 0.3048 m/ft = m/sec.
Mgal/day × 3785 = m^3/day.

Materials for sewer construction—cement, asbestos cement, and reinforced concrete are widely available. Materials such as PVC may have improved erosiveness characteristics and should be considered also (see Chapter 41 on Steep Slope Sewers). Also, possible health considerations for asbestos cement pipe should be taken into account.

10.5 OPERATION AND MAINTENANCE

Pipe material with special considerations are as follows:

Asbestos Cement: advantages include light weight, ease of handling, long laying lengths, and tight joints; disadvantages include corrosion where acids and hydrogen sulfide are present and potential adverse health impacts; diameters available from 4 to 42 in. (10.2 to 107 cm).

Clay Pipe: advantages include resistance to corrosion from acids and alkalies and also to erosion and scour; disadvantages are that clay pipe has a limited range of sizes and strengths, it is also brittle with short pipe lengths and many joints; diameters available from 4 to 42 in. (10.2 to 107 cm).

Concrete (both reinforced and non-reinforced): advantages are its strength especially in large diameters compared to other materials, availability of wide range of sizes and widespread use; disadvantages include a corrosion tendency when acid and hydrogen sulfide are present; relatively short pipe lengths and many joints are required; diameters available from 12 to

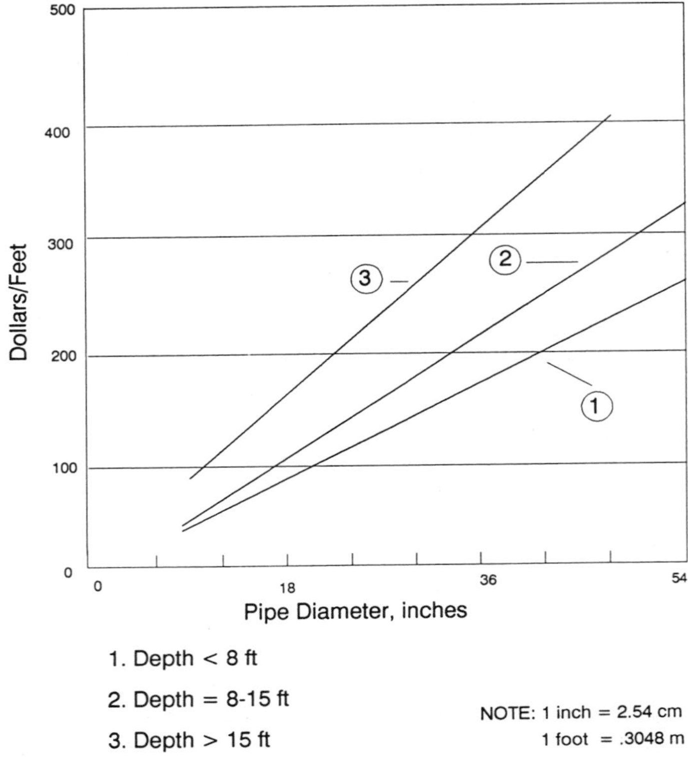

Figure 10.1. Capital cost of gravity sewer.

144 in. (30.5 to 366 cm) for reinforced and 4 to 24 in. (10.2 to 71 cm) for non-reinforced concrete pipe.

Cast Iron: advantages include long laying lengths and tight joints along with an ability to withstand high external loads and a corrosive resistant nature in neutral soils; disadvantages include potential for corrosion by acid, septic wastewater or corrosive soils; diameters available from 2 to 46 in. (5 to 122 cm).

Plastic Pipe: advantages include light weight, tight joints, long laying lengths and for some types, corrosion and erosion resistance; disadvantages include thin walls, susceptibility to sunlight and low temperature which affect shape and strength; diameters are available to 12 in. (30.5 cm) for solid wall and from 8 to 15 in (20.3 to 38 cm) for plastic or truss pipe. Larger diameters are available and may come into more popular use.

In general, gravity sewers are highly reliable, with long life expectancy.

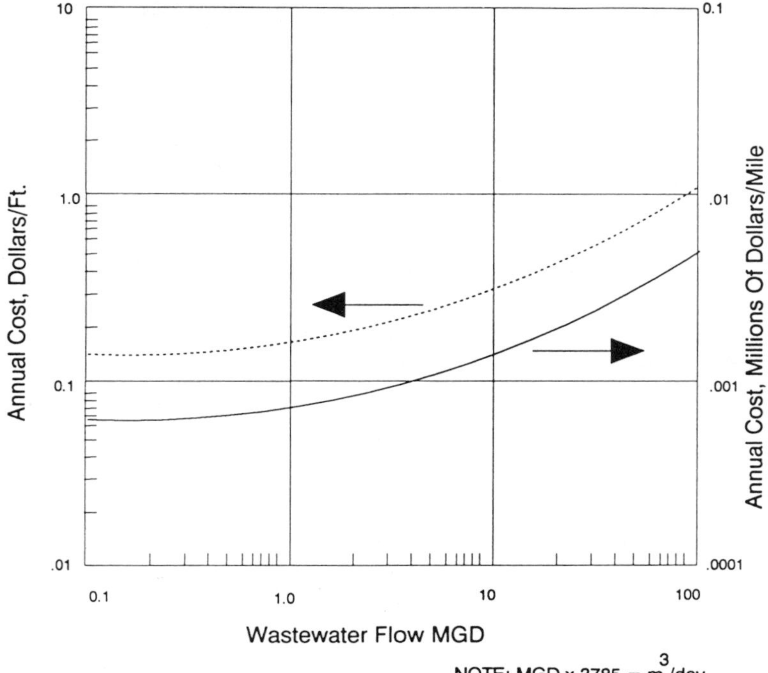

Figure 10.2. Operation and maintenance cost for gravity sewer.

10.6 SPECIAL FACTORS

Common modifications include addition of corrosion protection coatings (coal tar, PVC based tar) chemical grouting and slip-in liners (for pipes with diameters of less than 12 in., 131 cm) for rehabilitation of in-place sewers, inverted syphons, lift stations for hilly or excessively flat terrain and diversion regulators for combined sewers.

In rural communities where topography is favorable, small diameter gravity sewers which transport septic effluent to central treatment works have been employed in Australia and, to a limited extent in the U.S. Generally, these sewers have a minimum pipe diameter of 4 in. (10.2 cm). All installations to date have employed PVC pipe, owing to its light weight, long lengths and ease of laying. Curvilinear alignments in the vertical and horizontal planes are allowable, and manholes and meter boxes (depending on line depth) may be kept to a minimum (400 to 600 ft, 122 to 183 m spacing).

There is a considerable impact on land use during installation. Typically, the installation of sewers adjacent to vacant properties results in an increase in the rate of development of the land, especially when large diameter water and sewer lines are constructed.

11. PRESSURE SEWERS

11.1 DESCRIPTION

Pressure sewer systems use smaller (than gravity) pipe diameters and are operated with pumping instead of gravity. They result in lower construction costs relative to gravity sewer systems in less populated areas. Pressure sewers may be placed independent of slope. Pressure sewer systems have been developed and applied to reduce the high capital cost of sewer systems which have been designed in accordance with accepted design parameters, namely slope and velocity (generally sufficient slope to maintain a minimum of 2 ft/sec; 0.61 m/sec). There is a trade-off when compared to gravity systems, for more system operating complexity because of typically smaller diameters in pressure systems, operating and maintenance requirements for increased cleaning frequency and maintenance of pumps, and increased operating costs due to power requirements. Pressure sewer systems require a number of pressurizing inlet points and an outlet to a treatment facility or to a downstream gravity sewer, depending on the application.

Two major types of pressure sewer systems are the grinder pump (GP) system (see Figure 11.1) and the septic tank effluent pump (STEP) system shown in Figure 11.2. The major difference between the two systems is in the on-site equipment and layout. Neither pressure sewer system alternative requires any modification of household plumbing.

Design criteria are as follows: dendriform systems (irregular piping network) are used instead of rectangular grids. Pump requirements vary with the type of pump employed and its location in the system. Flushing/cleaning provisions are necessary. Pipe design is based on Hazen-Williams friction coefficient of 130 to 140. For GP systems a minimum design velocity of about 3 ft/sec (0.91 m/sec) occurring at least one time per day, is used to prevent deposition of solids. Meter boxes generally suffice in place of manholes.

Service connection lines between the pump and the pressure main are generally made of 1 to 2 in. (2.54 to 5 cm) PVC pipe with PVC drain, waste and vent fittings. Pressure mains are generally 2 to 12 in. (5 to 30.5 cm) diameter PVC pipe, depending on hydraulic requirements. Pipes need only be buried deep enough to avoid freezing of contents. Head loss due to pipe friction generally is in the range 1 to 4 ft water/100 ft of pipe m (0.3 to 1.2 m water/30.5 m of pipe).

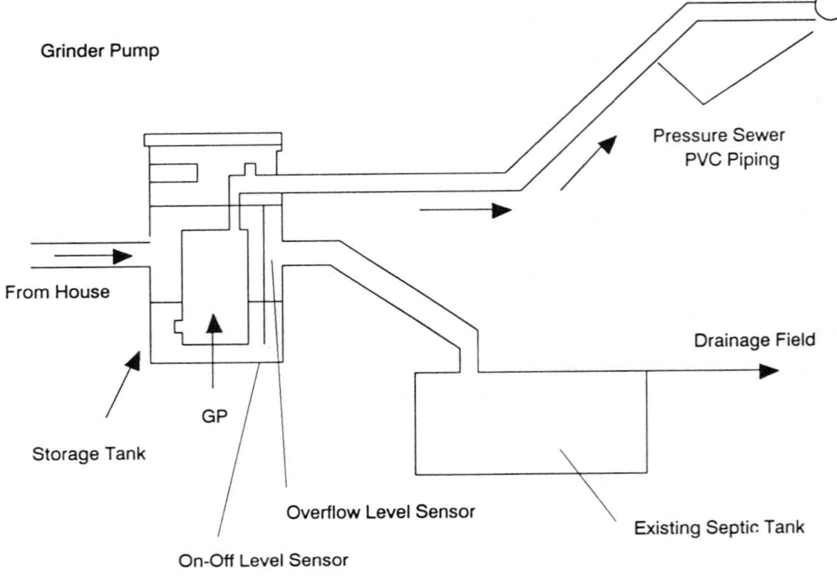

Figure 11.1. Pressure sewer connection (source: reference 2).

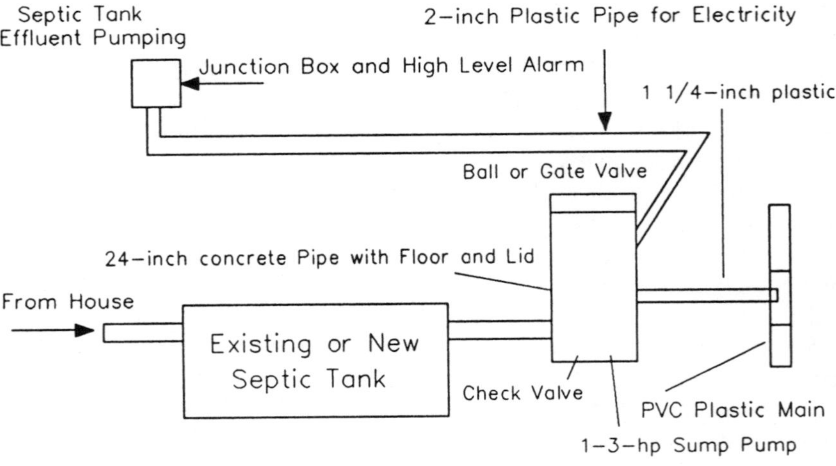

Figure 11.2. Pressure sewer system (source: reference 2).

11.2 ADVANTAGES AND DISADVANTAGES

Operation and maintenance costs are moderate to high because of the use of mechanical equipment at each point of entry to the system. In GP systems, the wastewater conveyed to the treatment facility may be more concentrated than normal wastewater. In STEP systems, a weaker, more septic waste is generated. Therefore, both systems require special considerations in design of the receiving station at the entry to the gravity system and/or at the treatment plant. It may be more difficult to monitor and maintain many small pumps than a few larger ones in gravity systems. While infiltration is virtually eliminated, exfiltration may be a problem in poorly constructed systems.

Advantages

- Low construction cost
- Adaptable to severe terrain conditions; no slope required
- No infiltration
- STEP systems reduce grit, grease, and solids present in wastewater flow
- Shallow sewer depths
- Cleanouts used instead of manholes which are the source of contaminants odor/corrosion

Disadvantages

- Multiple pumping units required
- Relatively high operation and maintenance requirements
- STEP systems require periodic cleaning of septic tanks and disposal of septage
- Individual services susceptible to power/pump failures (unless overflow storage is provided)
- Potential problems

11.3 COSTS

See Table 11.1 (2) for both construction and O&M costs. Local economics, cement and other material costs, distance to manufacturing and distribution centers, climate, geology and slope, soil type, and many other factors make efforts to give realistic costs extremely difficult. The costs given on Table 11.1 should be used for gross estimating purposes only.

11.4 AVAILABILITY

More than 100 pressure sewer systems are being operated in the U.S. to date. Availability of plastic small diameter piping should guide decisions about adopting such systems.

Table 11.1. Sewer Costs

COMPONENTS	CONSTRUCTION COST	OPERATION & MAINTENANCE COST
1. Mainline Piping (PVC)		
a. 1–3 in. diameter	$5.61/lin ft	$187–374/year/mile
b. 4 in.	$6.55/lin ft	
c. 6 in.	$8.70/lin ft	
TOTAL		$187–374/year/mile
2. On-lot Septic Tank Effluent Pumping (STEP)		
a. Pump, controls, etc.	$1700–2800	$75/year
b. Service line (100 ft @ $2/ft)	374	
c. Cocks, valves, etc.	94	
d. Septic tank (optional)	470	$19
e. Connection fee	94–187	
TOTAL	$2240–4000	$94/year
3. On-lot Grinder Pump (GP)		
a. GP unit, controls, etc.	$2400–3740	$140/year
b. Service line (100 ft @ $2/ft)	374	
c. Cocks, valves, etc.	94	
d. Connection fee	94–187	
TOTAL	$3000–4400	$140/year

Note: 1 ft = 0.3048 m
1 mile = 1.609 km
1 in. = 2.54 cm

11.5 OPERATION AND MAINTENANCE

In both designs, household wastes are collected in the sanitary drain and conveyed by gravity to the pressurization facility. The on-lot piping arrangement includes at least one check valve and one gate valve to permit isolation of each pressurization system from the main sewer. GP systems can be installed in the basement of a home to provide easier access for maintenance and greater protection from vandalism.

In STEP systems, wastewater receives primary treatment in a septic tank. The septic tank effluent then flows to a holding tank which houses the effluent pump, control sensors, and valves required for a STEP system.

Normally, small centrifugal pumps are employed for the STEP systems. These pumps are submersible and range in size from ¼ to 2 hp. Pump total head requirements generally range from 25 to 90 ft (7.6 to 27 m). Impellers can be made of plastic to reduce corrosion problems. Also included within the holding tank are level controls, valves and piping. The effluent holding tank can be made of properly cured precast concrete or cast-in-place reinforced concrete, molded fiberglass or reinforced polyester resin. Tank size considerations should provide accessibility for repairs and maintenance. Pump control switches are either a pressure sensing type or the mercury float type switch.

11.6 CONTROL

Severe corrosion can cause mechanical and/or electrical problems. Accumulations of grease and fiber can cause reduction in pipe cross sectional area of GP systems during partial flow conditions. Estimated life of current pump designs exceeds ten years and centralized maintenance is generally required for optimum service.

11.7 SPECIAL FACTORS

On GP systems an emergency (i.e., power failure, etc.) overflow tank may be provided. Measures such as standpipes and pressure control valves are sometimes used to maintain a positive pressure on the system. Air release valves are also provided to release gas pockets in the system. Polyethylene pipe, pneumatic ejectors, and mainline check valves have been used in some designs.

11.8 RECOMMENDATIONS

Pressure sewers are most applicable where population density is low, the service area is rocky, high ground water or unstable soils prevail, and also where terrain slope changes frequently. Pressure sewers and small diameter gravity systems should be considered for applications in mountainous areas and especially where service is to be provided for communities not currently sewered. Disruption of service because of construction activities is a minimum for these systems compared to installation of gravity sewers. The system is also appropriate in flat coastal flood plains where pump stations are necessary.

12. FACULTATIVE LAGOONS

12.1 DESCRIPTION

Facultative lagoons are low-cost, highly efficient alternatives for wastewater treatment in tropical and subtropical climates. Lagoons are of intermediate depth (3 to 8 ft; 0.91 to 2.4 m) ponds in which the wastewater is stratified into three zones (see Figure 12.1) (2). These zones consist of an anaerobic bottom layer, aerobic surface layer, and an intermediate zone. Stratification is a result of solids settling and temperature-water density variations. Oxygen in the surface stabilization zone is provided by reaeration from the atmosphere and photosynthesis. This is in contrast to aerated lagoons in which mechanical aeration is used to create aerobic surface conditions. In general, the aerobic surface layer serves to reduce odors while providing treatment of soluble organic by-products by means of anaerobic processes operating at the bottom.

Sludge at the bottom of the facultative lagoons will undergo anaerobic digestion producing largely carbon dioxide and methane. The photosynthetic activity in the aerobic lagoon surface produces oxygen diurnally, increasing the dissolved oxygen during daylight hours, while surface oxygen is depleted at night.

Design Criteria (see Table 12.1): at least three cells in series. Parallel trains of cells may be used for larger systems. Detention time: 20 to 180 days. Depth, ft: 3 to 8 (0.9 to 2.4 m), although a portion of the anaerobic zone of the first cell may be up to 12 ft (3.66 m) deep to accommodate large initial solids deposition. The pH: 6.5 to 9.0. Water temperature range: 35 to 90°F (2 to 32°C) for municipal waste applications. Optimum water temperature: 68°F (20°C). Organic loading: 10 to 100 lb BOD_5/acre/day, perhaps up to 300 lb/acre/day (approx. kg/ha/day) in tropical climates.

Performance: BOD_5 reductions of 60 to 95% have been reported. Effluent suspended solids concentrations of 20 to 150 mg/L can be expected, depending on the degree of algae separation achieved in the last cell. Efficiencies are strongly related to pond depth, detention time, and temperature.

Total containment ponds are a variant of facultative lagoons. In the case of total containment however, the design is based on the difference between evaporation and precipitation and the total expected flow. In areas where evaporation exceeds precipitation by a significant margin, and where wastewater flows are relatively small, containment ponds may be used. Large land areas are required. Thus, the option is a good one in regions where land costs

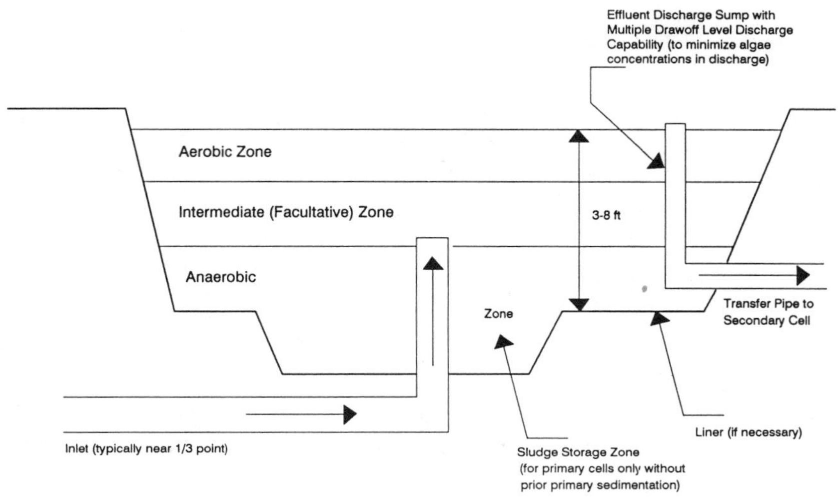

Figure 12.1. Typical facultative (aerobic-anaerobic) lagoon (*source:* reference 2).

are low. Moisture deficit is equal to annual evaporation minus annual precipitation. For various values, the design criteria are as follows (58):

45-inch moisture deficit:					
Flow (MGD)	0.25	0.50	1.00	1.25	1.50
Lagoon surface area (acres)	2	4	8	9	12
30-inch moisture deficit:					
Flow (MGD)	same				
Lagoon surface area (acres)	3.5	6	12	15	18
15-inch moisture deficit:					
Flow (MGD)	same				
Lagoon surface area (acres)	6	12	24	30	37

Note: 1 acre = 2.5 ha
1 MGD = 3785 m^3/day

12.2 LIMITATIONS

In very cold climates, facultative lagoons may experience reduced biological activity and treatment efficiency. Ice formation can also hamper operations. In overloading situations, odors can be a problem.

Bacteria, parasite, and virus removal are effective in multiple cell (mini-

Table 12.1. Typical Design Parameters for Anaerobic and Facultative Stabilization Ponds

PARAMETER	AEROBIC-ANAEROBIC (FACULTATIVE) POND	AEROBIC-ANAEROBIC (FACULTATIVE) POND	ANAEROBIC POND	AERATED LAGOONS
Flow regime	—	Mixed surface layer	—	Completely mixed
Pond size, ha	1–4 multiples	1–4 multiples	0.2–1 multiples	1–4 multiples
Operation[a]	Series or parallel	Series or parallel	Series	Series or parallel
Detention time, d[a]	7–30	7–20	20–50	3–10
Depth, m	1–2	1–2.5	2.5–5	2–6
pH	6.5–9.0	6.5–8.5	6.8–72.	6.5–8.0
Temperature range, °C	0–50	0–50	6–50	0–40
Optimum temperature, °C	20	20	30	20
BOD_5 loading, kg/ha·d	15–80	50–200	200–500	—
BOD_5 conversion %	80–95	80–95	50–85	80–95
Principal conversion products	Algae, CO_2, CH_4 bacterial cell tissue	Algae, CO_2, CH_4, bacterial cell tissue	CO_2, CH_4, bacterial cell tissue	CO_2, bacterial cell tissue
Algal concentration mg/L	20–80	5–20	0–5	—
Effluent suspended solids, mg/L[c]	40–100	40–60	80–160	80–250

[a] Depends on climatic conditions.
[b] Typical values (much higher values have been applied at various locations). Loading values are often specified by government agencies.
[c] Includes algae, microorganisms, and residual influent suspended solids. Values are based on an influent soluble BOD_5 of 200 mg/L and, with the exception of the aerobic ponds, an influent suspended-solids concentration of 200 mg/L.

Note:

ha × 2.4711 = acre
m × 3.2808 = ft
kg/ha · d × 0.8922 = lb/acre · d
mg/L = g/m^3

Source: Reference 4.

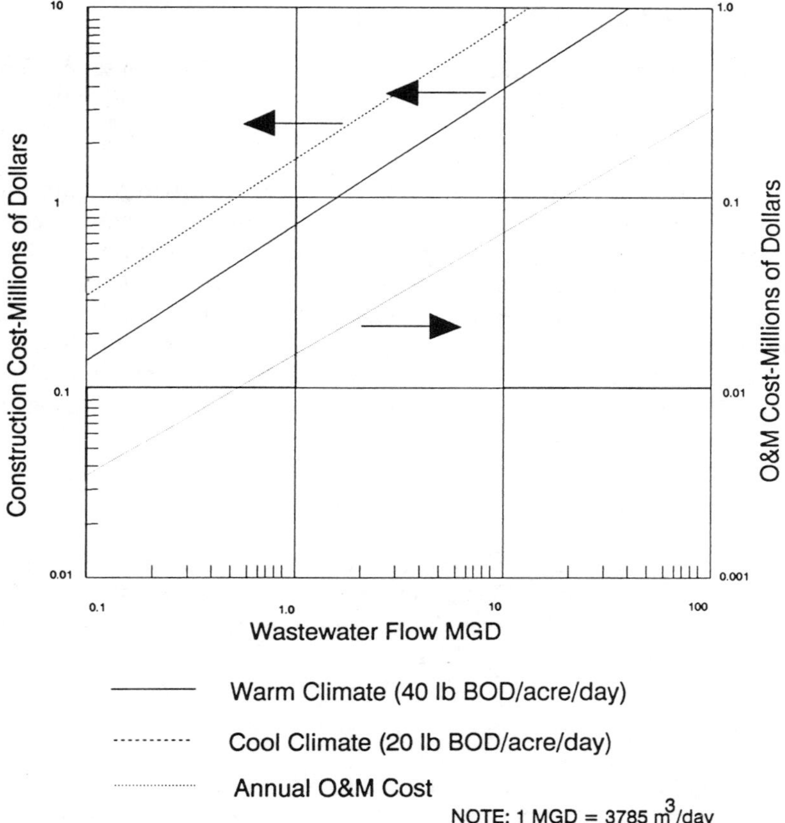

Figure 12.2. Construction, operation, and maintenance costs for facultative lagoons.

mum of 3) wastewater stabilization lagoons, as long as detention times are sufficient (minimum of 20 days).

12.3 COSTS (See Figure 12.2 (2,11).)

12.4 AVAILABILITY

Facultative lagoons have been fully demonstrated and are in moderate use especially for treatment of relatively weak municipal wastewater in areas where real estate costs are not a restricting factor. Such is the case in rural areas and some developing countries.

12.5 OPERATION AND MAINTENANCE

For best performance facultative lagoons are often operated in series. When three or more cells are linked, the effluent from either the second or third cell may be recirculated to the first. Recirculation rates of 0.5 to 2.0 times the plant flow have been used to improve overall performance.

Settled bottom solids may require removal once every 10 to 20 years; more frequent removal is necessary for highly contaminated wastes, especially some industrial wastewaters. Littoral pond areas must be maintained on a regular basis. This should include removal of peripheral and shallow area weeds, and accumulated debris. Significant reduction in available treatment volume and subsequent effluent quality is likely within short periods of time, several months to a few years, with poor maintenance. Maintenance should include the removal of canopy vegetation to facility exposure of the lagoon surface to ambient sunlight.

12.6 CONTROL

The service life of a lagoon is estimated to be 50 years, shorter if solids removal is not practiced. Operators need not be highly skilled, and the systems are highly reliable over the long term for intermediate level BOD removal efficiencies (see Table 12.1).

12.7 SPECIAL FACTORS

Facultative lagoons are customarily built within-ground or above ground using earthen dikes. Depending on soil characteristics, lining with various impervious materials, such as rubber, plastic, or clay may be desirable and/or necessary to control percolation to the groundwater and reduce weed growth. Uses of supplemental top layer aeration can improve overall treatment capacity, particularly in high elevations or cold climates where icing might occur.

If wastewater is nutrient deficient, a source of supplemental nitrogen and/or phosphorous may be needed. Compared to other secondary biological treatment processes, relatively small quantities of sludge requiring immediate disposal are produced. The pH may require adjustment to a vertical range by the addition of lime or acid.

12.8 RECOMMENDATIONS

Lagoons should be used for treating raw, screened, or primary settled domestic wastewaters and some biodegradable industrial wastewaters. Even high strength industrial wastewaters, such as cannery and dairy wastes may be lagooned. They are most suitable when land costs are not of concern, and operation and maintenance costs are to be minimized. The technology is preferable to mechanical systems for applications where average temperatures are relatively high, and high treatment efficiencies are not required. Lagoons may be used as pretreatment to other biological processes such as activated sludge, for very high strength industrial wastes. The technology is ideal for many locations in Latin America, Africa and India, and the southern and southwestern U.S.

13. AQUATIC PLANT—AQUACULTURE SYSTEM

13.1 DESCRIPTION

Aquaculture, or the production of aquatic organisms (both flora and fauna) under controlled conditions, has been practiced for centuries primarily for the generation of food, fiber, and fertilizer. The water hyacinth *(Eichhornia crassipes)* is a promising plant for wastewater treatment and has received the most attention. Other flora, however, are also being studied. Among them are duckweed, seaweed, midge larvae, and alligator weeds (see Chapter 14). Water hyacinths are large fast-growing floating aquatic plants with broad, glossy green leaves and light lavender flowers (see Figure 13.1) (2). Figure 13.2 (15) shows a plan and cross section with design suggestions. A native of South America, water hyacinths can also be found naturally in waterways, bayous, and other backwaters. Insects and disease apparently have little effect on the hyacinth, and they thrive in systems using either raw or partially treated wastewater.

Wastewater treatment by water hyacinths is accomplished by passing the wastewater through a hyacinth-covered basin, where the plants remove nutrients, BOD_5, suspended solids, metals, etc, during their normal growth cycle. Both batch treatment and flow-through systems, using single and multiple cell units, are possible. Hyacinths harvested from these systems have been investigated as a fertilizer/soil conditioner after composting, animal feed, and as a source of methane when anaerobically digested.

Design Criteria—experimental data vary widely from different experiences. Ranges herein refer to hyacinth treatment as a tertiary process on secondary effluent. Basin depth should be sufficient to maximize plant rooting and detention time sufficient to allow absorption by the plants. Detention time: depends on effluent requirements and flow; 4 to 15 days is the average; phosphorous reduction (TP): 10 to 75%; nitrogen reduction (TN): 40 to 75%; land requirement: 2 to 15 acres/Mgal/d (approx. $m^2/m^3/day$). Tables 13.1 (15), and 13.2 (2) present design criteria and performance expectations.

Performance—the pollutant removal results from five different wastewater streams are given in Table 13.2 (2). There is also some evidence that coliform, heavy metals, and refractory organics are removed to some extent, and neutralization may be achieved depending upon available alkalinity.

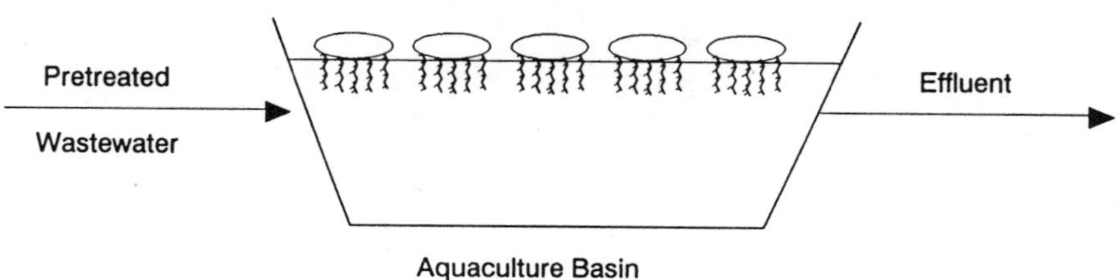

Figure 13.1. Generalized aquaculture basin with hyacinth plants (*source:* reference 2).

13.2 LIMITATIONS

Climate is the major limitation. Active growth begins when the water temperature rises above 10°C and flourishes when the water temperature is about 21°C. Plants die rapidly when the water temperature approaches the freezing point, therefore greenhouse structures may be necessary in cooler

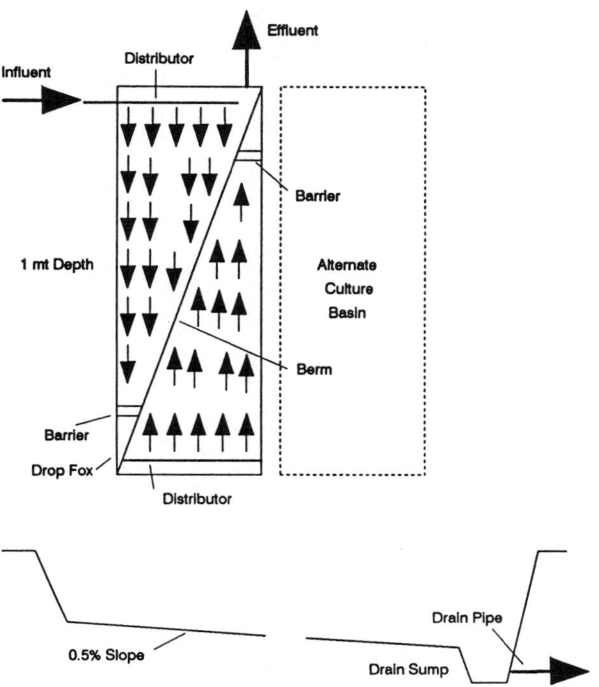

Figure 13.2. Suggested basic hyacinth culture basin design.

Table 13.1. Summary of Nutrient Loading Rates Applied to Water Hyacinths Wastewater Treatment Systems

LOCATION	ORGANIC LOADING RATE	NUTRIENT LOADING RATE TO FIRST UNIT		NUTRIENT REMOVAL %		COMMENTS
	kg BOD_5/HA · DAY	kg TN/HA · DAY	kg TP/HA · DAY	TN	TP	
Williams Creek, Texas						
Phase I (109 m^3/d)	43	15.3	—	70	—	Single Basin surface area = 0.0585 ha
Phase II (109 m^3/d)	89	18.5	—	64	—	Single Basin, surface area = 0.0585 ha
Coral Springs, Florida	31	19.5	4.8	96	67	Five Basins in Series Total surface area = 0.52 ha
National Space Technology Labs	26	2.9	0.9	72	57	Single Basin Receiving Raw Wastewater, surface area = 2 ha

Source: Reference 15.
Note: TN = Total Nitrogen.
TP = Total Phosphorus.

Table 13.2. Removal Performance of Five Wastewater Streams by Aquaculture Treatment System

Performance—In tests on five difference wastewater streams the following removals were reported:

FEED SOURCE	BOD_5 REDUCTION	COD REDUCTION	PHOSPHATE TSS REDUCATION	N REDUCTION	REDUCTION
Secondary Effluent	35%	—	—	44%	74%
Secondary Effluent	83%	61%	83%	72%	31%
Raw Wastewater	96%	—	75%	92%	60%
Secondary Effluent	60–79%	—	71%	47%	11%

Source: Reference 2.

climates. Water hyacinths are also sensitive to high salinity. Removal of potassium and phosphorous is restricted to the active growth period of the plants.

Metals such as arsenic, chromium, copper, mercury, lead, nickel, and zinc can accumulate in hyacinths and limit their suitability as a fertilizer or feed material. Dense growth areas of hyacinths may also result in small pools of stagnant surface water which can breed mosquitoes. Mosquito problems can be avoided by maintaining mosquito fish in the system. The spread of the hyacinth plant itself must be controlled by barriers since the plant will grow rapidly, can spread and clog nearby waterways. Hyacinth removal must be at regular intervals to avoid heavy intertwined growth. Evapotranspiration in treatment basins may be 2 to 7 times higher than evaporation from open ponds. Hyacinth treatment may prove impractical for large treatment plants because of land requirements.

Probably the biggest limitation is disposing of the plant mass produced during treatment. Where metals are not a problem, the plant may be used for fertilizer and mixed with soil as a conditioner. Harvest of the water hyacinth or duckweed plants is essential to maintain high levels of system performance and for high levels of nutrient removal. Harvesting equipment and maintenance procedures for accomplishing these tasks have been widely demonstrated. Disposal and/or reuse of the harvested materials is an important consideration in overall system design. A plan for disposal should be developed before systems are placed in operation. The water hyacinth plants have a moisture content similar to that of primary sludges. The amount of plant biomass produced in a water hyacinth pond system is about 4 times (dry basis) the quantity of waste sludge produced in conventional activated sludge secondary wastewater treatment. Composting, anaerobic digestion with methane production, and processing for animal feed are all technically feasible, however the techniques have not been demonstrated.

13.3 COSTS (See Figure 13.3 (2,5).)

13.4 AVAILABILITY

For widespread application to wastewater treatment this technology is still in the developmental stage. In the U.S. a number of full-scale demonstration systems are in operation, while in Mexico a few systems are in use.

13.5 OPERATION AND MAINTENANCE

The water hyacinth is a hardy, disease-resistant plant that thrives at above-freezing temperatures. While the water hyacinth system can successfully cope with a variety of stresses, health of the plants is essential for effective

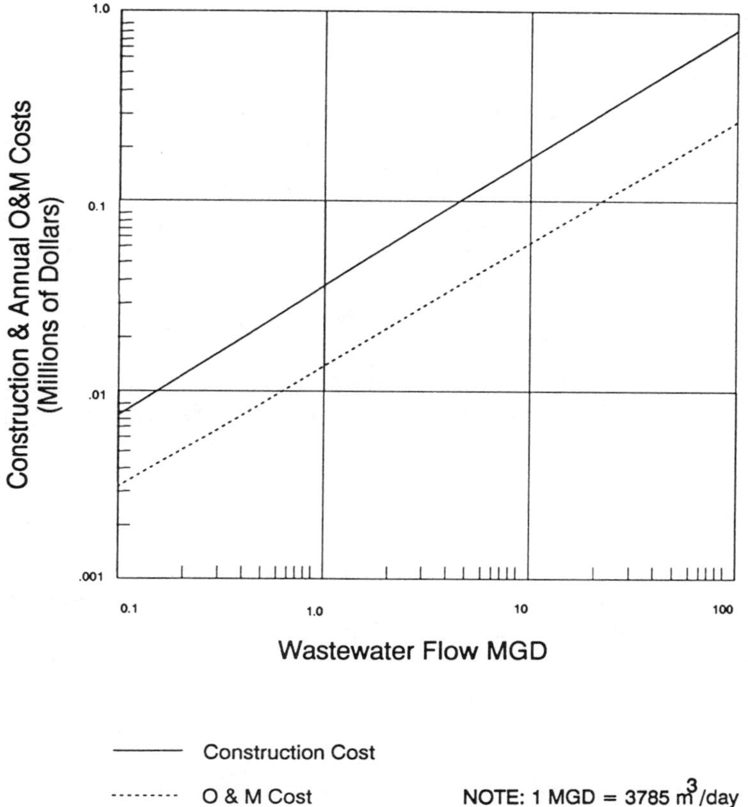

Figure 13.3. Operation, maintenance, and construction costs for aquatic plant-aquaculture system (*source:* references 2 and 5).

treatment. Presence of a high chlorine residual definitely inhibits plant growth. If possible, waste chlorination should be performed subsequent to hyacinth treatment. If local conditions dictate pre-hyacinth chlorination, chlorine residual in the influent to hyacinth treatment should not exceed 1 mg/L. Plant health is also adversely affected by high concentrations of chlorides.

Adequate nourishment is essential to plant health. The hyacinth has a voracious appetite, which if not satisfied also results in chlorosis and decreased uptake efficiency. Least efficient performance of the system has been obtained during periods of significantly reduced influent flow and when influent nitrogen concentration dropped below 10 mg/L. Plant health is also adversely affected by overcrowding.

The best indication of plant health is an abundant growth of dark green leaves. Any appearance of stunted leaf growth with yellowish-green leaves in immature plants, or of leaf yellowing in mature plants should be investigated immediately.

Intense sunlight with temperatures in the mid-nineties may cause some leaf browning and wilting. This is not a serious condition if new growth is evident beneath the brown wilted leaves. Wilted plants may be removed during the normal harvest cycle by selective harvesting.

The healthiest plant condition and best system performance have been obtained when ponds were maintained in a loosely packed condition using a four week harvest cycle. From 15 to 20% of the plants should be removed at each harvest. Uncovering more than 20% of pond surface area results in an algae problem.

Operation by gravity flow requires no energy. Operation and maintenance is relatively simple. Maintenance is largely associated with harvesting the plants on a regular basis as discussed above.

13.6 CONTROL

Hyacinth harvesting is continuous or intermittent. Studies indicate that average hyacinth production is on the order of 1,000 to 10,000 lb/d/acre—wet basis (approx. kg/ha/day). Basin cleaning at least once per year is suggested.

13.7 SPECIAL FACTORS

This technology is used in combination with other treatment, such as lagoons.

13.8 RECOMMENDATIONS

The process appears to be reliable from mechanical and process standpoints, but the system is subject to temperature constraints. This technology would be very useful in countries with warm/hot climates and where land costs are low.

14. AQUACULTURE—WETLANDS

14.1 DESCRIPTION

Aquaculture systems for wastewater treatment include natural and artificial wetlands as well as other aquatic systems which utilize the production of algae and higher plants (both submergent and emergent), invertebrates, fish and integrated polyculture food chains. Natural wetlands, both marine and freshwater, have incidentally served as natural waste treatment systems for centuries; however, in recent years marshes, swamps, bogs, and other wetland areas have been successfully utilized as managed natural "nutrient sinks" for polishing partially treated wastewater under relatively controlled conditions. Constructed artificial wetlands can be designed to meet specific treatment requirements while providing new wetland areas that also improve available wildlife habitats. Managed plantings of reeds (e.g., *Phragmites* spp.) and rushes (*Scirpus* spp. and *Schoenoplectus* spp.) as well as managed natural marshes, swamps, and bogs have been demonstrated to provide neutralization and some reduction of nutrients, heavy metals, organics, BOD_5, COD, TSS, fecal coliforms, and pathogenic bacteria. A system is shown schematically in Figure 14.1 (2).

Wastewater treatment by natural and constructed/managed artificial wetland systems is generally accomplished by sprinkling or flood irrigating the wastewater onto the wetland area. In managed systems the wastewater is passed through a constructed system of shallow ponds, channels, basins, or other devices in which aquatic vegetation has been planted or naturally grows and is actively growing (see Figure 13.1). Treatment performance by natural wetland areas may be enhanced by the addition of pre- and post-flow management devices. On the other hand, complete systems may be constructed in low-lying areas near streams or lakes.

In test units and constructed artificial marsh facilities using various wastewaters, the following percentage removals have been reported for secondary effluent treatment (10-day detention period): BOD_5, 80 to 95%; TSS 29 to 87%; COD, 43 to 87%; nitrogen, 42 to 94%; total phosphate, 94% (higher levels possible in warm climates and where plant harvesting is practiced); coliforms, 86 to 99%; heavy metals highly variable depending on the species. There is also evidence of reductions in concentrations of chlorinated organics and pathogens, as well as some neutralization without causing detectable harm to the wetland ecosystem.

Design parameters can be found in Table 14.1. By dividing the total area

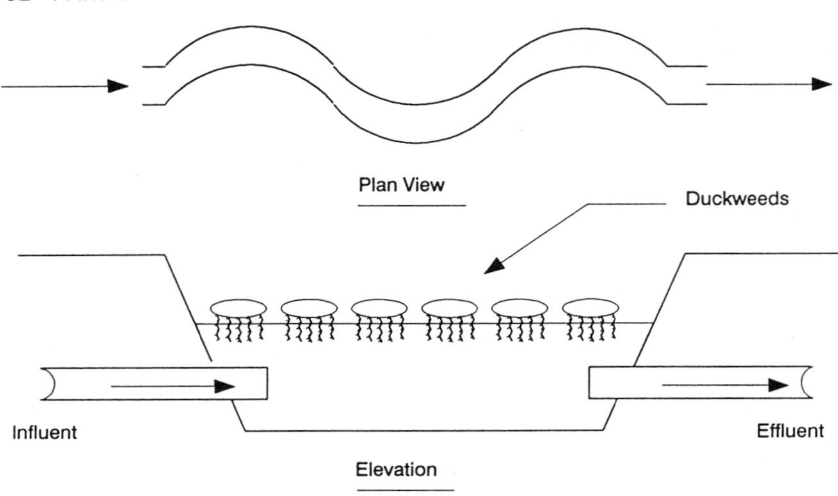

Figure 14.1. Generalized diagram of aquaculture wetland (*source:* reference 2).

designated for the wetlands into plots, a variety of habitats may be established. Flow variation during treatment is possible when a multiple plot system is used. Multiple plots allow for one plot to be isolated from the remainder of the system in case of major maintenance needs. Multiple plots also allow for depth variation. Depth is a key factor in habitat design; depth will determine whether or not emergent vegetation will be present and affect temperature and dissolved oxygen values. Plot shapes may vary, but small constricted areas should be avoided as they tend to promote stagnation and vector problems. Deciding which groups of organisms are desired in the wetlands and knowing which environmental growth conditions these organisms normally require, will determine the fundamental components of the design.

14.2 LIMITATIONS

Temperature is a major limitation since effective treatment is linked to the active growth phase of mergent (surface and above) vegetation. Herbicides and other materials toxic to the plants can adversely affect treatment performance. Wind may blow duckweeds to the shore if wind screens (e.g., deep trenches) are not employed. Small pools of stagnant surface water can promote mosquitoes. Wetland systems may prove impractical for large treatment needs because of the land requirements. Evapotranspiration is higher from wetland areas than from open ponds.

Table 14.1. Design Parameters for Artificial Wetland Wastewater Treatment Systems[a]

TYPE SYSTEM	FLOW REGIME[b]	CHARACTERISTIC/DESIGN PARAMETER					
		DETENTION TIME, DAYS		DEPTH OF FLOW, ft (m)		LOADING RATE G/ft² DAY (cm/DAY)	
		RANGE	TYP.	RANGE	TYP.	RANGE	TYP.
Trench (with reeds or rushes)	PF	6–15	10	1.0–1.5 (0.3–0.5)	1.3 (0.4)	0.8–2.0 (3.25–8.0)	1.0 (4.0)
Marsh (reeds, rushes, other)	AF	8–20	10	0.5–2.0 (0.15–0.6)	0.75 0.25	0.2–2.0 (0.8–8.0)	0.6 (2.5)
Marsh-pond							
1. Marsh	AF	4–12	6	0.5–2.0 (0.15–0.6)	0.75 (0.25)	0.3–3.8 (0.8–15.5)	1.0 (4.0)
2. Pond	AF	6–12	8	1.5–3.0 (0.5–1.0)	2.0 (0.6)	0.9–2.0 (4.2–18.0)	1.8 (7.5)

[a] Based on the application of primary or secondary effluent.
[b] PF = plug flow, AF = arbitrary flow.
Source: Reference 59.

14.3 COSTS

The generalized construction cost, and operation and maintenance costs are shown in Figure 14.2 (2). Case study examples of actual construction cost, operation, and maintenance costs for wetland treatment systems at Vermontville and Houghton Lake, Michigan are shown in Table 14.2 (16).

14.4 AVAILABILITY

This technology is in the development stage for general applicability to wastewater treatment. Several full scale, demonstration systems are in operation and/or under construction.

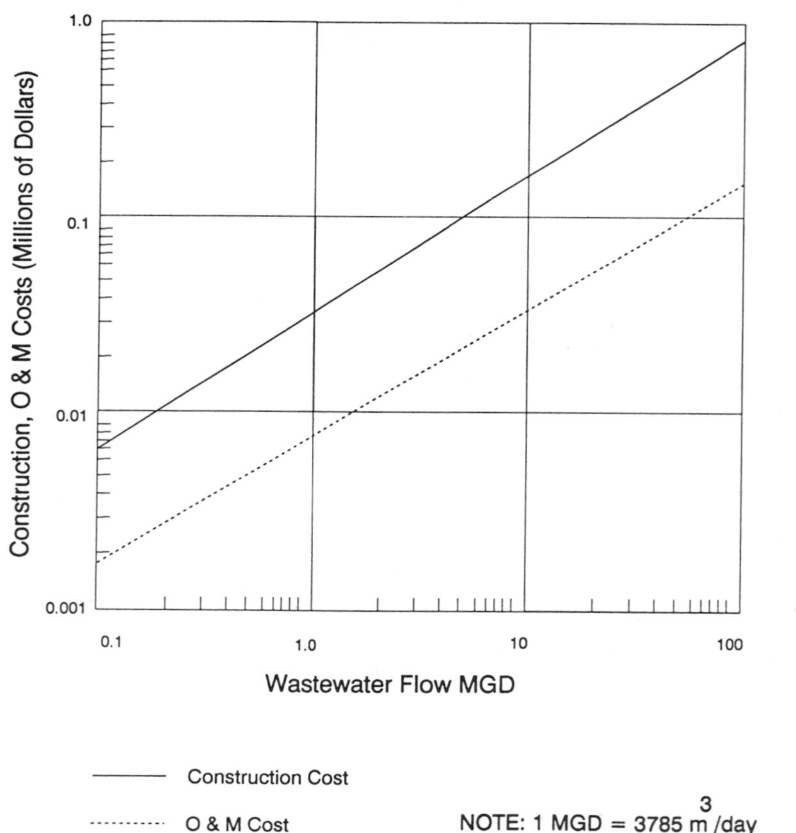

Figure 14.2. Construction, operation, and maintenance costs for aquaculture—wetlands.

Table 14.2. Wetlands Cost Summary

CATEGORY	VERMONTVILLE, MI 11-ACRE CONSTRUCTED WETLAND (0.07 MGD) COST	% OF TOTAL	HOUGHTON LAKE, MI NATURAL WETLAND (1.1 MGD SUMMER/ 0.42 MGD WINTER) COST	% OF TOTAL
Labor (including Overhead/ Administration)	$1,900	56.1	$4,750	53.9
Electrical Energy	372	11.2	2,400	27.4
Equipment Use (including fuel)	1,100	31.7	660	7.5
Repair/ Replacement	33	1.0	165	1.9
New Equipment	0	0	275	9.4
TOTALS	$3,300	100%	$8,800	100%
Cost breakdown per Million Gallons Treated	$ 128		$ 54	

Source: Reference 14.
Note: 1 MGD = 3785 m³/day.

14.5 OPERATION AND MAINTENANCE

Mosquito problems can be avoided by maintaining mosquito fish and an appropriate mix of aquatic flora in the system. Regular harvesting of plants and a range of plant types helps by limiting plant overcrowding—a major cause of stagnant pools.

Vegetation is the main form of erosion control in the wetland areas and works quite well once established. Without planting and cultivation, a minimum of one spring and summer are usually needed before the vegetation becomes well established. Vegetation is not sufficient to prevent erosion around weirs, gates, pipes, and other devices in the constructed portion(s) of the system. These areas may be fortified with stone or other available riprap material.

Clostridium botulinum is the cause of avian botulism and is deadly to waterfowl. Avoiding anaerobic conditions by keeping the water circulating and maintaining a depth below 3 feet are important factors in botulism avoidance. Removal of floating organic debris which collects behind weirs and in corners is important. Steep-sided levees, adjustable broad crested weirs for controlling water levels, conveying water by pipeline, and maintaining the ability to shunt a plot out of service for draining, are also factors in a botulism avoidance program.

Mosquitoes lay eggs in water, and the larva undergo metamorphosis to the adult form. In order to breathe, the larva must hang from the surface film of the water, piercing it with their respiratory tube to obtain oxygen. Knowledge of the mosquito life cycle and habitat needs helps the wetlands manager avoid mosquito breeding problems. Open water areas with circulation limit mosquito production because these areas are subject to wind action and provide easy access for predators.

The vegetation produced as a result of the system's operation may or may not be removed and can be utilized for various purposes (e.g., composted for use as a fertilizer/soil conditioner, dried or otherwise processed for use as animal feed supplements, or digested to produce methane). If vegetation disposal is desired, it is important to have a workable plan before the system goes into operation.

14.6 CONTROL

Limited operator attention is required if the system is properly designed.

14.7 SPECIAL FACTORS

Tie-ins with cooling water discharges from power plants to recover waste heat have potential for extending growing seasons in colder climates. Enclosed and covered systems are possible for small flows.

14.8 RECOMMENDATIONS

This technology is useful for polishing treated effluents. It has potential as a low cost, low energy consuming alternative or addition to conventional treatment systems, especially for small flows. It has been successfully used in combination with chemical addition and overland flow land pretreatment systems. Wetland systems may also be suitable for seasonal use in treating wastewaters from recreational facilities, some agricultural operations, or other wastes, in those cases where the necessary land area is available. Finally, it also has potential application as an alternative to lengthy outfalls extended into rivers, etc., and as a method of pretreatment of surface waters for domestic supply, storm waters, and for other purposes.

15. PRELIMINARY TREATMENT

15.1 DESCRIPTION

Preliminary treatment may be defined as consisting of those processes used to remove grit, heavy solids and, floatable material from wastewater by using grit settling, coarse screening (bar racks), medium screening, and comminution/grinding. In conventional wastewater treatment or raw water supply treatment, a preliminary treatment system is used to protect pumps, valves, pipelines, and other appurtenances from damage or clogging by large solids or high density materials. Preliminary treatment will also remove large particulate material, thus reducing loadings on following processes.

Preliminary treatment typically consists of bar screens and grit chambers. These two processes are designed in varying sizes with several maintenance options, such as hand-cleaned, and/or mechanically cleaned. A typical flow diagram of preliminary treatment is shown in Figure 15.1 (2).

Design Criteria—the arrangement of preliminary treatment units varies depending on sewage/wastewater characteristics and/or subsequent treatment processes. The preliminary system may also include flow measurement devices such as flumes. Low lift pumping may be included to adjust for operating head losses in the subsequent treatment processes. A few general design rules are (17): Bar Screen—bar size ¼ to ⅝ in. (0.6 to 1.6 cm) width, by 1 to 3 in. (2.54 to 7.2 cm) depth; spacing 0.75 to 3 in. (1.9 to 7.2 cm). Slope—from vertical to 45°; velocity 1.5 to 3 ft/sec (0.5 to 0.9 m/sec). Typical grit removal chambers may operate with horizontal velocities of from 0.5 to 1.25 ft/sec (0.15 to 0.4 m/sec). Sufficiently long retention times should be provided in the grit chambers to settle the lightest and smallest grit particles, generally between 10 and 30 minutes.

15.2 LIMITATIONS

Bars and screens require regular cleaning. If the cleaning is by mechanical means, preventive maintenance is required on a regular basis, especially in those cases where solids are heavy and solids concentrations are high. Accumulations are highest during periods of rainfall for both water and wastewater treatment systems. Operational problems have been experienced if comminutors are used in certain installations with a heavy influx of plastic or high density objects.

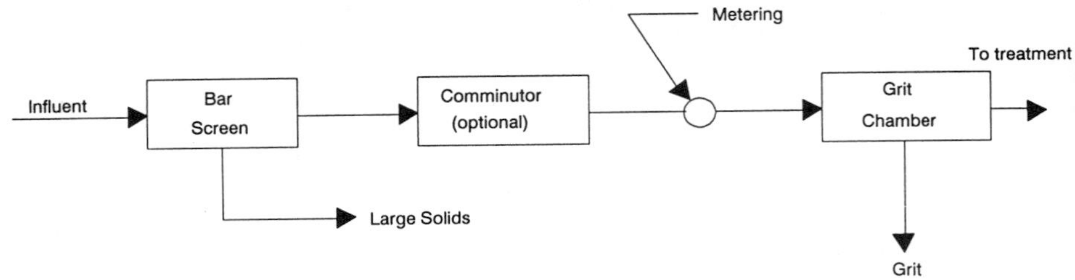

Figure 15.1. Flow diagram of preliminary treatment (*source:* reference 2).

15.3 COSTS

The construction costs are shown in Figure 14.2 (2,11), for flow channels and superstructures, bar screen (mechanical), horizontal grit chamber with mechanical grit handling equipment, and Parshall flume and flow recording equipment. Figure 15.2 also shows the operation and maintenance cost (cost for grit disposal not included).

15.4 AVAILABILITY

Preliminary treatment has been used since the earliest days of municipal wastewater and water supply treatment in virtually every type of application.

15.5 OPERATION AND MAINTENANCE

Preliminary treatment consists of two separate and distinct unit operations —bar screening and grit removal. There are two types of bar screens or racks. The most commonly used and oldest technology, consists of hand-cleaned bar racks. Hand cleaning is generally used in smaller treatment plants but requires regular operator attention. The mechanically cleaned type of bar screen is commonly used in larger facilities.

Grit is most commonly removed in chambers, which are capable of settling out high density solid materials, such as sand and gravel. There are two types of grit chambers: horizontal flow and aerated. In both types the settleable solids collect at the bottom of the unit. The horizontal units are designed to maintain a relatively constant velocity by use of proportional weirs or flumes. Constant velocity prevents settling of organic solids while simultaneously resulting in relatively complete removal of the inorganic grit.

The aerated type produces spiral action within the settling chamber. The heavier particles to be removed remain at the bottom of the tank, while lighter and organic particles are maintained in suspension by rising air bub-

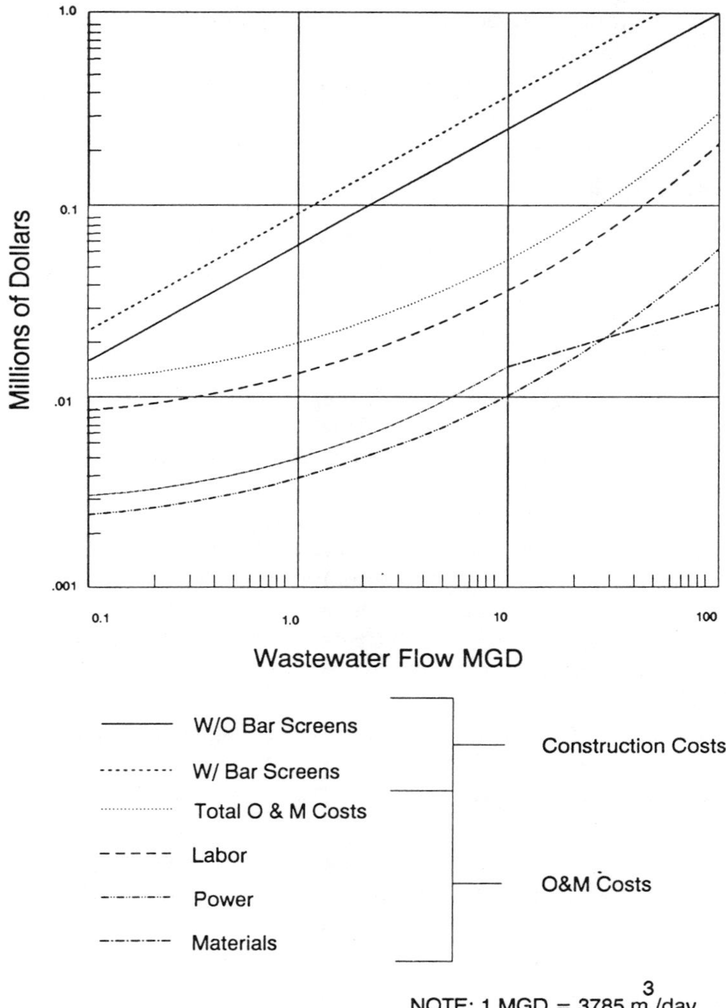

Figure 15.2. Construction, operation, and maintenance costs for preliminary treatment.

bles. The organic material is then treated in subsequent treatment processes. Advantages of aerated units are that the amount of air can be regulated to control the amount of grit/organic solids separation, and offensive odors are controlled. Aeration also facilitates cleaning of the grit. The grit and other solids are often landfilled. The grit removed may require additional treatment prior to disposal depending on organic content.

All unit operations in preliminary treatment except for comminutors generate solids requiring disposal. Screens remove up to 1 yd^3 of 12 to 15% solids/Mgal (202 m^3/Mm^3). Removal however is very much related to the character of the water being treated.

15.6 CONTROL

Preliminary treatment systems are extremely reliable and, in fact, are designed to improve the reliability of downstream treatment systems.

15.7 SPECIAL FACTORS

Many wastewater treatment plants use comminutors. These are mechanical devices that grind up the material normally not removed in the screening process. Therefore, these solids remain in the wastewater to be removed in downstream unit operations.

In recent years, the use of static or rotating wedge-wire screens has increased for preliminary treatment. These remove large organic particles prior to degritting. The units are superior to comminutors in that they remove the screenable material from the waste immediately instead of creating additional loads on downstream treatment processes. On the other hand the screened material is more likely to contain a significant organic fraction in the case of wastewater. Odors are common for stored grit when it contains excess organic solids, in which case short term disposal is suggested.

15.8 RECOMMENDATIONS

This technology should be used at all municipal wastewater treatment plants and water treatment plants with the potential for high solids in the influent. It may be used prior to water and wastewater pumping.

16. HORIZONTAL SHAFT ROTARY SCREEN

16.1 DESCRIPTION

The rotating drum filter operates intermittently or continuously and can be used as changes in the influent quality characteristics dictate. The rotating drum is covered with a plastic or stainless steel screen of uniform sized openings and is partially submerged in a chamber (see Figure 16.1) (2). The chamber is designed to permit the entry of water to the interior of the drum and collection of filtered (or screened) water from the exterior side of the drum. Coarse screens have openings of ¼ in. (0.64 cm) or more; fine screens have openings of less than ¼ in. Screens with openings of 20 to 70 microns are called microscreens or microstrainers. Drum diameters are 3 to 5 ft (0.91 to 1.5 m) and drums are 4 to 12 ft (1.2 to 3.7 m) in length.

Design Criteria—screens are submerged 70 to 80% in the water to be treated. Loading rate: 2 to 10 gal/min/ft^2 (0.82 to 4.1 L/min/m^2) of submerged area depending on pretreatment and mesh size (2). Flow rates may be as high as 10 to 30 gpm/ft^2 (4.1 to 12.3 L/min/ft^2), at head losses around 12 to 18 in. (0.3 to 0.46 m) of water through the filter system (55). Screen openings: 150 microns to 10 mm for pretreatment; 20 to 70 microns for fine particle removal. Drum rotations/min: 1 to 7. Screen materials: stainless steel or plastic cloth. Washwater requirement is 2 to 5% of flow being treated. Performance of fine screen device varies considerably depending on influent solids type, concentration and loading patterns, mesh size, and hydraulic head.

Typical removal rates for some pollutants are as follows: BOD_5, 40 to 60%; and SS, 50 to 70%. Head loss is usually from 0.3 to 2 ft (0.09 to 0.61 m).

16.2 LIMITATIONS

Reducing the speed of rotation of the drum and less frequent flushing of the screen has resulted in increased removal efficiencies but reduced capacities. Very high influent solids concentrations may block or blind the screen media. Thus pretreatment may be required in these cases. Wide variations in influent solids concentrations may result in more frequent cleaning or uneven cleaning frequencies.

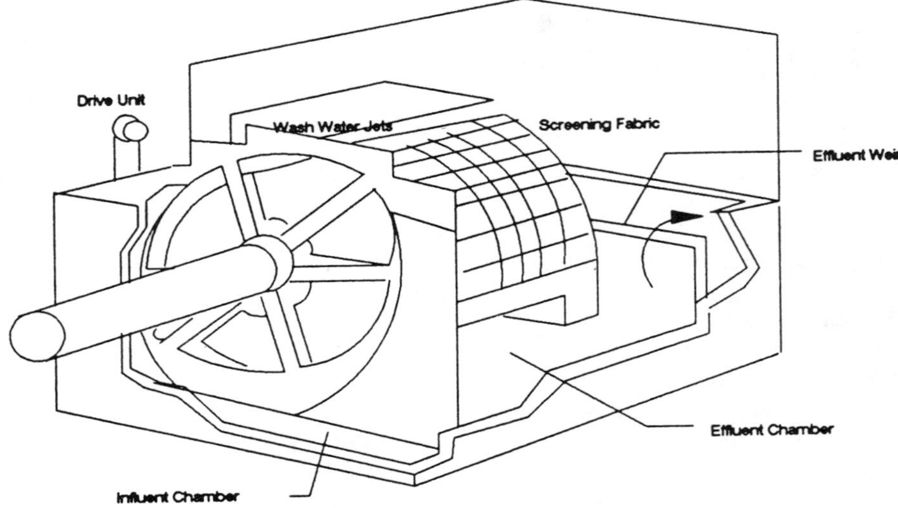

Figure 16.1. Horizontal shaft rotary screen (*source:* reference 2).

16.3 COSTS (See Figure 16.2 (2).)

16.4 AVAILABILITY

Rotating drums are in widespread use for pretreatment and for secondary biological treatment plant effluent polishing.

16.5 OPERATION AND MAINTENANCE

With each revolution of the drum, the solids are flushed from the exposed screen surface into a collecting trough, by water sprays. Blinding by grease is a problem when the screens are used in some pretreatment applications. Effective cleaning is important for efficient removal or suspended solids on a consistent basis. Hand cleaning with acid solution is required for stainless steel cloths under difficult treatment conditions.

Odor problems around equipment is created if solids are not flushed frequently enough from the screen. This is particularly true if the screens are used on influents containing suspended solids with significant organic fractions.

16.6 CONTROL

Rotating drums possess a high degree of reliability from both the process and mechanical viewpoints. The process is simple to operate and associated

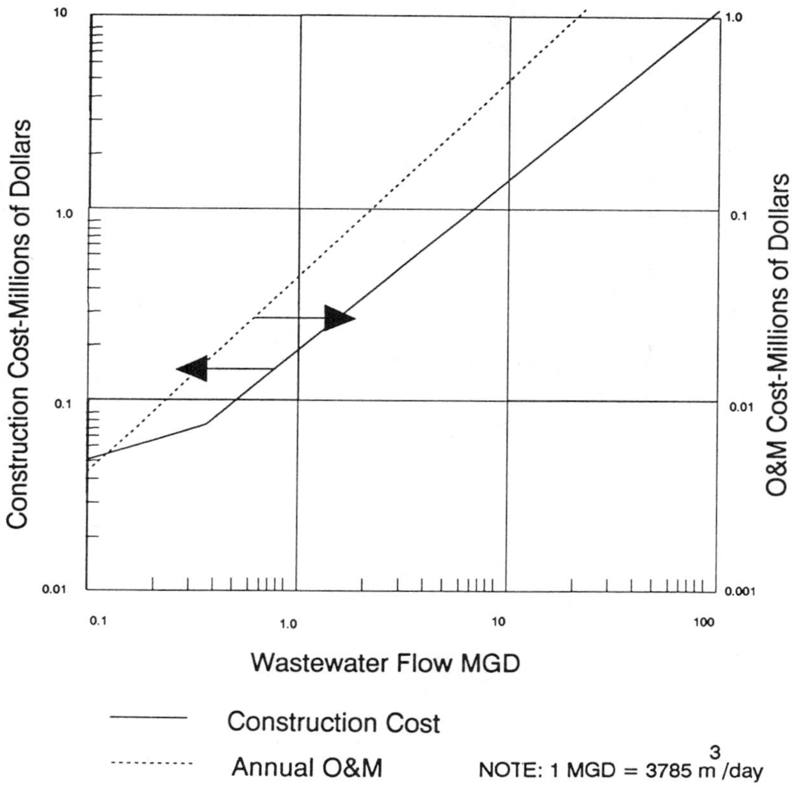
10
1.0
0.1
0.01
1.0
0.1
0.01
0.001
0.1
1.0
10
100
Construction Cost-Millions of Dollars
O&M Cost-Millions of Dollars
Wastewater Flow MGD
Construction Cost
Annual O&M
NOTE: 1 MGD = 3785 m3/day

Figure 16.2. Construction, operation, and maintenance costs for horizontal shaft rotary screen.

mechanical equipment is generally straightforward to maintain. Occasional problems may arise because of incomplete solids removal by flushing.

16.7 SPECIAL FACTORS

Modifications to available hardware may help achieve improved consistency in TSS removals and/or easier operation and maintenance. Some common modifications which are in use include:

- Tile treatment chamber, reinforced concrete chamber, steel chamber, for special waste applications.
- Variable speed drive for drum, useful in situations where influent TSS concentrations vary.
- Addition of backwash storage and pumping facilities.

- Addition of ultraviolet light slime growth control equipment.
- Addition of chlorinating equipment.

16.8 RECOMMENDATIONS

This technology is most useful in the removal of coarse wastewater solids from the wastewater treatment plant influent after bar screen treatment (150 microns to 10 mm). Also for polishing activated sludge effluent (screen openings from 20 to 70 microns).

Screens are used in water supply source applications to protect against such things as leaves. Travelling, rather than rotary screens, have been used successfully (48) with not less than 2 and sometimes as many as 8 meshes to the in. (79 to 315/m). Travelling screens should have a velocity of 3-½ in./sec (8.89 mm/sec). The best operating region for screens is in the 50 to 100 micron size range (55).

17. WEDGEWIRE SCREEN

17.1 DESCRIPTION

Screening is used to remove coarse and/or gross solids from water or wastewater before subsequent treatment. A wedgewire screen is a device onto which water to be treated is directed across an inclined stationary screen or a rotary drum screen. In both cases the screen openings are of uniform size. The solids are trapped on the screen surface and subsequently moved either by gravity (stationary) or by mechanical means (rotating drum, also see Chapter 16) to a solids collection zone for discharge (see Figure 17.1) (2). In stationary screen devices the water to be treated is introduced as a thin film flowing downward across the wedgewire screen with a minimum of turbulence. The stationary screen is generally in three sections of progressively flatter slope downstream from the inlet device.

The design criteria for wedgewire screens for the range of water flow, 0.05 to 36 MGD (0.0022 to 1.6 m^3/sec), are:

STATIONARY	PARAMETER	ROTARY DRUM
0.01 to 0.06 in.	Screen opening	0.01 to 0.06 in.
4 to 7 ft	Head required	1.5 to 4.5 ft
10 to 750 ft^2	Surface required	10 to 100 ft^2
—	Motor size	0.5 to 3 hp

Source: Reference 13
Note: 1 in. = 2.54 cm; 1 m = 3.28 ft; 746 W = 1 hp

Expected pollutant removals are: BOD_5, 5 to 20%; suspended solids, 5 to 25%.

17.2 LIMITATIONS

Wedgewire screens require regular cleaning and prompt residuals disposal, especially in cases of high organic content to prevent odors. High efficiency TSS removal is generally not possible with this technology, i.e., the process is not applicable as a polishing step. Small particle size TSS may resist treatment.

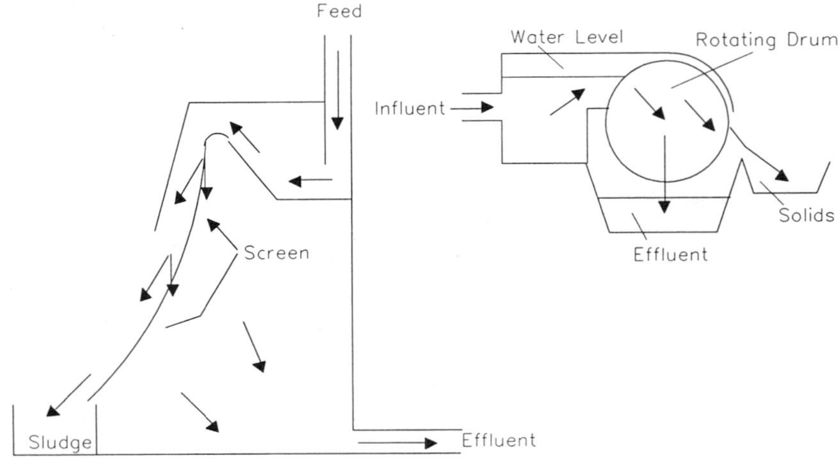

Figure 17.1. Typical wedgewire screen (*source:* reference 13).

17.3 COSTS

Figure 17.2 (2) shows the construction and O&M costs for wedgewire screens. The costs are based on wedgewire stainless steel screen with 0.06 in. (1.5 mm) openings, including equipment and installation. Pumping equipment and piping for effluent or sludge are included. Operation and maintenance costs are based on labor costs at $7.50/hour, power at $0.02/Kwh, and pumping head (for stationary screen) at 4.5 ft (1.4 m).

17.4 AVAILABILITY

This technology has been in general use since 1965 and in municipal wastewater treatment since 1967. There are over 100 installations to date in the U.S. and many more worldwide.

17.5 OPERATION AND MAINTENANCE

These screens, as with all other screening devices, require effective cleaning on a regular basis to maintain treatment effectiveness. Cleaning schedules should be based on the results of the early runs using the devices. Where screen blinding due to grease is a problem, increased frequency of cleaning is required.

Pretesting of the screen media may be required for some waste applications. Highly acid wastes, strongly corrosive wastes, or those with strong base

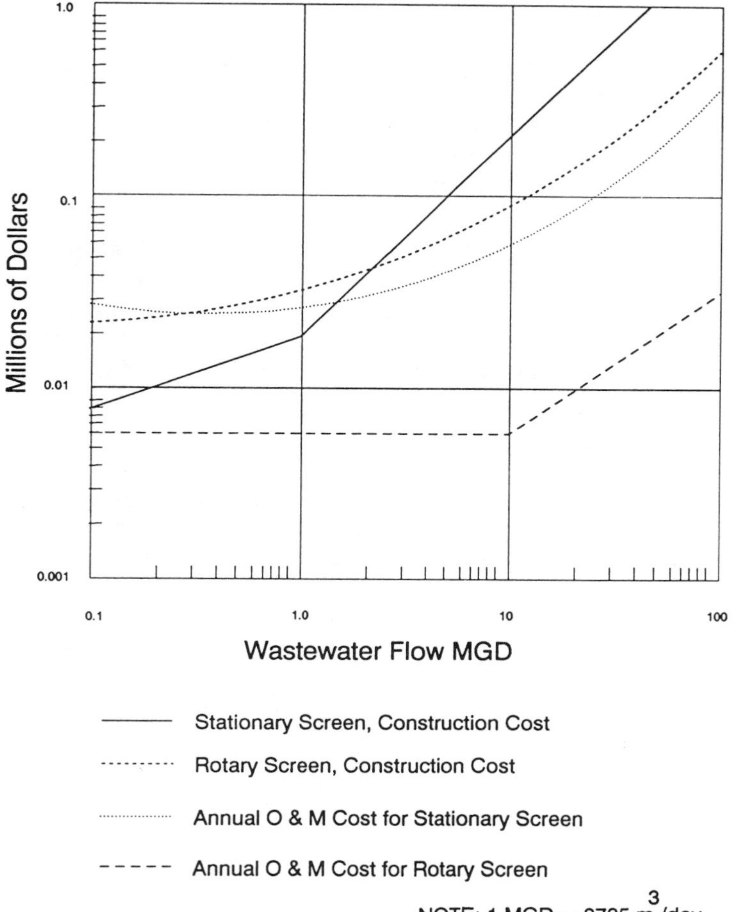

Figure 17.2. Construction, operation, and maintenance costs for wedgewire screen.

characteristics may react with some screen metals or fabrics. Pretreatment is desirable in these cases.

17.6 CONTROL

Screening devices have a relatively high mechanical reliability for most applications. The stationary wedgewire screens are highly reliable. High performance in the expected ranges of TSS characteristics is likely for well maintained and cleaned systems.

17.7 SPECIAL FACTORS

Wedgewire screen spacing should be selected based on specific treatment applications, especially TSS characteristics. The preferred way to obtain expected performance data is by testing the actual water to be treated with various filter fabrics and filter configurations. For domestic wastewater treatment applications, spacings are generally between 0.01 and 0.06 in. (0.25 and 1.5 mm).

Inclined stationary screens can be housed in stainless steel or fiberglass structures for special applications; wedgewire screens may be curved or straight; the screen face may be of a single- or multi-angle unit design, three separate multi-angle segments, or a single-curved unit design. Rotary screens may have a single rotation speed drive or a variable speed drive.

17.8 RECOMMENDATIONS

Stationary and rotary screens are ideally suited to use following bar screens and prior to grit chambers. They have also been used for primary treatment, scum dewatering, dilute sludge screening, treatment of digester supernatant and cleaning residuals, and for storm water and combined sewer overflow treatment. Generally, the rotary drum type unit is preferred.

18. TRICKLING FILTER—PLASTIC MEDIA

18.1 DESCRIPTION

The process utilizes a fixed bed of plastic media over which wastewater is applied to achieve aerobic biological treatment. Treatment is accomplished by fixed growths of microorganisms attached to the media. The bed is dosed by a distributor system, and the treated wastewater is collected by an underdrain system. Pretreatment, e.g., settling normally results in satisfactory operation and performance. The process is shown schematically in Figure 18.1. Recirculation may be accomplished with trickling filter or settled effluent as shown but is usually not necessary except during periods of replacement of the fixed growth on the filter media or during seasonal sloughing.

The rotating mechanical distribution system for the wastewater on the filter medium has become standard practice because of its reliability and ease of maintenance, but mostly because it provides for even distribution of the wastewater and, accordingly, improved performance. However, fixed nozzles are often used in roughing filters.

Plastic media is comparatively light with a specific weight 10 to 30 times less than rock media. Its high void space (about 95%) promotes better oxygen transfer than rock media to fixed organisms during treatment in the filter. Rock media has approximately 50% void space. Because of its light weight, containment structures for plastic media may be built as elevated towers 20 to 30 ft high. The construction cost is lower than other options. Existing containment structures for rock media (normally of concrete in excavations) can sometimes serve as a foundation for elevated towers when converting an existing facility to plastic media and when capacity expansion is desired.

Plastic media trickling filters are employed to provide independent secondary treatment or roughing ahead of a second-stage biological process. Operation of trickling filtration in the roughing mode is useful when pretreatment of high strength wastewater is required for subsequent application of the wastewater to another downstream process or when only low level treatment is desired. When used for secondary treatment, the media bed is circular in plan and dosed by a rotary distributor. Roughing applications often utilize rectangular media beds with fixed nozzles for distribution.

Design criteria are presented on Table 18.1 (2). Using a single-stage configuration with filter effluent recirculation and pre- and post-clarification,

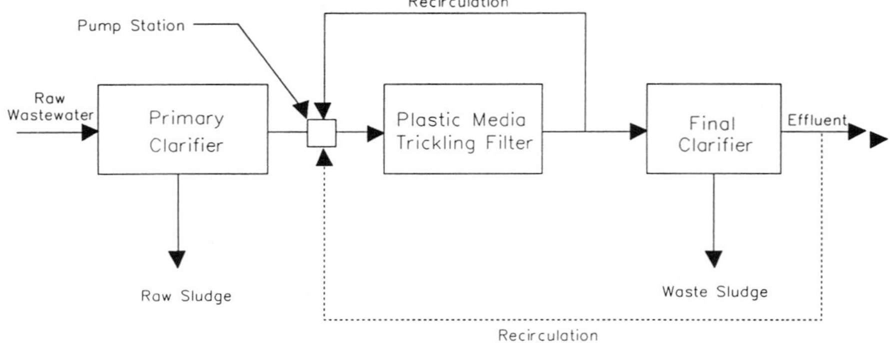

Figure 18.1. A flow diagram of trickling filter plastic media (*source:* reference 2).

removal rates for some pollutants are as follows: BOD_5, 80 to 90%; phosphorous, 10 to 30%; ammonia nitrogen (NH_4-N), 20 to 30%; TSS, 80 to 90%.

18.2 LIMITATIONS

This technology provides low level treatment capability in single stage operation. It is less effective in treatment of wastewater containing high concentrations of soluble organics. It also has limited flexibility and control in comparison with competing biological treatment processes, particularly the activated sludge options. There is a potential for vector and odor problems, although high rate systems have less problems in this regard than low rate trickling filters.

18.3 COSTS (See Figure 18.2 (2, 11).)

18.4 AVAILABILITY

This technology (plastic media) has been used as an alternative to rock media filters for 20 to 30 years.

18.5 OPERATION AND MAINTENANCE

The organic material present in the wastewater is degraded by a population of microorganisms attached to the filter media. As the microorganisms grow, the thickness of the slime layer increases. Periodically the liquid will wash some of the slime off the media, and a new slime layer will start to grow. This

Table 18.1. Trickling Filter Design Criteria

HYDRAULIC LOADING (WITH RECIRCULATION)

Secondary treatment—15 to 90 Mgal/acre/d
350 to 2050 $gal/d/ft^2$
Roughing—60 to 200 Mgal/acre/d
1400 to 4600 $gal/d/ft^2$

ORGANIC LOADING

Secondary treatment—450 to 1750 lb BOD_5/d/acre ft
10 to 40 lb BOD_5/d/1000 ft^3
Roughing—4500 to 22,000 lb BOD_5/d/acre ft
100 to 500 lb BOD_5/d/1000 ft^3

OTHER

Bed Depth—20 to 30 ft
Power requirements—10 to 50 hp/Mgal
Underdrain minimum slope = 1%
Recirculation ratio—0.5:1 to 5:1
Dosing interval—Not more than 15 sec (continuous)
Sloughing—continuous

Note: 1 MGD = 3785 m^3/day
1 ft = 0.3048 m
Mgal/acre/day × 9462 = m^3/ha/day
gal/day/ft^2 × 0.041 = m^3/day/m^2
lb BOD_5/day/acre ft × 8.2 = lb BOD_5/day/ha m
lb BOD_5/day/1000 ft^2 × 33 = lb BOD_5/day/1000 m^2
hp/Mgal × 0.2 = W/m^3

phenomenon of losing the slime layer (sloughing) is primarily a function of the organic and hydraulic loadings on the filter. Sloughing is more likely to occur when seasonal changes are more pronounced, e.g., in less temperate climates. Recirculation may be important in low flow situations. With all media, but especially with plastic media trickling filters, recirculation ensures continuous wetting of the attached media growth and is useful in sloughing control.

18.6 CONTROL

Trickling filters have a high degree of reliability if operated properly, waste flow and characteristic variability is a minimum, and climate is favorable (wastewater temperatures above 13°C). Mechanical reliability is high and the process is simple to operate.

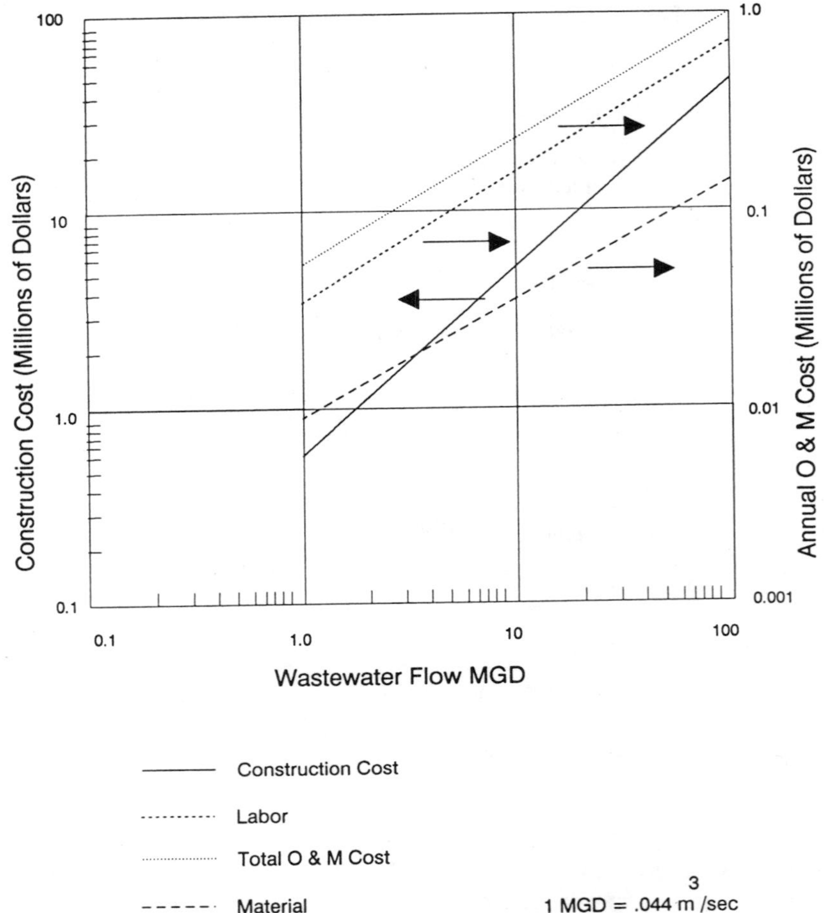

Figure 18.2. Construction, operation, and maintenance costs for trickling filter and plastic media.

18.7 SPECIAL FACTORS

Some common modifications of the system include recirculation flow schemes, multistaging, electrically powered distributors, forced ventilation, filter covers, and use of various methods of pre- and post-treatment of wastewater. The process may also be used as a roughing filter at flow rates above 1400 gal/d/ft^2 (57.4 m^3/m^2/day) and also as a separate stage nitrification process.

Sludge is withdrawn from the secondary clarifier at a rate of 3000 to 4000 gal/Mgal (m^3/Mm^3) of wastewater, containing 500 to 700 lb (225 to 315 kg) of dry solids.

18.8 RECOMMENDATIONS

This technology is best suited for treatment of domestic and compatible industrial wastewaters amenable to aerobic biological treatment. Industrial and joint (mixed industrial and domestic) wastewater treatment facilities may use the process as a roughing filter prior to activated sludge or other unit processes. Existing rock filter facilities can be upgraded by elevation of the containment structure and conversion to plastic media. Finally, it can be used for nitrification, often with follow-on biological treatment.

19. TRICKLING FILTER, HIGH RATE, ROCK MEDIA

19.1 DESCRIPTION

The process consists of a fixed bed of rock media over which wastewater is applied for aerobic biological treatment. Refer to the description in Chapter 18. Zoogleal slimes (collections of microorganisms) form on the media which assimilate and oxidize substances from the wastewater. The bed is dosed by a distributor system and the treated wastewater is collected by an underdrain system (see Figure 19.1). Presettling is normally required beforehand. Post-treatment is often necessary to achieve comparable effluent quality from activated sludge or to meet receiving water quality limitations. The primary difference between rock and plastic media systems is that lower density plastic medium requires less support, thus the plastic media filter can generally be above ground. Notice the pump station in Figure 18.1 (page 120).

The rotating arms are mounted on a pivot in the center of the filter. Nozzles distribute the wastewater as the arms rotate. Continuous recirculation of filter effluent is used to maintain a constant hydraulic loading to the distributor arms and, subsequently to the filter.

Underdrains are manufactured from specially designed vitrified-clay blocks that support the filter media and pass the treated wastewater to a collection sump for transfer to the final clarifier.

The filter media consists of 1 to 5 in. (2.5 to 12.7 cm) diameter stone. The high rate trickling filter bed is typically circular in plan, with a depth of 3 to 6 ft (0.91 to 1.8 m). Containment structures are normally made of reinforced concrete and installed in the ground to support the weight of the media.

Design criteria for the high rate systems are given on Table 19.1 (2). Using a single-stage configuration with filter effluent recirculation and primary and secondary clarification, the removal rates for some pollutants are as follows: BOD_5, 60 to 80%; phosphorous, 10 to 30%; ammonia nitrogen (NH_4-N) 20 to 30%; TSS, 60 to 80%.

19.2 LIMITATIONS

See the discussion for Chapter 18.

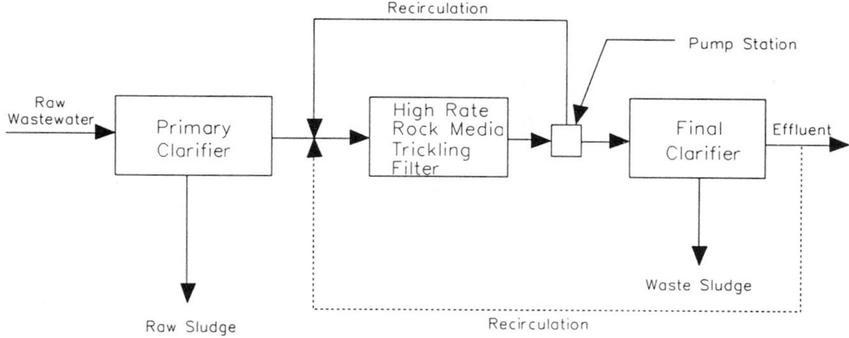

Figure 19.1. Flow diagram of trickling filter high rate, rock media (*source:* reference 2).

19.3 COSTS (See Figure 19.2 (2, 11).)

19.4 AVAILABILITY

This technology has been in widespread use since the 1930s as an improved alternative to the low rate trickling filter process.

Table 19.1. High Rate Trickling Filter Design Criteria

HYDRAULIC LOADING (WITH RECIRCULATION—10 TO 50 MGAL/ACRE/D 230 TO 1150 GAL/D/FT^2
Recirculation ratio—0.5:1 to 4:1 Dosing interval—not more than 15 sec (continuous)
ORGANIC LOADING—900 TO 2600 LB BOD_5/D/ACRE FT 20 TO 60 LB BOD_5/D/1000 FT^3
Other Bed Depth—3 to 6 ft Power requirements—10 to 50 hp/Mgal Underdrain minimum slope = 1 percent Sloughing—continuous Rock, 1 in. to 5 in. diameter

Note: 1 MGD = 3786 m^3/day
1 ft = 0.3048 m
Mgal/acre/day × 9462 = m^3/ha/day
gal/day/ft^2 × 0.041 = m^3/day/m^2
lb BOD_5/day/acre ft × 8.2 = lb BOD_5/day/ha
lb BOD_5/day/1000 ft^2 × 33 = lb BOD_5/day/1000 m^3
hp/Mgal × 0.2 = W/m^3

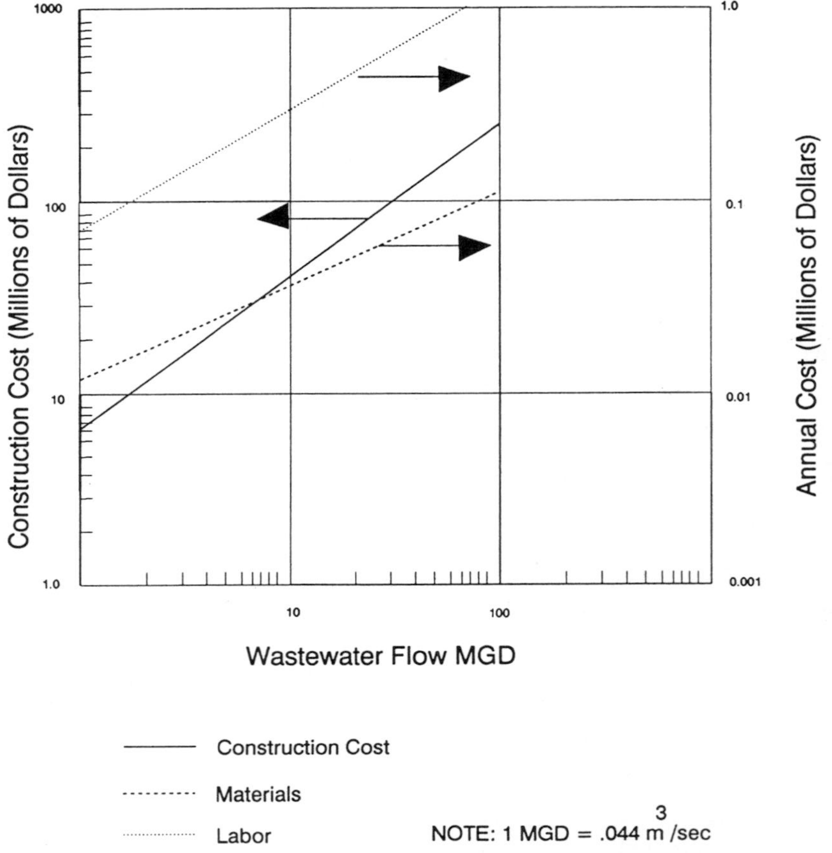

Figure 19.2. Construction, operation, and maintenance costs for trickling filter, high rate, rock media.

19.5 OPERATION AND MAINTENANCE

The organic material present in the wastewater is degraded by a population of microorganisms attached to the filter media. As the microorganisms grow, the thickness of the slime layer increases. As the slime layer increases in thickness, the absorbed organic matter is metabolized before it can reach the microorganisms near the media face. As a result, the microorganisms near the media face enter into an endogenous phase of growth. In this phase, the microorganisms lose their ability to cling to the media surface. The liquid then washes the slime off of the media, and a new slime layer begins to grow.

Sloughing is primarily a function of the organic and hydraulic loadings on the filter. Filter effluent recirculation may be important in high rate trickling filters to promote the flushing action necessary for effective sloughing control. Without sloughing control media clogging and anaerobic conditions could develop, and sloughing tends to increase the solids content in the effluent.

19.6 CONTROL

This process has a high degree of reliability if operating conditions minimize variability and the installation is in a climate where wastewater temperatures do not fall below 13°C for prolonged periods. Sloughing tends to occur during seasonal climatic changes. Mechanical reliability is high, and the process is simple to operate.

19.7 SPECIAL FACTORS

Common modifications that can be applied to the system include various recirculation methods, multistaging, electrically powered distributors, forced ventilation to enhance oxygen supply to the microorganisms, and filter covers to reduce odor and insect problems.

19.8 RECOMMENDATIONS

This technology is best suited for treatment of domestic and compatible industrial wastewaters amenable to aerobic biological treatment. Industrial and joint wastewater treatment facilities may use the process as a roughing filter prior to activated sludge or other unit processes. The process is effective for the removal of suspended or colloidal materials and is less effective for removal of high concentrations of soluble organics. The rock media high rate option should only be used when plastic media is not available or in those cases where plastic media replacement supply may be uncertain.

Trickling filters have been used effectively as satellite treatment systems within sewer collection networks in Guatemala (60). Although this application is with the expectation of further treatment—perhaps at a treatment plant at the end of the collection network—the effluent quality is suitable for sand filtration and for groundwater replenishment with the intent of indirect reuse.

20. TRICKLING FILTER, LOW RATE, ROCK MEDIA

20.1 DESCRIPTION

The process consists of a fixed bed of rock media over which wastewater is applied for aerobic biological treatment. Zoogleal slimes form on the media which assimilate and oxidize substances in the wastewater. The reader should refer to Chapters 18 and 19 for discussions of process description. The bed is dosed by a distributor system, and the treated wastewater is collected by an underdrain system. Preliminary settling is normally required [see Figure 20.1 (2)]. The Figure 20.1 schematic should be compared to Figures 18.1 and 19.1. A summary of design criteria for various trickling configurations is given in Table 20.1 (4).

In contrast to the high rate trickling filter, which may use continuous recirculation of filter effluent to maintain a constant hydraulic loading to the distributor arms, either a suction-level controlled pump or a dosing siphon is employed for that purpose with a low rate filter. Nevertheless, programmed rest periods may be necessary at times because of inadequate influent flow.

Underdrains are manufactured from vitrified-clay blocks that support the filter media and pass the treated wastewater to a collection sump for transfer to the final clarifier. The filter medium consists of 1 to 5 in. (2.5 to 12.7 cm) diameter stone. Containment structures are normally made of reinforced concrete and installed in the ground to support the weight of the media.

The low rate media trickling filter bed generally is circular in plan, with a depth of 5 to 10 ft (1.5 to 3 m). Although filter effluent recirculation is generally not utilized, it can be provided as a standby tool to keep filter media wet during low flow periods.

Design Criteria for low rate filters are given in Table 20.2 (2,4). Filter media material properties are given in Table 20.3 (4). For a low rate trickling filter with a single-stage configuration, primary and secondary clarification and no recirculation, the expected pollutant removal rates are as follows: BOD_5, 75 to 90%; phosphorous, 10 to 30%; ammonia nitrogen (NH_4-N), 20 to 40%; TSS, 75 to 90%.

20.2 LIMITATIONS

Filter flies and odors are common in low rate systems; periods of inadequate moisture for slimes can be common; low rate systems are less effective for

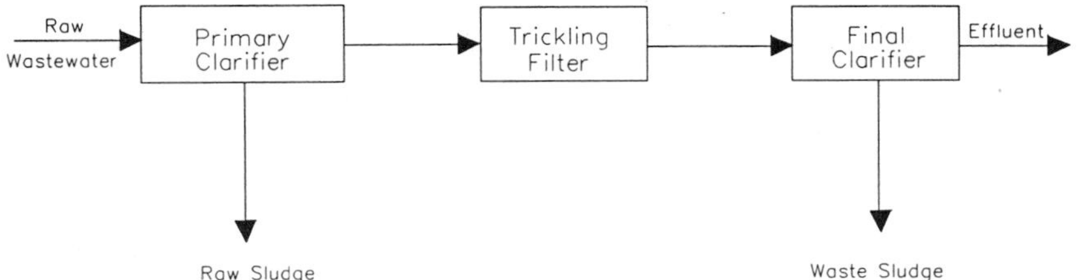

Figure 20.1. Typical flowsheet for trickling filter, low rate, rock media (*source:* reference 2).

treatment of wastewater containing high concentrations of soluble organics. Low rate filters possess limited flexibility and process control in comparison with competing processes; require higher land and capital cost investments than activated sludge for similar capacity and effectiveness.

20.3 COSTS (See Figure 20.2 (2, 11).)

20.4 AVAILABILITY

This process works well in moderate climates and is in widespread use all over the world. Use of post-treatment and multistaging has been found necessary to ensure uniform compliance with effluent limitations. It is presently being superseded more and more by plastic media systems and high rate configurations.

20.5 OPERATION AND MAINTENANCE

The organic material present in the wastewater is degraded by a population of microorganisms attached to the filter media, as described in Chapters 18, 19, and 20. The organisms must be preserved by dosing at regular intervals, especially in intermittent flow situations. When the dosing period is longer than about 2 hours, the efficiency of the process deteriorates because the slime becomes dry. When sloughing has occurred, the operator must pay attention until regrowth has occurred.

Typically, only the top 2 to 4 ft (0.6 to 1.2 m) of the filter medium will have appreciable attached slime growth. The lower segments of the media may have attached growths which are populated largely by nitrifying bacteria oxidizing ammonia nitrogen to nitrite and nitrate forms. Thus, it is possible to produce both a well nitrified, well treated effluent from the point of view of BOD removal.

Table 20.1. Typical Design Information for Trickling Filters

ITEM	LOW-RATE FILTER FILTER	INTERMEDIATE-RATE	HIGH-RATE FILTER (ROUGHING FILTER)	SUPER-RATE
Hydraulic, loading, $m^3/m^2 \cdot d$	1–4	4–10	10–40	40–200
Organic loading $kg/m^3 \cdot d$	0.08–0.32	0.24–0.48	0.32–1.0	0.80–6.0
Depth, m	1.5–3.0	1.25–2.5	1.0–2.0	4.5–12
Recirculation ratio	0	0–1	1–3	1–4
Filter media	Rock, slag, etc.	Rock, slag, etc. synthetic materials	Rock, slag, materials, redwood	Synthetic
Power requirements, $kW/10^3\ m^3$	2–4	2–8	6–10	10–20
Filter flies	Many	Intermediate	Few	Few
Sloughing	Intermittent	Intermittent	Continuous	Continuous
Dosing intervals	<5 min	15 to 60 sec	<15 sec	Continuous
Effluent	Usually fully nitrified	Partially nitrified loadings	Nitrified at low loadings	Nitrified at low loadings

Note: $m \times 3.2808 = ft$
$m^3/m^2/d \times 1.0691 = mgal/acre/d$
$m^3/m^2/d \times 0.0170 = gal/ft^2/min$
$kg/m^3/d \times 62.4280 = lb/1000\ ft^3/d$
$kW \times 1.3410 = hp$

Source: Reference 4.

Table 20.2. Low Rate Trickling Filter Design Criteria

Hydraulic Loading—1 to 4 Mgal/acre/day; 25 to 90 gal/d/ft^2
Organic Loading—200 to 900 lb BOD_5/day/acre ft;
5 to 20 lb BOD_5/day/1000 ft^3
Dosing Interval—Continuous for majority of daily operating schedule, but may become intermittent (not more than 5 min.) during low flow periods
Effluent Channel Minimum Velocity = 2 ft/sec average daily flow
Media—Rock, 1 in. to 5 in., must meet sodium sulfate soundness test
Recirculation Ratio = 0
Depth = 5 to 10 ft
Sloughing—Intermittent
Underdrain Minimum Slope = 1%

Note: m × 3.2808 = ft
in. × 2.54 = cm

20.6 CONTROL

This process is highly reliable under moderate climate conditions, and process operation requires little skill.

20.7 SPECIAL FACTORS

Some common modifications of the system include recirculation flow schemes, multistaging, electrically powered influent distributors, forced ventilation, filter covers, and use of various methods of pre- and post-treat-

Table 20.3. Physical Properties of Trickling-Filter Media

MEDIUM	NORMAL SIZE, MM	DENSITY KG/M^3	SPECIFIC SURFACE AREA, IN.^{2}/IN.^{3}	VOID SPACE, %
River rock (diameter)				
Small	25–65	1250–1450	55–70	40–50
Large	100–120	800–1000	40–50	50–60
Blast-furnace slag (diameter)				
Small	50–80	900–1200	55–70	40–50
Large	75–125	800–1000	45–60	50–60
Plastic				
Conventional	600 × 600 × 1200	30–100	80–100	94–97
High rate	600 × 600 × 1200	30–100	100–200	94–97
Redwood	1200 × 1200 × 500	150–175	40–50	70–80

Note: mm × 0.03937 = in.
kg/m^3 × 62.4280 = lb/10^3 ft^3
m^2/m^3 × 0.3048 = ft^2/ft^3
Source: Reference 4.

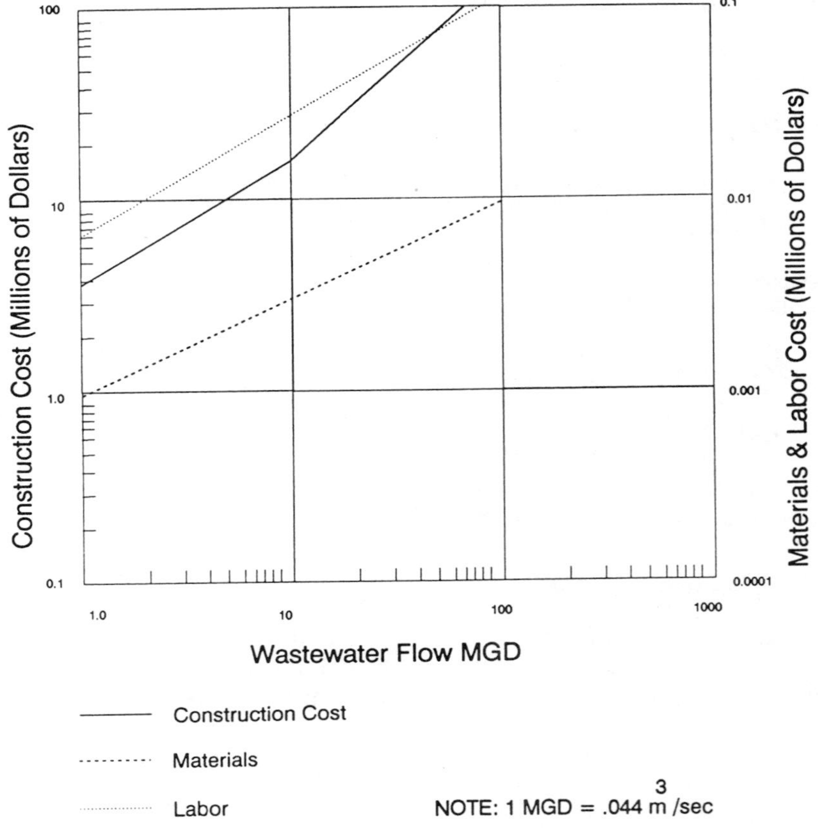

Figure 20.2. Construction, operation, and maintenance costs for trickling filter, low rate, rock media.

ment of wastewater. High head losses—5 to 10 ft (1.5 to 3 m) may be possible.

20.8 RECOMMENDATIONS

This technology is best used for treatment of domestic and compatible industrial wastewaters amenable to aerobic biological treatment. Industrial and joint (mixed domestic and industrial) wastewater treatment facilities may use the process as a roughing filter prior to activated sludge or other processes. Existing rock filter facilities can be upgraded via elevation of the containment structure and conversion to plastic media. Finally, it can be used for nitrification following biological treatment.

21. AERATED/AEROBIC LAGOONS

21.1 DESCRIPTION

This chapter is presented on the basis that *aerobic* lagoons operate naturally, while *aerated* lagoons are artificially oxygenated mechanically.

Aerated

Aerated lagoons are medium-depth basins designed for biological treatment of wastewater on a continuous basis. In contrast to stabilization ponds, which obtain oxygen from photosynthesis and surface reaeration, aerated lagoons typically employ devices which supply supplemental oxygen to the system (2). Aerated lagoons evolved from stabilization ponds when aeration devices were added to counteract odors arising from anaerobic conditions. The aeration devices may be mechanical (i.e., surface aerator) or diffused air systems using submerged pipes. Surface aerators are divided into two types: caged aerators and the more common turbine and vertical shaft aerators. Diffused air systems utilized in lagoons consist of plastic pipes supported near the bottom of the lagoon cells with regularly spaced sparger holes drilled in the tops of the pipes. Because aerated lagoons are normally designed to achieve partial mixing only, aerobic/anaerobic stratification may occur, and a large fraction of incoming solids and a large fraction of the biological solids produced from waste conversion settle to the bottom of the lagoon cells.

Removal rates for certain pollutants are as follows: BOD removal, for an influent concentration of 200 to 500 mg/L, 60 to 90%; COD, 70 to 90%; TSS, for an influent concentration of 200 to 500 mg/L, 70 to 90%.

Design criteria: one or more aerated cells, followed by a settling (typically unaerated) cell; detention time, 3 to 10 days; depth, 6 to 20 ft (1.8 to 6.1 m); pH, 6.5 to 8.0; water temperature range, 0 to 40°C; optimum water temperature, 20°C; oxygen requirement, 0.7 to 1.4 times the weight of BOD_5 removed; organic loading, 10 to 300 lb BOD_5/acre/d (approx. 10 to 300 kg/ha/day). Table 21.1 presents design criteria for several lagoon configurations (4). The distinction among the various types of lagoons and ponds is not clear in many cases. This is especially true since the systems are often operated differently than the original design intended.

Energy requirements are as follows: for aeration, 6 to 10 hp/million gal

Table 21.1. Typical Design Parameters for Aerobic Stabilization Ponds and Lagoons

PARAMETER	AEROBIC (HIGH-RATE) POND	AEROBIC POND[a]	AEROBIC POND (MATURATION)	AERATED LAGOONS
Flow regime	Intermittently mixed	Intermittently mixed	Intermittently mixed	Completely mixed
Pond size, ha	25–1	<4 multiples	1–4	1–4 multiples
Operation[b]	Series	Series or parallel	Series or parallel	Series or parallel
Detention time, d[b]	4–6	10–40	5–20	3–10
Depth, m	0.30–0.45	1–1.5	1–1.5	2–6
pH	6.5–10.5	6.5–10.5	6.5–8.0	6.5–8.0
Temperature range, °C	5–30	0–30	0–30	0–30
Opt. temp. °C	20	20	20	20
BOD_5 loading, kg/ha/d[c]	10–300	40–120	<15	
BOD_5 conversion, %	80–95	80–95	60–80	80–95
Principal conversion products	Algae, CO_2, cells	Algae, CO_2, cells	Algae, CO_2, cells, NO_3	CO_2, cells
Algal concentration, mg/L	100–260	40–100	5–10	—
Effluent TSS mg/L[d]	150–300	80–140	10–30	80–250

[a] Long detention time, conventional aerobic ponds designed to maximize the amount of oxygen produced rather than the amount of algae produced.
[b] Depends on climatic conditions.
[c] Typical values (much higher values have been applied at various locations). Loading values are often specified by state control agencies.
[d] Includes algae, microorganisms, and residual influent suspended solids. Values are based on an influent soluble BOD_5 of 200 mg/L and, with the exception of the aerobic ponds, an influent suspended-solids concentration of 200 mg/L.

Note: ha × 2.4711 = acre
m × 3.2808 = ft
kg/ha · d × 0.8922 = lb/acre · d

capacity (2 to 1.2 W/m^3); to maintain all solids in suspension, 60 to 100 hp/million gal capacity (12 to 20 W/m^3); to maintain some solids in suspension, 30 to 40 hp/million gals capacity (6 to 8 W/m^3).

Aerobic

Aerobic stabilization ponds are typically large, shallow earthen basins that are used for the treatment of wastewater by natural processes involving the use of both algae and bacteria. These ponds are also called high rate aerobic ponds, or maturation ponds when dissolved oxygen concentration levels are maintained throughout their entire depth. They are usually 12 to 18 in. (30.5 to 46 cm) deep, and it is desirable to allow light to penetrate to the full depth. Artificially aerated ponds may be deeper. Mixing is primarily provided by gases produced from the photosynthesis process of algae.

There are two basic types of aerobic ponds. The first type is shallow with a limited depth of 6 to 18 in. (15 to 46 cm), containing high populations of algae. The second type is about 5 ft (1.5 m) deep with large amounts of bacteria present. To achieve best results with aerobic ponds, both types are mixed periodically using pumps or surface aerators. The oxygen released by the algae through the process of photosynthesis is used by the bacteria in the aerobic degradation of organic matter. The nutrients and carbon dioxide released by bacterial degradation are used by the algae. Typical design parameters for aerobic ponds are shown in Table 21.2 (7, 15, 24).

There is considerable variation in design approaches to ponds and lagoons. This is probably because the actual operating conditions lay somewhere among the aerobic, anaerobic, and facultative modes. The various design approaches are based on studies during which the concentration of dissolved oxygen was uncertain. At best, the dissolved oxygen concentration was undoubtedly different at different points within the same pond. The design for any technology should be based on specific site related studies, but this is especially true for ponds and lagoons. The operating conditions can be fixed given the waste quantity and quality characteristics during the site specific studies. Potential cost savings from smaller sizes alone may pay for pilot plant studies. Table 21.3 (64) is presented to illustrate the results of several design approaches for facultative ponds. In warm climates loading rates may be higher. The loading rates in Table 21.3 should be compared to those in Table 21.2 for anaerobic lagoons.

21.2 LIMITATIONS

Lagoon systems are simpler to operate and maintain than trickling filter or activated sludge biological systems but are less controllable. Thus, it is more

Table 21.2. Typical Design Parameters for Aerobic, Anaerobic, and Faculative Stabilization Ponds

	PONDS/LAGOONS		
PARAMETER	AEROBIC	ANAEROBIC	FACULATIVE
Flow	Complete, mixed	—	Mixed surface layer
Size (acre)	2.5 to 10	0.5 to 2.5	2.5 to 10
Operation	Series	Series	Series
Detention Time (day)	3 to 10	20 to 50	7 to 20
Depth (ft)	10 to 33	8 to 16	3 to 8
pH	6.5 to 8	6.8 to 7.2	6.5 to 8.5
Temperature range (°F)	32 to 86	43 to 122	32 to 122
Optimum temp. (°F)	68	86	68
BOD Loading Rate lb/acre/day	10 to 300	178 to 446	45 to 178
Algal conc. (mg/L)	—	0 to 5	5 to 20
Effluent TSS (mg/L)	80 to 250	80 to 160	40 to 60
Principal conv. products	CO_2 bacteria	CO_2, CH_4, bacteria	Algae, CO_2, CH_4
BOD_5 conversion, %	80 to 95	50 to 85	80 to 95

Source: Reference 7.

difficult to treat highly variable waste streams. Toxicity impacts may affect the lagoon contents and long periods of readjustment may be required.

The dominant algae group or bacterial species present in any section of aerobic ponds depends on several factors, such as organic loading, degree of pond mixing, nutrients, sunlight, pH and temperature. Temperature is the key limiting factor to high loading rates for aerobic biological treatment processes.

21.3 COSTS

See Figures 21.1, and 21.2 (2, 11). Operation and maintenance costs for anaerobic lagoons should be lower than the costs shown for aerated lagoons because of the absence of mechanical aerators. On the other hand, aerated lagoons may be significantly smaller in size than facultative or anaerobic lagoons because of improved oxygen transfer allowing higher loadings. Thus, construction costs for aerated lagoons may be lower.

21.4 AVAILABILITY

While not widely used when compared with the large number of stabilization ponds in common use throughout Latin America, the Caribbean, and the U.S. for example, the technology has been fully demonstrated and in widespread use for many years.

Table 21.3. Summary of Results from Various Design Methods for Facultative Ponds

DESIGN METHOD	DETENTION TIME, d		VOLUME, m^3		SURFACE AREA, ha		DEPTH M	NO. CELLS IN SERIES	SURFACE LOADING RATE kg BOD_5/ha/d	
	PRIMARY POND	TOTAL SYSTEM	PRIMARY POND	TOTAL SYSTEM	PRIMARY POND	TOTAL SYSTEM			PRIMARY POND	TOTAL SYSTEM
Areal loading rate	66[a]	180	125,600[a]	386,800	9.5	22.3	2 (1.4)[c]	4	40	17
Gloyna	—	140	125,600[a]	265,000	—	26.5	2 (1)[c]	—	—	14
Marais & Shaw	37[b]	74	69,300[b]	138,600	2.8	5.6	2.4	2	135	68
Plug Flow	66[a]	180	125,600[a]	386,800	9.5	22.3	2 (1.4)[c]	4	40	25
Wehner & Wilhelm	66[a]	80–132	125,600[a]	151,400–249,900	9.5	10.8–17.9	2 (1.4)[c]	4	—	30–48

[a] Controlled by state standards and is equal to value calculated for an areal loading rate of 40 kg/ha/d and an effective depth of 1.4 m.
[b] Also would be controlled by state standards for areal loading rate; however, the method includes a provision for calculating a value and this calculated value is shown.
[c] Effective depth.
Source: Reference 64.

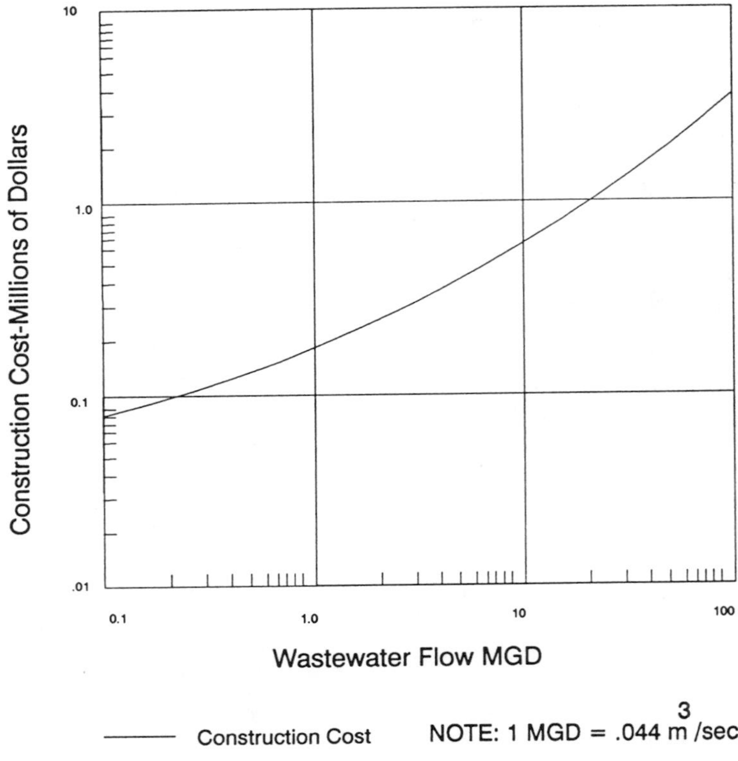

Figure 21.1. Construction cost of aerated lagoons.

Ponds are the most economical method of sewage treatment wherever sufficient land is available at relatively low cost. Aerobic ponds have found only limited application due to higher cost and somewhat more complex operation than facultative or anaerobic ponds.

21.5 OPERATION AND MAINTENANCE

Aerated

As the solids begin to build up, a portion will undergo aerobic decomposition. Some volatile organic toxic compounds are removed by the aeration process in aerated lagoons (more recent work shows that volatile organics may be adsorbed onto the biomass and later degraded or become part of the sludge solids). Incidental removal of toxic organics are similar to an activated sludge system.

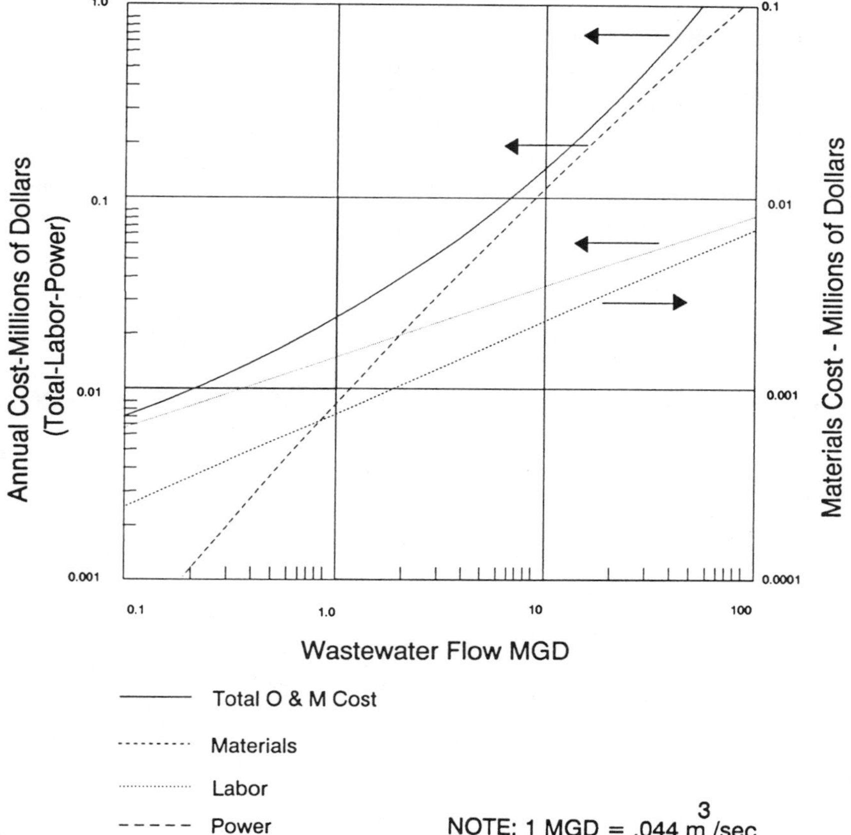

Figure 21.2. Operation and maintenance cost of aerated lagoons.

Several smaller aerated lagoon cells in series are more effective than one large cell. Tapering aeration intensity downward in the direction of flow (through successive cells) appears to promote enhanced settling of solids in the last cell. A non-aerated polishing cell following the last aerated cell is an optional, but recommended design technique to enhance suspended solids removal prior to discharge.

Aerobic

In aerobic ponds the microbiological decomposition of organic wastes is accomplished by algae and bacteria in a combination of cyclic-symbiotic relationships. The principal advantage of this process is that it removes pathogens at a much lower cost than any other form of treatment. Ponds should be cleaned and maintained regularly. It is particularly important to remove grass and other plant growth from the periphery and in the pond itself. Some pond installations in Latin America are known to have such significant growth that capacity is significantly reduced. High solids containing wastes should be pretreated before pond treatment. Solids accumulation in ponds can significantly reduce capacity and lower treatment efficiency. Floating scum mats should be removed, or oxygen transfer is impaired.

Aerated and Aerobic

Abnormal operation occurs when lagoons are overloaded because the BOD loading rate is too high. Excessive effluent BOD concentrations can occur when influent loads exceed design capacity due to population or industrial waste increases. Under these conditions expanded facilities should be considered. Large amounts of brown or black scum on the surface of a pond is an indication that the pond is overloaded. Scum on the surface of a pond often leads to odor problems.

Hydraulic overloading can occur when too much flow is diverted to one lagoon. This can happen when an operator accidentally feeds one lagoon more than another or when a pipe opening is blocked by rags, solids, or grit. Once this happens and the overloaded lagoon starts producing odors, it must be taken out of service and the flow diverted to the other lagoons until the overloaded lagoon recovers. Hydraulic overloading may occur during storms and periods of high runoff.

During winter conditions the pond can become covered with ice and snow. Sunlight is no longer available to the algae, and oxygen cannot enter the water from the atmosphere. Without dissolved oxygen available for aerobic decomposition, anaerobic decomposition of the solids occurs but slowly because of the low temperatures. By keeping the pond surface at a high level, a longer detention time and lower temperature differential will be obtained.

During periods of ice cover, odorous gases formed by anaerobic decomposition accumulate under the ice and are dissolved into the wastewater being treated. Some odors may be observed in the spring just after the ice cover breaks up because the lagoon is still in an anaerobic state, and some of the trapped dissolved gases are being released. Ice melting in the spring provides dilution water with a high oxygen content, thus the lagoons usually become facultative in a few days after ice breakup if they are not organically (BOD) overloaded.

21.6 CONTROL

Aerated

The service life of an aerated lagoon is estimated at 30 years or more with consistent cleaning. The reliability of equipment and the process is high. Little operator expertise is required, more important is that aeration equipment requires regular maintenance, cleaning and attention.

Aerobic

Aerobic ponds are designed for high BOD removal (95%) and may be used as post-treatment for facultative and anaerobic ponds. They may also be used for destruction and removal of highly concentrated organic matter (especially some industrial wastes) and pathogens. Consideration should always be given to designing and operating ponds in series configurations for added control, and also for combining aerobic with anaerobic and facultative ponds, especially for high strength wastes.

Aerated and Aerobic

Aerated/aerobic pond operation is somewhat more complex than facultative and anaerobic ponds because the effluent may contain high suspended solids and algae. Also, the relatively shorter detention time may result in higher amounts of coliform in the effluent. Because of their shallow depth, covering/paving the bottom may be required to prevent weed growth.

21.7 SPECIAL FACTORS

Both aerated and aerobic lagoons may be lined with concrete or an impervious flexible lining, depending on soil conditions. When high-intensity aeration produces completely mixed (all aerobic) conditions a final settling tank is generally required. Solids may be recycled to maintain about 800 mg/L mixed liquor volatile suspended solids (MLVSS) in this mode.

There is an opportunity for volatile organic material and pathogens to enter the air with any aerated wastewater treatment process. The amount depends on air/water contact afforded by the aeration system. There is also potential for seepage of wastewater into groundwater unless a lagoon is lined. Compared to other biological treatment processes, aerated lagoons generate less solid residue for immediate disposal, since solids retention times are very long, and organic degradation is more complete as a result.

21.8 RECOMMENDATIONS

These processes are used for domestic and industrial wastewater of low and medium organic strength. Combinations may be used to treat higher strength wastes. They are commonly used when land is inexpensive and costs and operational control are to be minimized. It is relatively simple to upgrade existing oxidation ponds, lagoons, and natural bodies of water to this type of treatment.

Mechanical aeration increases the oxidation capacity of the pond(s) and is useful in overloaded anaerobic and/or facultative ponds that generate odors. It is also useful when supplemental oxygen requirements are high or when wastewater treatment requirements are either seasonal or intermittent, such as for canning or other food industries.

22. SLUDGE DRYING BEDS

22.1 DESCRIPTION

Drying beds are used to dewater sludge by two mechanisms: drainage through the sludge mass to the underlying soil; and evaporation from the surface exposed to the air. Collected filtrate is returned to the treatment plant by means of the underdrains (spaced from 8 to 20 ft, 2.44 to 6.1 m, apart) shown in Figure 22.1. Sludge drying beds usually consist of 4 to 9 in. (10.2 to 23 cm) of sand which is placed over 8 to 18 in. (20 to 46 cm) of graded gravel or stone. The sand has an effective size of 0.3 to 1.2 mm and a uniformity coefficient less than 5. Gravel is normally graded from ⅛ to 1.0 in. (3 to 25.4 mm). Underdrain piping is often vitrified clay laid with open joints, has a minimum diameter of 4 in. (10.1 cm), and is laid on a minimum slope of about 1%.

Drying beds achieve a cake of 40 to 45% solids in 2 to 6 weeks in good weather, with a well digested waste activated, primary or mixed sludge. Dewatering time may be reduced by 50% or more with chemical conditioning. Conditioning is accomplished by addition of chemicals and even with water treatment plant sludges. Also, sludges from both wastewater treatment and water treatment plants possess considerable soil conditioning properties (see Chapter 23). Solids content as high as 85 to 90% have been achieved on sand beds, but generally the time required (and thus land requirements) to achieve such high concentrations is impractical.

Design Criteria: for area requirements for open bed drying, 1.0 to 1.5 ft^2/capita (0.093 to 0.135 m^2/capita) for primary digested sludge; 1.75 to 2.5 ft^2/capita (0.163 to 0.23 m^2/capita) for primary and activated sludge; 2.0 to 2.5 ft^2/capita (0.19 to 0.23 m^2/capita) for alum or iron precipitated sludge. Enclosed beds require only 60 to 75% of the area for open beds. Solids loading rates vary from 10 to 28 lb/ft^2/yr (49 to 137 kg/m^2/yr) for open beds, to 12 to 40 lb/ft^2/yr (59 to 196 kg/m^2/yr) for closed beds. Sludge beds should be located at least 200 ft (61 m) from dwellings to avoid odor complaints which may result from application of poorly digested sludges.

Sludge is placed on the bed in an 8 to 12 in. (20 to 30 cm) layer. The drying area is partitioned into individual cells, approximately 20 ft (6.1 m) wide by 20 to 100 ft (6.1 to 30 m) long. Individual cell size is such that one or two cells are filled with a normal withdrawal quantity of sludge from digesters, Imhoff tanks, or other treatment systems. The interior partitions of cells consists of two or three creosoted planks, one on top of the other, 15 to 18 in. high (38 to

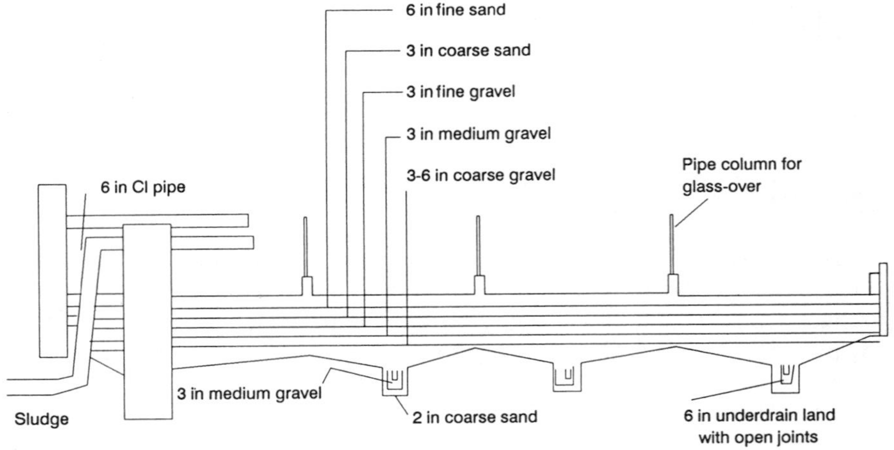

Figure 22.1. Diagram of sludge drying beds (*source:* reference 2).

46 cm). These may stretch between slots in precast concrete posts or other convenient construction techniques may be used. The outer boundaries of the cells may be concrete, or earthen embankments may be used for open beds. Concrete foundation walls are required to support bed enclosure structures.

Piping to the sludge beds is generally of cast iron and designed for a minimum velocity of 2.5 ft/sec (0.75 m/sec). Piping arrangements allow drainage into the beds from sludge treatment processes in the plant. Provisions are made to flush the lines in order to remove blockages and prevent freezing effects on piping in cold climates. Distribution boxes are provided to divert sludge influent flow to the selected cells; splash plates are used at the sludge inlets to distribute the sludge evenly over the bed and to prevent erosion of the sand.

22.2 LIMITATIONS

Air drying on beds is restricted to well digested or stabilized wastewater sludges. Raw biological sludge is malodorous, attracts insects, and does not dry satisfactorily when applied at reasonable depths. Oil and grease clog sandbed pores and may retard drainage. The design of drying beds is affected by weather conditions, sludge characteristics (especially dewatering characteristics), land values, and proximity of nearby residences. Operation is significantly restricted during periods of prolonged freezing and rain. Land requirements may also be significant.

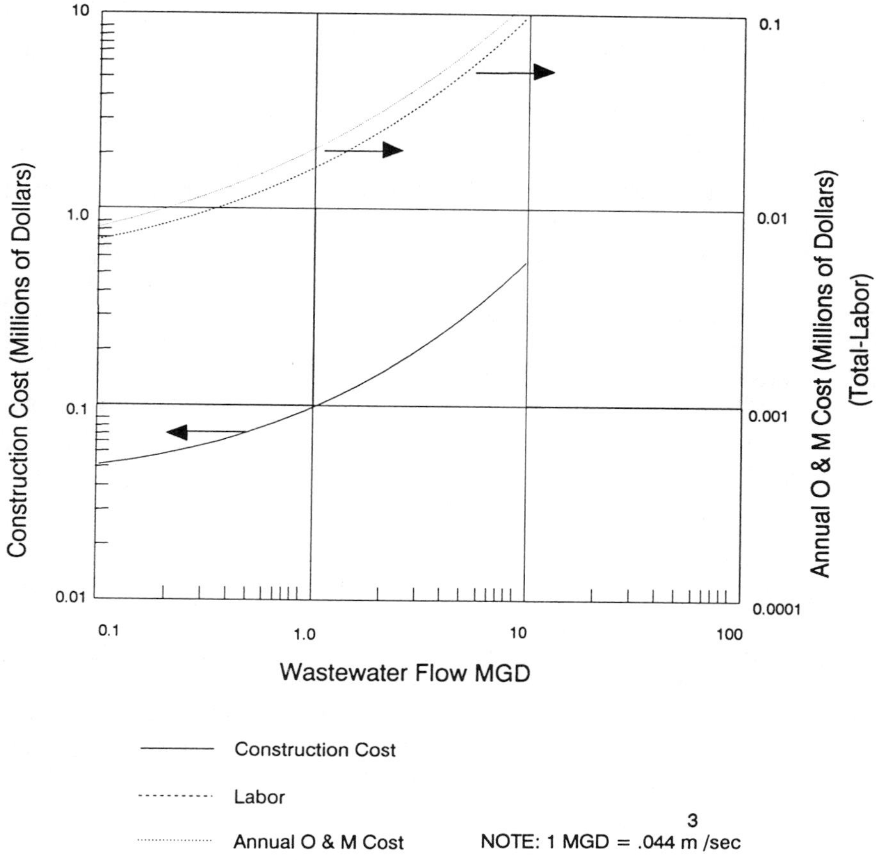

Figure 22.2. Construction, operation, and maintenance costs for sludge drying beds.

22.3 COSTS (See Figure 22.2 (2, 11).)

22.4 AVAILABILITY

Over 6000 wastewater treatment plants in the U.S. use open or covered sandbeds. Also, brick underlain beds are used in Uruguay for drying Imhoff tank sludges. Sand, gravel, and bricks are usually available almost everywhere.

22.5 OPERATION AND MAINTENANCE

Sludge is removed from the drying bed after it has drained and dried sufficiently to be spadable. Sludge removal is accomplished by manual shoveling into wheelbarrows or trucks or by a scraper or front-end loader. Mechanical devices remove sludges of 20 to 30% solids while cakes of 30 to 40% require hand removal. Provisions should be made for driving a truck onto or alongside the bed to facilitate loading.

Paved drying beds permit the direct use of mechanical equipment for cleaning but may possess limited drainage capability. Paved drying beds result in more economical operation when compared with conventional sandbeds because, as indicated above, the use of mechanical equipment for cleaning allows the removal of sludge with a higher moisture content than in the case of hand cleaning. Shorter drying times may be possible. Paved beds have worked successfully with anaerobically digested sludges but are less desirable than open sandbeds for aerobically digested activated sludge.

22.6 CONTROL

Sandbeds are used to dewater sludges in small plants although if land is available, large quantities of sludge may be treated. Beds require little operator attention or skill. Drying beds should be cleaned regularly. Indeed, the material from the beds may be reused for local soil conditioning; for example, in Peru, there has been a demand by the local farmers for treated sludge from water treatment plants. Most localities can use chemical and biological sludges. Some effort is required on the part of the utility or the plant operator to find disposal opportunities nearby.

22.7 SPECIAL FACTORS

Sand beds may be enclosed by glass allowing higher solids loading rates. Glass enclosures protect the drying sludge from rain, control odors and insects, reduce the drying periods during cold weather, and can improve the appearance of a wastewater treatment plant using beds.

Wedgewire drying beds have been used successfully in England. This approach prevents the rising of water by capillary action through the media; the construction approach lends itself well to mechanical cleaning. The first U.S. installations of this type have been made at Rollingsford, New Hampshire and in Florida. It is possible in these small plants to place the entire dewatering bed in a platform rotating about a horizontal axis from which sludge may be removed by tilting the entire unit mechanically. This approach is applicable to all small plants and even larger plants where the drying units are made small enough for easy hand tilting.

22.8 RECOMMENDATIONS

Sand drying beds are ideal for dewatering sludge in small or large plants. Some mechanization, such as the use of screens and/or a tilting apparatus results in a more easily operable process. The soil conditioning potential for most sludges is considerable. The possibility of using sludges in this manner should be actively explored with local agricultural persons.

Codisposal of wastewater treatment sludge with water treatment plants sludges has been suggested (48) and should be practiced more often in order to utilize the water treatment sludge chemical conditioning capability.

23. LAND APPLICATION OF SLUDGE

23.1 DESCRIPTION

Techniques for applying liquid sludge, dried sludge, and sludge cake to the land include direct application from trucks, injection, ridge and furrow spreading, and spray irrigation. Sludge can be incorporated into the soil by plowing, discing, or similar methods.

Municipal sludge contains most of the essential plant nutrients but not necessarily in sufficient quantities to act as a complete fertilizer. It is considered a fertilizer supplement and in almost all cases a soil conditioner—its strongest potential use. The application of water treatment plant sludge to the land has been virtually ignored. Both water and wastewater sludges should be actively considered for land application in all areas where they are generated but especially in poor soil areas.

Wastewater treatment plant sludge can be applied at rates that will supply all the nitrogen and phosphorous needed by most crops. When application rates are high however, the concentration in plants of certain elements, especially metals, may increase. Animal diets are often deficient in trace elements, such as zinc, copper, nickel, chromium and selenium. Thus sludge application improves the quality of feeds and forages used for animal consumption. On the other hand, applications of some industrial wastes may result in animal and even human toxicity if food chain crops are harvested.

Design Criteria—application rates of wastewater sludge depend on sludge composition (nitrogen, phosphorus, metals), soil characteristics, climate, vegetation, and cropping practices. Annual application rates have varied from 0.5 to 10 tons (0.45 to 9.1 mt) per acre (2.5 acres = 1 ha). Application rates based on phosphorous needs are lower. A pH of 6.5 or greater will minimize heavy metal uptake by most crops. Table 23.1 presents sludge fertilizer values which may be used for design.

There are eight basic sludge disposal options for water treatment plant sludge (48):

- discharge to a waterway,
- discharge to sanitary sewers (not a good option in countries where sewers are not connected to treatment plants),
- codisposal with wastewater treatment plant sludge,

Table 23.1. Wastewater Sludge Fertilizer Value

SLUDGE	NITROGEN	PHOSPHATE	POTASH
1 ton dry sludge provides	60 lbs (50% avail)	40 lbs	5 lbs
Typical corn fertilizer provides	180	50	60
If 6 tons dry sludge/acre were applied, would provide (lb/acre)	180 (avail)	240 (avail)	30 (avail)

Source: Reference 2.
Note: 2.2 lbs = 1 kg
lb/acre × 1.14 = kg/ha

- lagooning—requiring ultimate disposal of the residue,
- mechanical dewatering followed by landfilling (soil conditioning value not recovered),
- coagulant recovery from sludge and disposal of residue,
- land application, especially of softening sludge, and
- use as building or fill material.

Land application is an especially attractive option for water treatment plant sludges. As time passes, the land disposal option may be viewed as an increasingly attractive option, and more data will be developed. The data in this section may be applied to arrive at only very preliminary estimates of cost and for sludge application rates in the case of water treatment plant sludge. Land application of water treatment sludges has not been adequately utilized anywhere as partially illustrated in Table 23.2 (48). The basis for the

Table 23.2. Methods for Disposal of Water Treatment Plant Waste

	PERCENT OF PLANTS USING INDICATED DISPOSAL METHOD	
	SOFTENING SLUDGE	COAGULATION SLUDGE
Sludge lagoon	34*	43
Sanitary sewer	8	27
River or lake	13	20
Recalcination	5	—
Direct land application	5	—
Other	—	10

* Fifty-six percent of plants surveyed had sludge lagoons, 60% of which were considered "permanent lagoons," thus 34% of plants used sludge lagoons for disposal.
Source: Reference 48.

Table 23.3. Major Site Considerations for Land Application of Sludge

- Soil type
- Site susceptibility to flooding
- Slope
- Depth to seasonal ground water table
- Permeability of the most restrictive soil layer
- Cropping patterns and vegetative cover
- Soil nutrient and organic matter content

Source: Reference 61.

data shown is a survey of U.S. facilities in 1981. But two earlier surveys within the previous twenty years had shown similar results.

Design Approach for All Types of Sludges—the approach to be used for land application of virtually any sludge is based on the limiting constituent concept. The maximum rates of sludge application are determined for EACH contaminant (e.g., nitrogen, water, metals, organics, etc.). That contaminant which possesses the limiting concentration (and thus limits the mass application rate) limits the application rate of the sludge for the particular site (62). This is discussed further below.

Once a suitable application site (see Table 23.3) has been selected and the process objectives defined, proper sludge loading rates are determined. This determination often involves characterizing the waste for a number of constituents of interest or concern, e.g., nitrogen and heavy metals. The following constituents are of most concern for municipal sludges (61): pathogens, phosphorus, nitrogen, cadmium, copper, nickel, lead, and zinc. Often nitrogen is the limiting constituent in the case of wastewater sludges. When sludge is applied at rates to meet the nitrogen requirements of the crops being grown, other constituent application rates are usually not limiting. Nitrogen losses in excess of those expected from commercial fertilizer use should not be expected.

Regarding heavy metals, the useful life of land application sites can be based on the cumulative amounts of the five metals previously listed. The recommended limits, shown in Tables 23.4 and 23.5, should not result in interference with crop growth or use of the crops at any future time, while serving to protect human and animal health. In addition to the values given, the application rate of cadmium is limited to 0.45 lbs/acre/yr.

After the allowable loading rate of each constituent has been calculated, the actual sludge loading rate is based on the most limiting constituent of those being considered. Higher loading rates than those calculated can be used if nonfood chain vegetation is used, and the site is well monitored. General guidelines are:

Table 23.4. Maximum Amount of Metal (lb/acre, cumulative) suggested for application to agricultural soils

METAL	SOIL CATION EXCHANGE CAPACITY (MEG/100 G)		
	<5	5 TO 15	>15
Lead	500	1000	2000
Zinc	250	500	1000
Copper	125	250	500
Nickel	125	250	500
Cadmium	5	10	20

Source: Reference 61, see also 40 CFR Part 503, 9 February 1989 *Federal Register.*
Note: lb/acre × 1.14 = kg/ha

- sludges should be used that are well stabilized and that have acceptable concentrations of critical contaminants,
- rates of application which meet crop nitrogen or phosphorus requirements are desirable,
- soil pH should be maintained near neutral, and
- adequate monitoring of sludge and soil quality should be provided.

For high application rate disposal practices or when sludges contain relatively high concentrations of critical contaminants and/or sludge stabilization is limited, one or more of the following additional precautions can be helpful:

- remote application sites should be selected or access to the sites by food chain animals limited,

Table 23.5. Maximum Amount of Cadmium (Cumulative) Suggested for Application to Agricultural Soils

SOIL CATION EXCHANGE CAPACITY (MEQ/100 G)	IF BACKGROUND SOIL pH IS BELOW 6.5 (LB CD/ACRE)	IF BACKGROUND SOIL pH IS ABOVE 6.5 OR MAINTAINED AT 6.5 BY LIMING (LB CD/ACRE)
<5	5	5
5 to 15	5	10
>15	5	20

Source: Reference 61.
Note: lb/acre × 1.14 = kg/ha

- sludge should be applied to areas used for the production of animal feed or nonfood chain crops, and
- adequate monitoring of vegetation, nearby surface and groundwater should be provided.

23.2 LIMITATIONS

Constituents of sludge may limit the rate of application, the crop that can be grown, or the management or location of the site. Trace elements added to soil may accumulate to levels that are toxic to plants or may be taken up and concentrated in edible portions of plants to toxic levels. Trace element problems are prevented by limiting the amount of sludge applied; pretreatment of industrial wastes which may contribute to objectionable wastewater sludge concentrations; selection of tolerant or nonaccumulating crops; se-

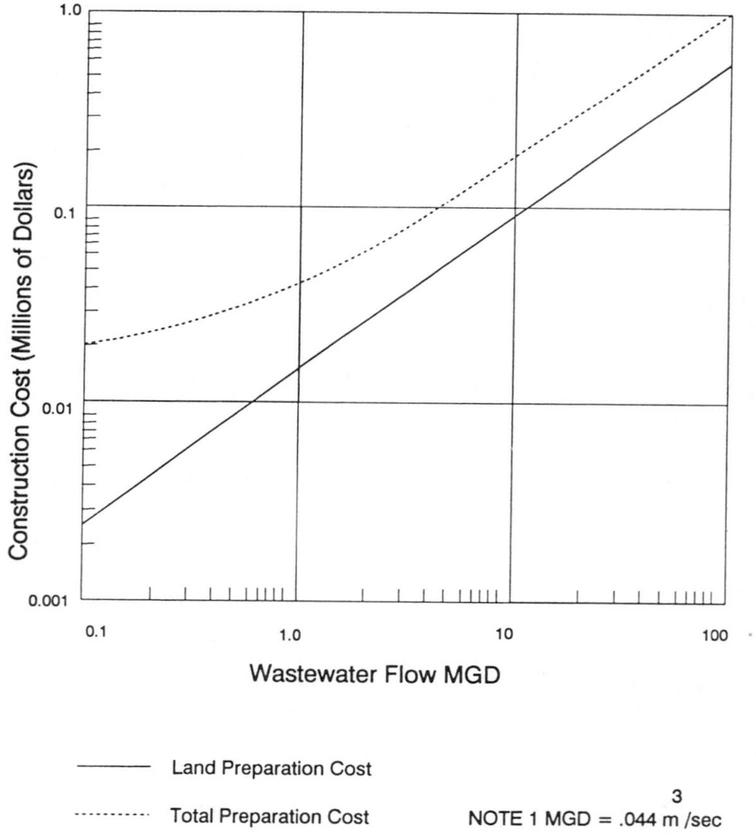

Figure 23.1. Construction cost of land application of sludge.

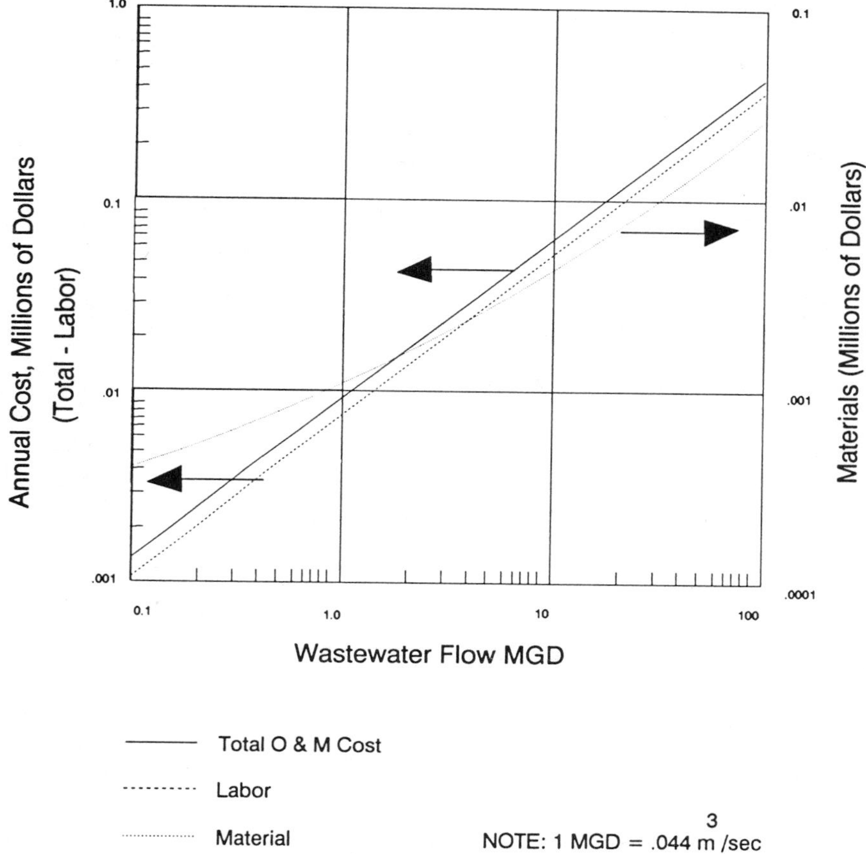

Figure 23.2. Operation and maintenance cost of land application of sludge.

lection of crops not used in the human food chain; and adopting appropriate agronomic practices, such as liming of the soil application area. Where population is concentrated and agricultural land limited, sufficient land for sludge application may not be available. Terrain must be properly selected; steep slopes and low lying fields are less suitable and require more active management. Equipment with standard tires can cause ruts, compacted soil and crop damage, or get stuck in muddy terrain.

23.3 COSTS

See Figures 23.1 and 23.2 (2, 11). Additional cost information is presented in Figures 23.3 and 23.4 for application to marginal lands (lands which have been disturbed). Table 23.6 presents the assumptions for the cost estimates

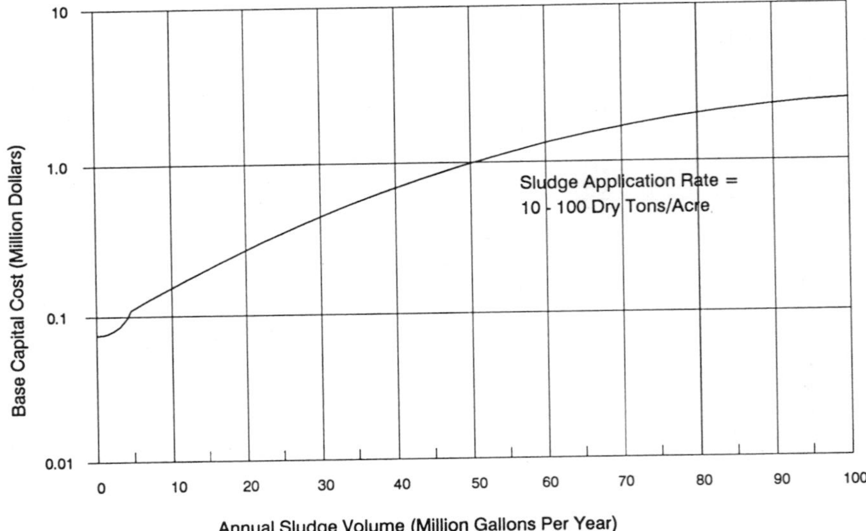

Figure 23.3. Base capital cost of applying sludge to marginal land for reclamation as a function of annual sludge volume applied.

in Figures 23.3 and 23.4. Figure 23.5 may be used to derive multiplication factors for the capital costs in Figure 23.3 based on number of application days per year.

23.4 AVAILABILITY

Land application of wastewater sludge is practiced throughout the U.S. Perhaps as much as 40% of the sludge produced finds its way eventually to the land (61). This practice could be used immediately in developing countries with little or no additional technology required. Equipment needs are simple. Once the nutrient requirements of the land are understood, the technology may be used.

Land application of softening sludge is not a new method of disposal but has never been widely applied. Farmers were allowed to remove dewatered softening sludge from a plant in Ohio approximately 30 to 40 years ago (48).

The solids content of softening sludge discharged from clarifiers is between 1 and 5%. If land application of softening sludge is employed, this sludge could be thickened further and applied for soil conditioning as a liquid at 8 to 10% solids or as a solid after dewatering to about 40% solids. If

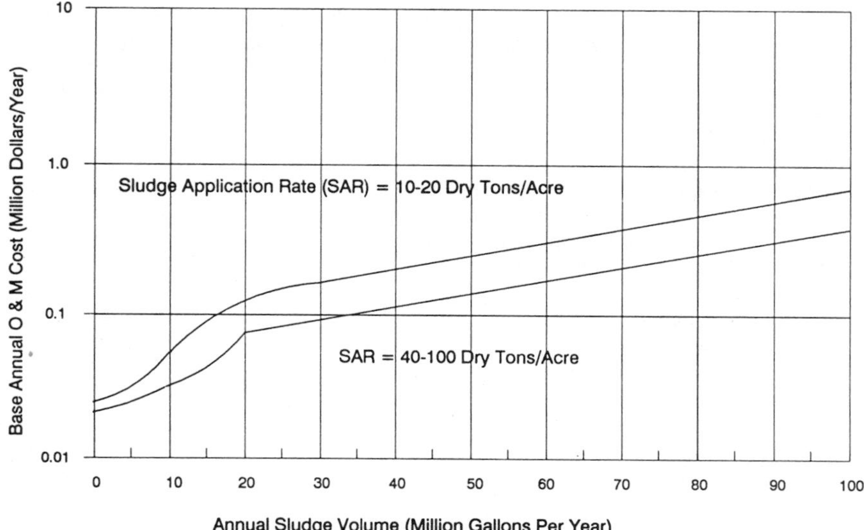

Figure 23.4. Base annual operation and maintenance costs for applying sludge to marginal land for reclamation as a function of annual sludge volume applied and dry solids application rate.

the solids content results in sludge between pumpable liquid consistency and a solid (a semi-solid), and if conventional farming equipment is used, handling problems may be encountered.

In farming regions, the application of commercial nitrogen fertilizers sometimes causes a reduction in soil pH. If optimum pH conditions do not exist, crop yields are reduced. Therefore, farmers must apply sufficient quantities of calcium carbonate as a means of counteracting the effects of fertilizer applications. For each 100 lb (45.4 kg) of ammonia fertilizer, 3 to 4 lb (1.4 to 1.8 kg) of limestone must be applied.

In 1969, the Ohio Department of Health reported that the total neutralizing power of lime sludge was greater than that of marketed liming materials. To bring the soil pH into a desirable range, 3 tons/acre (0.67 kg/m^2) lime, or about 10 tons (9.07 metric tons) of lime sludge at a 30% solids concentration, are required. Subsequent lime applications may be required to maintain the desired pH. In Illinois, a calcium carbonate equivalent test performed on several softening sludges indicated that the softening sludges were superior to agricultural limestones available locally. Because softening sludges contain a high quantity of calcium carbonate and offer a high degree of neutralization,

Table 23.6. Assumptions Used in Developing Cost Requirement Curves for Land Application of Sludge to Marginal Land

PARAMETER	ASSUMED VALUE
Sludge Solids Concentration	5%
Daily Application Period	7 hr/day
Annual Application Period	140 days/yr
Fraction of Land Required in Addition to Application Area	0.3
Fraction of Land Area Requiring Lime Addtion	0
Fraction of Land Area Requiring Grading	0
Cost of Land	0
Cost of Lime Addition	0
Cost of Grading Earthwork	0
Cost of Operation Labor	$14.90/hr
Cost of Diesel Fuel	$1.50/gal
Cost of Monitoring Wells	$5,700 each

Source: Reference 51.
Note: 3.785 L = 1 gal.

this resource should be used when it is practical for soil conditioning. The addition of softening sludge also increased the porosity of tight soils, making them more workable for agricultural purposes (48).

Both alum and lime water treatment sludges increase the cohesiveness of soils, but more importantly, both types also widen the moisture content range over which soil remains cohesive. Alum sludge increases soil cohesiveness at high moisture contents, whereas lime sludge increases soil cohesiveness at lower moisture contents.

Use of lime sludge on agricultural land has had limited success because farmers are unfamiliar with its use as a soil conditioner and with the logistics of its transportation and applications to their lands. In addition, the availability of dewatered sludge should be scheduled to coincide with farmers' demands. Agricultural use of dewatered sludge would offset some portion of sludge disposal costs. Alum sludge has little soil conditioner value.

23.5 OPERATION AND MAINTENANCE

Ridge and furrow methods of sludge application to the land involve spreading sludge in the furrows and planting crops on the ridges. Utilization of this technique is best suited to relatively flat land and is well suited to certain row crops. Spray irrigation systems are more flexible, require less soil preparation and can be used with a wider variety of crops. High application rates are commonly used to reclaim strip mine spoils or other low quality land. Sludge spreading in forests has been limited, but offers opportunities for improved soil fertility and increased tree growth.

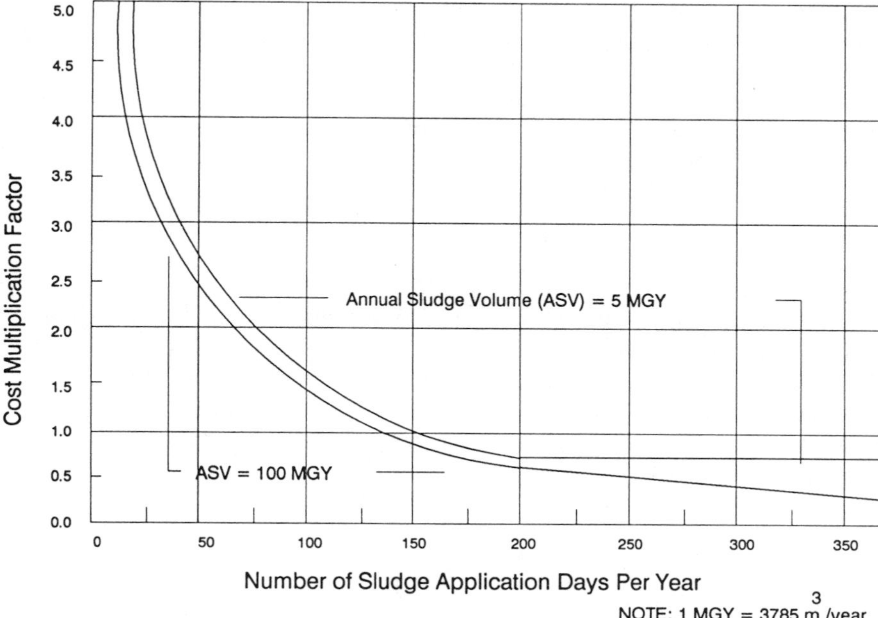

Figure 23.5. Multiplication factor to adjust sludge application to marginal land costs in Figure 23.4 for variations in days of application per year.

23.6 CONTROL

As a disposal process, land application is very reliable. As a utilization process, careful control should be exercised as previously described.

23.7 SPECIAL FACTORS

There is always a potential for toxic chemicals and pathogens from sludge to contaminate soil, water, air, vegetation, adversely affect animal life, and ultimately cause hazards to humans. Accumulation of toxic compounds in the soil may cause phytotoxic effects, the degree of which varies with the tolerance level of the particular plant species and variety. Toxic substances, such as cadmium that accumulate in plant tissues can subsequently enter the food chain, reaching humans directly through ingestion or indirectly through animals. If available nitrogen exceeds plant requirements it may reach groundwater in the nitrate form. Toxic materials and pathogens can contaminate groundwater supplies or can be transported by runoff or erosion to surface waters if overloading is practiced. Aerosols which contain pathogenic organisms may be present in the air over a landspreading site, particularly where spray irrigation is the means of sludge application. Some pathogens remain viable in the soil and on plants for periods of several months; some parasitic ova can survive for a number of years. Another potential impact is odor, and the public may find land application unacceptable. A good long term monitoring program for soil and groundwater quality is essential for an effective system.

23.8 RECOMMENDATIONS

This technology is a popular method of disposal because it is simple. It also serves as a resource reuse measure since sludge is beneficial as a soil conditioner for agricultural, marginal, or drastically disturbed (e.g., mined) land. Finally, wastewater treatment sludges contain considerable quantities of organic matter, most of the essential plant nutrients, and a capacity to produce humus exists.

24. CHLORINATION DISINFECTION

24.1 DESCRIPTION

Chlorination is the most commonly used water and wastewater disinfection process worldwide. This process involves the addition of elemental chlorine or hypochlorite, either calcium or sodium (the calcium form most frequently used is high test hypochlorite, HTH), to the wastewater. Figure 24.1 shows the gas chlorination option schematically. Figure 24.2 shows the hypochlorination option for a well water supply.

Chlorine is supplied as a liquified gas under high pressure in containers varying in size from 100 to 150 lb (45 to 68 kg) up to 1 ton (0.91 mt), as well as tank cars of larger sizes. Precautions should be taken when handling chlorine gas: 1) chlorine gas is both very poisonous and very corrosive; adequate exhaust ventilation at floor level should be provided since chlorine gas is heavier than air, 2) chlorine-containing liquid and gas can be handled in black wrought-iron piping, but chlorine solution is highly corrosive and should be handled in rubber-lined or resistant plastic piping with hard rubber parts where necessary, 3) cylinders in use are set on platform scales flush with the floor, and the loss of weight is used as a positive record of chlorine dosage (4, 5, 6, 7) see Figure 24.3 (3).

Calcium hypochlorite is available commercially in either dry or wet form. High-test calcium hypochlorite (HTH) contains about 60% available chlorine. Because calcium hypochlorite granules or pellets are readily soluble in water and under proper storage conditions are relatively stable, they are often favored over other available forms. Because of its oxidizing potential, calcium hypochlorite should be stored in a cool, dry location away from other chemicals in corrosion-resistant containers (4, 6). Figure 24.3 shows a typical hypochlorite installation (7).

Sodium hypochlorite is available in strengths from 1.5 to 15% with 3% the typical maximum strength used. Therefore, transportation costs may limit application. The solution decomposes more readily at high concentrations and is affected by exposure to light and heat. It must therefore be stored in a cool location in a corrosion-resistant tank.

Disinfection is used to kill harmful organisms, and generally does not result in sterile water (free of all microorganisms). Table 24.1 (2) shows some wastewater dosages and results. Typically, 30 min of chlorine contact time is required with good mixing. Water supply treatment dosages are established on the basis of maintaining a residual in the treated water.

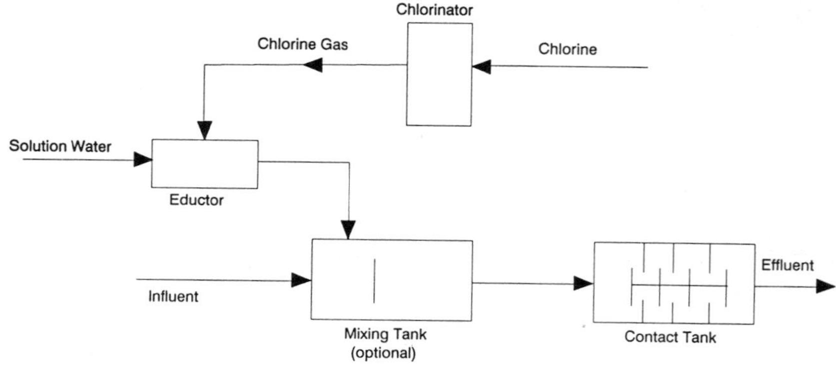

Figure 24.1. A flow diagram of chlorination (*source:* reference 2).

Design Criteria—generally a contact period of 15 to 30 min at peak flow is required. Detention/contact tanks should be designed to prevent short-circuiting. Baffling in contact tanks (as shown schematically in Figure 24.1) improves efficiency. Baffles can either be of the over-and-under type or of the end-around variety. Residual concentrations of at least 0.5 mg/L in water

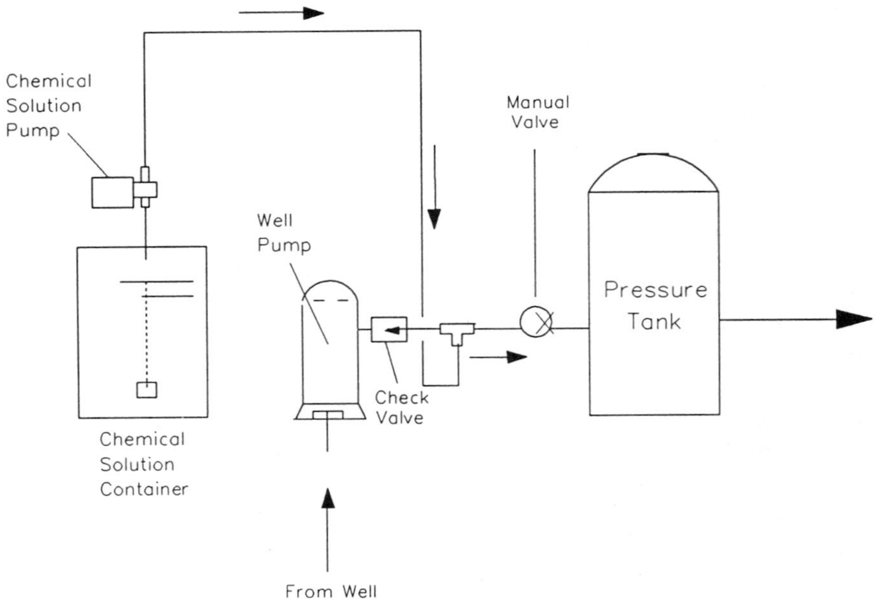

Figure 24-2. Electrical pump connection to hypochlorinator with on-off control (*source:* reference 7).

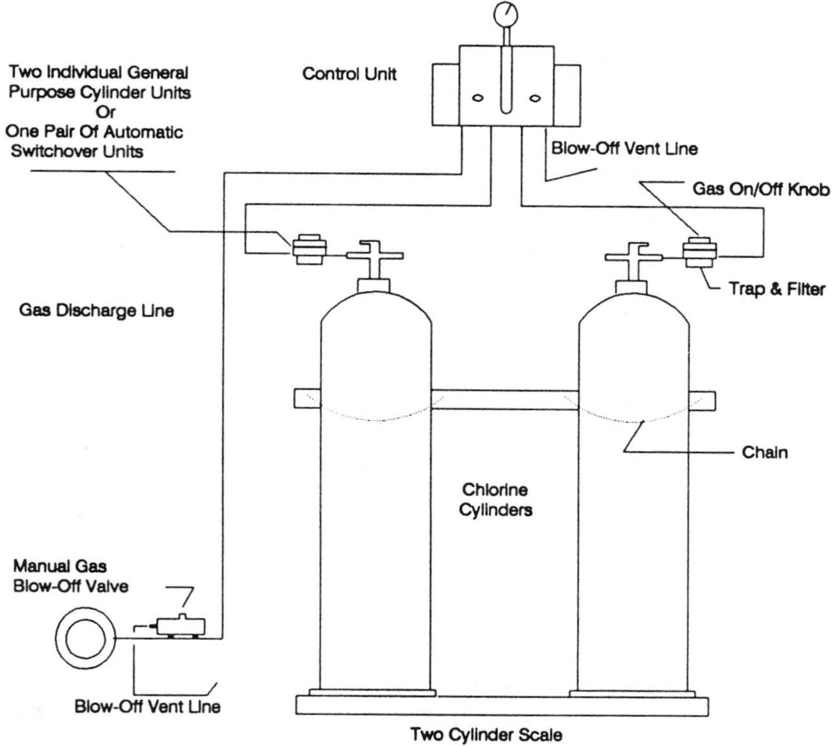

Figure 24-3. A typical chlorine setup for chlorination treatment (*source:* reference 3).

supplies are necessary. In the case of chlorination of wastewater discharges residuals often result in halomethanes or other toxic compounds which are undesirable in receiving waters, especially those used as water supply downstream. Recently, dechlorination before discharge is being practiced to eliminate this problem.

Chlorine may be applied by two basic methods: 1) gas chlorination employing compressed chlorine gas or 2) hypochlorination employing a chemical feed pump to inject a water solution of chlorine compounds.

Gas Chlorination—a gas chlorinator must be employed to meter the gas flow and mix it with water. The mixture is then injected for treatment as a water solution. The 100 or 150 lb (48 or 65 kg) gas container may be handled manually, but the larger containers require special hoists and cradles. Chlorine gas is a highly toxic oxidant, and lung irritant and special facilities are required for storing and housing gas containers and chlorinators. The ad-

Table 24.1. Typical Chlorine Dosages and Results

CHLORINE RESIDUAL, mg/L	TOTAL COLIFORM MPN/100 mL PRIMARY EFFLUENT	SECONDARY EFFLUENT
0.5–1.5	24,000–400,000	1,000–12,000
1.5–2.5	6,000–24,000	200–1,000
2.5–3.5	2,000–6,000	60–200
3.5–4.5	1,000–2,000	30–60

In normal low dose disinfection treatment, the COD, BOD_5, and TOC of the treated wastewater are not measurably changed.

EFFLUENT FROM	DOSAGE RANGE, mg/L
Untreated wastewater (prechlorination)	6–25
Primary sedimentation	5–20
Chemical-precipitation plant	3–10
Trickling-filter plant	3–10
Activated-sludge plant	2–8
Multimedia filter following activated-sludge plant	1–5

Source: Reference 2.

vantage of this method is the convenience afforded by a relatively large quantity available for continuous operation for several days or weeks without mixing chemicals. Gas chlorinators have an advantage where variable influent flow rates are encountered, as feed rates may be synchronized to inject variable quantities of chlorine. Capital costs are somewhat greater for gas chlorination, but chemical costs may be less.

Hypochlorination—water solutions of either the liquid or dry forms are prepared in predetermined stock solution strengths. Solutions are injected into the water supply using special chemical metering pumps, hypochlorinators. The positive displacement meters and pumps are accurate and reliable and are preferred over hypochlorinators employing other feed principals (usually based on a suction principal, eductors). Positive displacement type hypochlorinators are available at modest costs. These small chemical feed pumps are designed to pump (or inject under pressure) an aqueous solution of chlorine into the water to be treated. They are designed to operate against pressures as high as 100 psi (7 kg/cm^2) but may also be used to inject chlorine solutions at atmospheric or negative head (suction side of water pump) conditions. Hypochlorinators come in various capacities ranging from 1.0 to 60 gal/day (3.8 to 227 L/day). The pumping rate is manually adjusted by varying the stroke of the piston or diaphragm. Once the stroke is set, the hypochlorinator feeds accurately at that rate, maintaining a constant dose.

This works effectively if the influent flow rate is fairly constant, as with pumped influent. If the influent flow rate varies considerably, a metering device is used to vary the hypochlorinator feed rate synchronized with flow rate. In the case of a pumped well supply, the hypochlorinator is connected electrically with the on-off controls of the pump.

24.2 LIMITATIONS

Chlorination may cause the formation of chlorinated hydrocarbon compounds, some of which are known to be carcinogenic (e.g., halomethanes). The effectiveness of chlorination is greatly dependent on pH and temperature of the wastewater. Chlorine gas is a hazardous material and requires sophisticated handling procedures implemented by trained personnel. Chlorine will react with organic matter in the wastewater and maintaining residuals in water supplies is important. It will oxidize ammonia and hydrogen sulfide, as well as metals present in their reduced states.

24.3 COSTS (See Figure 24.4 (2, 11).

24.4 AVAILABILITY

Chlorination of water supplies on an emergency basis was practiced as early as about 1850. Presently, chlorination of both water supplies and wastewaters is a widespread practice throughout the world. System designers should be cognizant of the need to continuously and consistently supply chlorine for the process to be effective. This is especially true for systems where chlorine or hypochlorite are not produced on-site. Inconsistent disinfection of wastewaters may adversely impact the downstream water quality and potentially downstream users. Chlorination systems are often not operational, especially in small and medium sized communities, and where operation budgets are restrictive.

24.5 OPERATION AND MAINTENANCE

Gas Chlorinator

Normal operation of a gas chlorinator requires routine observation and preventive maintenance. The operator on a daily basis should:

- read the chlorinator rotameter,
- record the reading time, date and initial the entries,
- read the meters and record the number of gallons of water pumped,
- check the chlorine residual; if the residual is too low in the distribution

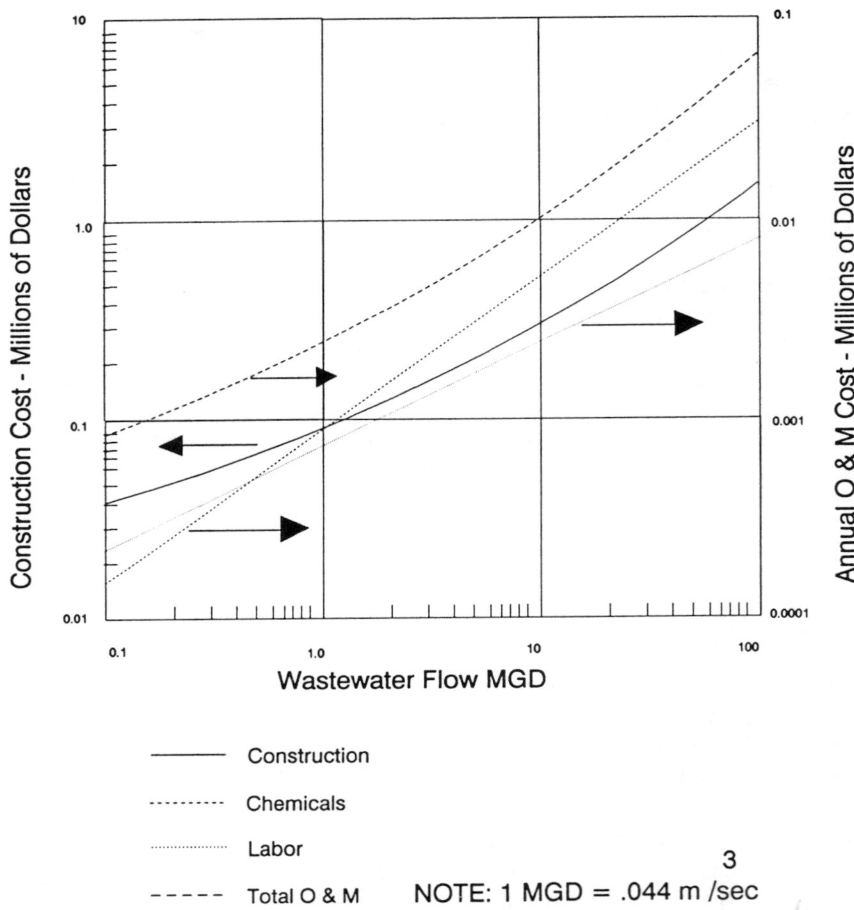

Figure 24-4. Construction, operation, and maintenance costs for chlorination (disinfection).

system the feed rate should be increased; if the residual is too high the feed rate should be decreased,

- calculate and record chlorine usage.

On a weekly basis the operator should:

- clean the equipment and the building,
- perform preventive maintenance on the equipment.

Hypochlorinator

Operators should follow the daily maintenance schedule:

- read and record the level of the solution tank at the same time every day,
- read the meters and record the amount of water pumped,
- check the chlorine residual (0.2 mg/L) in the system and adjust the chlorine feed rate as necessary; maintain this residual at the most remote point in the distribution system; suggested free chlorine residual for treated water or well water is 0.5 mg/L at the point of chlorine application provided the 0.2 mg/L is maintained throughout the distribution system,
- check the chemical feed pump operation; most hypochlorinators have a dial which indicates the chlorine feed rate; operate the pump in the upper ranges of the dial; require the frequency of the strokes or pulses from the pump to be frequent enough so that the chlorine will be fed continuously to the water being treated; adjust feed rate after testing chlorine residual levels.

An operator should on a weekly basis:

- clean the equipment and building,
- replace the chemicals and wash the chemical storage tank; try to have a 15- to 30-day supply of chlorine in storage for future needs; when preparing hypochlorite solutions, prepare only enough for a 2- or 3-day supply.

An operator should on a monthly basis:

- check the operation of the check valve,
- perform any required preventive maintenance,
- cleaning; commercial sodium hypochlorite solutions (such as household bleaches) contain an excess of caustic (sodium hydroxide or NaOH); when this solution is diluted with water containing alkalinity, the resulting solution may become supersaturated with calcium carbonate which tends to form a coating on the poppet valves in the solution feeder, preventing sealing and proper feeding.

Hypochlorinators on small systems are small sealed systems that cannot be repaired, so replacement of the entire unit is required at failure. Maintenance requirements are normally minor, such as changing the oil and lubricating the moving parts.

24.6 CONTROL

These systems are reliable with regular maintenance. The hypochlorite system is somewhat easier to operate than the gas system, and operators need not be as skilled or as cautious.

24.7 SPECIAL FACTORS

Chlorine Hazards

Chlorine is a gas, heavier than air, extremely toxic, and corrosive in moist, humid atmospheres. Dry chlorine can be safely handled in steel containers and piping, but in humid environments corrosion-resisting materials such as silver, glass, teflon®, and certain other plastics are necessary. Chlorine gas at container pressure must be piped in appropriate systems. Even in dry atmospheres, chlorine combines with the moisture in the mucous membranes of the nose and throat, and with the fluids in the eyes and lungs; a very small percentage in the air can be very irritating and can cause severe coughing. Heavy exposure can be fatal.

Hypochlorite Safety

Hypochlorite compounds are nonflammable; however, they can cause fires when they come in contact with organics or other easily oxidizable substances. Calcium hypochlorite is used commonly by small water supply systems. Incrustation may occur when sodium fluoride is injected at the same point as the hypochlorite. Hypochlorite does not present many of the hazards that gaseous chlorine does and, therefore, is safer to handle. When spills occur, it is necessary to wash with large volumes of water. Hypochlorite may cause damage to eyes and skin upon contact. Hypochlorite solutions are very corrosive. Skin burns are possible from contact with concentrated solutions.

Dechlorination may be used for wastewater treatment, which involves the addition of sulfur dioxide, aeration, or even activated carbon.

24.8 RECOMMENDATIONS

Chlorination for disinfection is used to prevent the spread of waterborne diseases and to control algae growth and odor. Chlorination is the most popular method of disinfection used in developing countries. Economics, ease of operation, and convenience are the qualities used for evaluation.

For reasons of consistency of supply it is useful to consider on-site generation of chlorine (6). Most commercially available equipment for chlorine generation will operate on seawater as well as brine solutions prepared for the purpose. About 2.5 kWh/lb of available chlorine is required for most commercial equipment. Hypochlorite solutions prepared from seawater are usually limited to about 1800 mg/L available chlorine, and those produced form brine to about 8000 mg/L. Heavy metal ions present in seawater interfere with stability of hypochlorite solutions.

Hypochlorite can also be made on-site by using common salt, manganese dioxide, low grade slaked lime, and sulfuric acid. Often these products are available nearby. Manganese dioxide and common salt are mixed and placed in a reaction tank which is suspended above an open type water boiler. Above the chemical reaction tank is a sulfuric acid tank with a control valve which allows regulation of the flow of acid to the reaction tank. The rate of chlorine gas generation is regulated by the flow of sulfuric acid and the temperature of the water. The chlorine gas generated is passed through a foam trap and desiccator, where it is dried before going to an absorption chamber. This chamber contains slaked lime. Chlorine reacts, and a bleaching powder is formed. It has been shown that this method will produce about 30 pounds of bleaching powder with 35 percent available chlorine in 12 hours of operation (67).

25. ULTRAVIOLET DISINFECTION

25.1 Description

Ultraviolet (UV) radiation is used to disinfect drinking water as well as wastewater on an increasingly frequent basis as reliable equipment becomes more available. Commercially, UV light is generated artificially by a wide variety of arcs and incandescent lamps. For disinfection, the UV radiation is generated from special low pressure mercury-vapor lamps that produce UV radiation as a result of an electron flow between the electrodes in an ionized mercury vapor. UV may also be laser generated although the necessary generators are not commonly available. The inactivation of microorganisms by UV radiation is based on photochemical reactions in the DNA molecule that produce reproductive system errors. Figure 25.1 (20) is a schematic representation of a UV disinfection unit showing a submerged lamp (in a quartz sleeve) placed perpendicular to the direction of wastewater flow (18, 20).

Ultraviolet (UV) radiation has been considered as an alternative means of disinfecting small drinking water supplies. A major impetus for this was the increase in reported waterborne disease outbreaks caused by Giardia lamblia, an organism that forms cysts which are highly resistant to conventional chlorination. Other advantages include:

1. No chemical consumption—eliminates large scale storage, transportation and handling, and potential safety hazards.
2. Low contact time—no chemical contact basin is necessary and space requirements are reduced.
3. No harmful by-products are formed.
4. A minimum of, or no moving parts—high reliability.
5. Low energy requirements.

The UV lamp is submerged in or suspended above the water to be treated. A relatively new configuration marketed by a firm in Brazil uses a series of vertical flow-through cylinders with the UV bulbs located in an annular tube. Thus, the flow is longitudinal (parallel to) rather than perpendicular to the UV source, see Figure 24.1 (also see Figure 25.2).

Testing of UV disinfection has been done by the U.S. EPA for small water supplies, and was found effective on Giardia, Yersinia, and *G. muris* cysts, as well as the lesser resistant *E. coli* and Yersinia enterocolitica cysts.

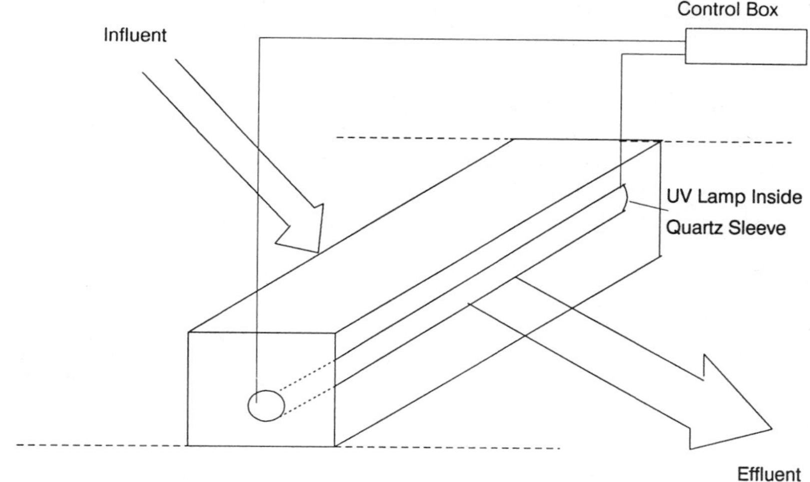

Figure 25.1. Ultraviolet design schematic (*source:* reference 20).

Reactor design should include isolatable modules so that maintenance can be easily done. Lamps and ballast must be accessible. Operators must wear UV filtering goggles and must be made aware of the potential hazards. Design should include redundant modules. Given the lack of actual plant operating experience, perhaps complete redundancy should be considered.

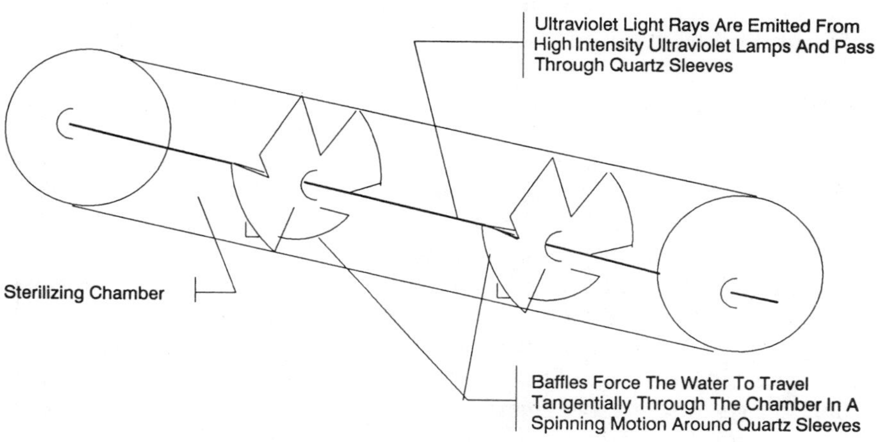

Figure 25.2. Typical UV sterilizing chamber (*source:* reference 19).

The evaluation process leading to possible adoption of UV should include consideration of:

- power supply—continuity,
- current UV equipment reliability as of the time of installation,
- safety tradeoffs vs. other disinfection systems,
- achievable UV exposure indices under actual plant conditions,
- impact of TSS, color and turbidity variations on effectiveness.

25.2 LIMITATIONS

The wavelength of useful UV light for disinfection is 4 to 400 nm. The maximum microbiological kill is in the narrow band between 200 and 310 nm. Apparently the most efficient wavelength is around 260 nm. This is a limitation of UV radiation as an effective disinfection process, although mercury vapor lamps can now be commercially manufactured to emit near this wavelength. UV intensity of the lamps and consistency of output at required wavelengths have been a problem. Establishing manufacturing standards, equivalent to pump manufacturers criteria for example, could help to ensure equipment manufacturing consistency, output, and effectiveness.

The lethal effect of UV radiation (lethal dose: 110 MW · sec/cm^2 @ 20 min duration) occurs because of a photochemical reaction initiated by absorption of a photon by the molecular structure of the microorganism, rather than by the presence of an oxidizing agent or toxic chemical (as in the case of chlorine or ozone). Thus, another limitation is that the germicidal action depends on the absorption of UV radiation in the lethal wavelength range. The organisms must be exposed to the UV wavelengths. Process effectiveness is interfered with by the presence of suspended solids (which must be kept in the range of 10 to 15 mg/L and lower). Turbidity and color are as important, perhaps more important than SS concentration and alkalinity. Initial studies on *G. muris* cysts have shown that color increases cyst survival at 254 nm. Other studies on *G. muris* cysts indicated that storage time and temperature affected the viability of the cysts and that the rate of decrease in viability approximately doubled with each 10°C increase in temperature above freezing. Because of this, cyst viability was shortened to hours rather than to days for the above freezing conditions. Physical stress produced by pressure and alum addition in water treatment processes appears to damage and even destroy cysts. All of these factors impact the effectiveness of UV disinfection of water and should be considered if UV is to be adopted.

The efficiency of a UV system is related to a dispersion index, which in turn is related to exposure time and lamp configuration (largely spacing of lamps). An ideal dispersion index is zero. Exposure time is the order of 10 to 30 sec, similar to the requirements for chlorination. However, lamp aging affects the required exposure time. The deterioration of the effectiveness for germicidal lamps at 8000 hr of use is as high as 40%. Thus, corrections for treatment capacity must be made to correct for deterioration with use.

The disadvantages of UV disinfection are: lack of a persisting residual in the treated water, low level of equipment reliability, inconsistent operating reliability in the "typical" small plant, a previous industry-wide commitment to chlorination (momentum), and lack of efficient process designs.

Frequently, it is not possible to replace chlorination at a treatment plant with other disinfection technology such as UV. Chemical disinfection is still required for pretreatment of wastewater for odor control and treatment of bypassed flows, and may be required for prefiltration and preadsorption applications.

25.3 COSTS

Cost data for application of UV and other disinfection technologies in developing countries and for small systems is very limited. Table 25.1 contains comparative costs for a range of disinfection applications to wastewater, mostly in the U.S. (6).

25.4 AVAILABILITY

The sun is the most important natural source of UV light. Since the maximum UV sensitivity of microorganisms and the UV emission of the low-pressure mercury vapor lamp are well matched, the nearly monochromatic low-pressure mercury lamp has prevailed as the dominant radiation source in research and practical applications.

UV has significant disinfection qualities. It has even been used successfully where sterilization is required, in the pharmaceutical, cosmetic, and beverage industries for example. UV application for disinfection of potable water supplies however, has received little attention.

The use of UV for disinfecting wastewater has had greater success largely as a result of EPA funded development under plant conditions. The use of UV for highly treated effluents and the use of ozone together with UV appear promising.

There are 51 facilities in operation in the U.S., most below 1.5 MGD (19,000 m^3/day) wastewater plant size. About 30 are under construction, and the remainder under design (18).

Table 25.1. Disinfection Cost Summary

	PLANT SIZE		
	1 MGD	10 MGD	100 MGD
Capital Cost		$1,000	
Process			
Chlorine	60	190	840
Chlorine/SO_2	70	220	930
Chlorine/SO_2/aeration[b]	120	360	1,580
Chlorine/carbon	640	2,800	8,400
Ozone/air[a]	190	1,070	6,880
Ozone/oxygen[a]	160	700	4,210
Ultraviolet[a]	70	360	1,780
Bromine chloride	50	130	410
Operation Cost		cents/1000 gal	
Process			
Chlorine	3.49	1.42	0.70
Chlorine/SO_2	4.37	1.75	0.89
Chlorine/SO_2/aeration[b]	7.66	2.39	1.19
Chlorine/carbon	19.00	8.60	3.28
Ozone/air	7.31	4.02	2.84
Ozone/oxygen[a]	7.15	3.49	2.36
Ultraviolet[a]	4.19	2.70	2.27
Bromine chloride	4.52	3.04	2.65

[a] Additional treatment is not included in these costs.
[b] Aeration is not required following dechlorination by SO_2 because a properly designed system should not remove DO from the effluent.
Source: Reference 6.
Note: 1 gal × 3.785 × 10^{-3} = m^3

25.5 OPERATION AND MAINTENANCE

Operation depends on continuous availability of power. An alternate power supply for the UV system must be provided to ensure dependable continuous disinfection of a water supply.

Proportioning dose to flow is accomplished by changing the number of lights exposed to the quantity to be treated. Water absorbance should be checked continuously (best done with a "slave" UV lamp so that comparisons with the operating system are appropriate).

Unless the quartz-ultraviolet system is disinfecting a highly treated water, the sleeves quickly foul with suspended and dissolved matter in the water. Therefore, it is necessary to clean the lamps on a regular basis to preserve the high UV transmittance of the quartz and the disinfection capability of the overall system.

Two nonchemical cleaning methods have been used with limited success.

The first is a mechanical wiper system, in which a wiper periodically scrapes fouling deposits off of the outer surface of the quartz sleeves. For this technique to work effectively, very close tolerances are required on quartz sleeve outer diameter and alignment. Close tolerances are also required on the wiper system. These severe tolerance requirements add to manufacturing expense and are very difficult to achieve with large tube bundles.

Misalignment or improper tolerance control will cause reduced ultraviolet intensity in some portions of the unit and a lower than expected delivered dose. Supplemental chemical cleaning is often required on wastewater disinfection systems.

Another nonchemical cleaning system used on quartz ultraviolet units is high frequency ultrasound. Ultrasonic cleaning is a new technique and has not been used on large capacity systems over an extended time period. This method is based on the same principle that is used for cleaning laboratory glassware. There may be potential deleterious effects of ultrasound on the quartz sleeves, the lamps, and the seals. Also penetration of ultrasound into the inner portions of large bundles must be demonstrated. Field testing has shown that ultrasonic cleaning has had to be supplemented with frequent chemical cleaning.

Expected lives of lamps are variable, normally ranging from 7,000 to 12,500 hours. It is good practice, however, to replace lamps on a regular basis or when metered UV intensity falls below acceptable values. A complete cleaning of quartz glass enclosures with alcohol is required during lamp replacement. Power requirements for the UV system for design flow rates up to 4 gpm (0.25 L/sec) are approximately 1.5 KWh/day.

25.6 CONTROL

Giardia cyst inactivation is a function of UV energy absorption and thus depends on the amount of UV light that reaches the cyst and the time of exposure. If there is short-circuiting, the time for 100% inactivation can be expected to be protracted.

25.7 SPECIAL FACTORS

Size and morphological characteristics of organisms and particles appear to be very important factors in shielding them from UV radiation.

Though laser-generated UV radiation has a considerably greater intensity than the mercury-vapor UV lamps, the detention time for the laser pulse is on the order of 10 sec. Thus, equivalent dose ranges can be obtained from both sources. Data suggest that the commercially available UV lamp units are more effective than some laser units in inactivation of Giardia cysts.

25.8 RECOMMENDATIONS

If power supplies are expected to be more reliable than chemical supplies, UV and other similar disinfection technologies are preferred. Most operating treatment processes require power; power supplies may therefore be more consistent than chemical deliveries in remote areas. Operating UV systems are in operation in Brazil for both wastewater and water supply applications.

26. PLATE/LAMELLA SETTLING

26.1 INTRODUCTION TO CHAPTERS 26 AND 27

Treatment process technologies are often combined in practice in the same or similar ways to achieve treated water supply or wastewater quality requirements, e.g., chemical coagulation and settling. Simplified water treatment plants are considered for analysis and costing as single combined units rather than individual unit treatment process technologies. Package plants are often cost effective when designed using the combined approach. Clariflocculators are an example of actually combining unit process technologies into a single operating hardware package.

Combined simplified plants have been in use in Latin America, the Caribbean, and other developing countries. Detailed reviews have been made of such plants (21, 22). A review of process technology has been prepared by the U.S. EPA (80). The technologies presented in this book may be used in conjunction with the unit processes in the "simplified treatment plant" approach or as replacements thereto. The combined water treatment plant approach was also used in the cost model developed for Latin America and the Caribbean (11). The processes that are suggested for a simplified approach are:

- CHEMICAL DOSING, MIXING, AND FLOCCULATION—See Chapters 3, 9, and 27.
- SEDIMENTATION—Using largely plate settlers because of improved efficiency over classical designs for small plants.
- DUAL MEDIA FILTRATION—See Chapter 4. Declining rate filtration is suggested. Filtration rate is adjusted as the filtration process proceeds, rather than remaining constant. In actual practice the rate declines as the head loss increases. Declining rate operation allows the filtration rate to adjust to the degree of cleanliness of individual filter units in an operational array. The ratio of maximum to average flow may be as high as 2. Brazil requires 1.3, and practice in the U.S. is around 1.5.

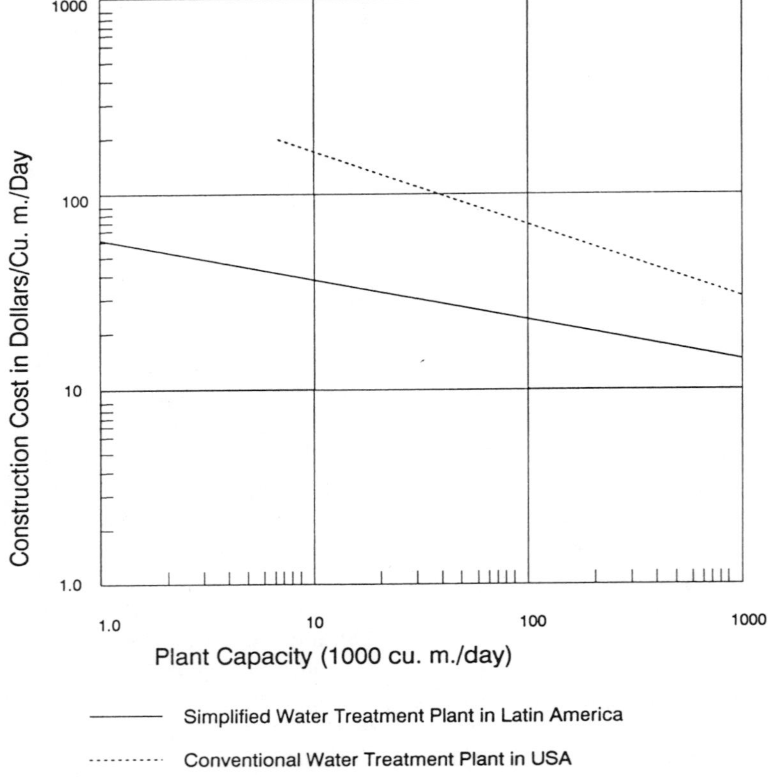

Figure 26.1. Construction cost comparison between conventional water treatment plant and simplified water treatment plant and simplified water treatment plant in Latin America (*source:* references 21 and 22).

26.2 COSTS

Costs for simplified plants are presented in Figure 26.1. Simplified plants consist of dosing and mixing of chemicals, flocculation, sedimentation (with plate settlers), and filtration with dual media. The data on which Figure 26.1 are based are given in Table 26.1. Table 26.1 includes operating and maintenance costs on a unit flow basis. Other small system costs are given in Table 26.2.

26.3 DESCRIPTION

Plate settlers are used in developing countries, especially Latin America not only for new plants, but also for upgrading existing facilities. They are more

Table 26.1. Cost of Simplified Water Treatment Plants in Latin America

LOCATION	RATED CAPACITY ×1000 GPD	RATED CAPACITY ×1000 m^3/D	(×1000) CONSTRUCTION COST (ADJ. TO $U.S. 1988)	O&M UNIT COST IN $U.S. GPD	O&M UNIT COST IN $U.S. m^3/D
COLOMBIA					
Filandia	571	2.16	188	14500	55
Rio de Oro	571	2.16	68	8300	32
Paz del Rio	571	2.16	160	20000	74
Manaure	571	2.16	127	16000	60
Circasia	914	3.46	126	9600	36
Pailitas	914	3.46	200	15000	57
Becerril	914	3.46	140	11000	41
Santa Fe	914	3.46	178	13600	52
Girardota	1028	3.89	290	20000	75
Abrego	1028	3.89	115	7700	29
Mompos	1598	6.05	165	7100	27
Andalucia	1963	7.48	290	10000	39
Chiquinquira	2830	10.71	459	11400	43
LaPaz SnDgo.	3197	12.10	445	9600	36
Zarzal	3424	12.96	592	12000	46
Florencia	5478	20.74	452	5900	22
Pereira	13696	51.84	1000	5200	20
Manizales	18261	69.12	1700	6500	25
Barranquilla	22827	86.40	3400	10500	40
Cali	22287	86.40	2800	8300	32
BRAZIL					
Prudentopolis	264	1.00	70	18000	69
Parana	528	2.00	123	16000	61
Parana	661	2.50	84	9000	34
Parana	1585	6.00	250	11000	41
Parana	3170	12.00	400	8700	33
Parana	4491	17.00	374	5900	22
Parana	6605	25.00	425	4700	18
Parana	8454	32.00	510	4300	16
Parana	11361	43.00	760	4600	18
Parana	11889	45.00	636	3700	14
Aracayu	17120	64.80	1600	6500	25
Parana	26420	100.00	1275	3400	13
BOLIVIA					
Cochabamba	5284	20.00	515	6800	26
CHILE					
Santiago	91307	345.60	5900	4600	18

Source: Reference 21.

Table 26.2. Estimated Capital, Operations and Maintenance, and Total Annual Costs for Various Unit Processes, Showing Cost Per 1,000 Gal*

	CAPITAL COST			O&M COST		
UNIT PROCESS	TOTAL $	ANNUAL $	$/1,000 GAL	ANNUAL $	$/1,000 GAL	TOTAL COST $/1,000 GAL
0.10-mgd coagulation-filtration package plant with tube settlers	176,000	20,670	1.13	11,000	0.60	1.73
0.10-mgd pressure depth clarifier and pressure filter	206,000	24,200	1.33	10,400	0.57	1.90
0.10-mgd pressure depth clarifier and pressure filter plus GAC adsorber	246,000	28,900	1.58	16,300	0.89	2.47
0.25-mgd adsorption clarifier and mixed-media filter	262,000	30,800	0.67	11,700	0.26	0.93
0.10-mgd lime-softening package plant, one stage with recarbonation	160,000	18,800	1.03	17,300	0.95	1.98
0.10-mgd lime-softening package plant, two stages with recarbonation	265,000	31,100	1.70	25,300	1.39	3.09
0.10-mgd DE vacuum filter	103,000	12,100	0.66	11,100	0.61	1.27
0.10-mgd DE pressure filter	106,000	12,400	0.68	10,600	0.58	1.26
0.10-mgd covered slow sand filter	580,000	68,100	3.73	7,700	0.42	4.15

0.10-mgd covered slow sand filter (field data)	100,000	11,800	0.64			
0.10-mgd open slow sand filter	335,000	39,300	2.16	7,100	0.39	2.55
0.29-mgd open slow sand filter (field data)	Approximately 150,000	17,600	0.33			
0.10-mgd high-pressure RO	275,000	32,300	1.77	41,300	2.26	4.03
0.10-mgd low-pressure RO	275,000	32,300	1.77	29,800	1.63	3.40
0.010-mgd high-pressure RO	84,000	9,900	5.42	8,200	4.49	9.91
0.010-mgd low-pressure RO	84,000	9,900	5.42	7,200	3.94	9.36
0.10-mgd cation exchange	151,000	17,700	0.97	8,500	0.47	1.44
0.10-mgd anion exchange	115,000	13,500	0.90	10,300	0.56	1.46
0.10-mgd activated alumina	104,000	12,200	0.67	14,600	0.80	1.47
0.10-mgd GAC in pressure vessel	175,000	20,600	1.13	14,400, six-month replacement	0.79	1.92
				9,800, twelve-month replacement	0.54	1.67
0.10-mgd packed-tower aerator	45,100	5,300	0.29	2,900	0.16	0.45
0.029-mgd packed-tower aerator	35,000	4,100	0.77	1,800	0.34	1.11
0.34-mgd packed-tower aerator	62,000	7,300	0.12	4,300	0.07	0.19

* Amortization of capital at 10 percent interest for 20 years; average flow—50 percent of design flow

Source: Reference 79.

economical to build, produce better effluent quality than horizontal sedimentation tanks, and are more stable and reliable in operation. Plate settlers consist of a series of parallel plates (in small plants, even sheets of plywood have been used) inclined at a steep enough angle to allow self-cleaning of the trays (see Figures 26.2 and 26.3). The water flows upward while the sludge slides toward the bottom of the tank where it is concentrated and then manually or mechanically removed. Because the flow velocity near the plates is almost zero, the particles that fall on them are not subject to drag forces and can easily move in an opposite direction toward collection.

The design of the sedimentation basin with plates is governed by three basic criteria: 1) the quantity of water to be treated, 2) selected detention time, and 3) the selected surface loading rate. These three criteria and applicability ranges are given in Tables 26.3 and 26.4. The asbestos cement and wood plates are most commonly used in South and Central America, because of low cost. The standard plate is 4 ft × 8 ft (1.2 × 2.4 m), and 0.2 to 0.3 in. (5 to 7.6 mm) thick. Plates of this type can handle without damage, a concentrated load of 176 lb (80 kg) at the center. Plates are installed about 7.5 ft (2.3 m) apart.

The following relationship, which is derived from geometrical considerations and laboratory performance studies, can be used for designing inclined-plate and tube settlers. Refer to Figure 26.3 (Q in appropriate units).

$$V_{sc} = Q/A = Q/A_0 f$$

where

V_{sc} = critical surface loading rate or settling velocity (m/day)
$f = \sin\theta - (L - 0.013\ R_N)\cos\theta$
A = surface area of conventional horizontal settling tank (m^2)
A_0 = surface area of high rate settling (m^2)
L_u = effective relative depth, $L_u = L - 0.013\ R_N$, (m)
L = measured from inlet edge of plate, m
R_N = Reynold's Number, probably should not exceed 400 for plate settlers but in any case should be between 280 and 1800.

Hence, for a flow Q:

$$A_0 = A/f$$

The f factor then becomes the number of times the area of a horizontal settling tank must be divided to obtain the area of an inclined-plate or tube settler. For example, for an effective relative depth $L_u = 40$ ft, and angle $\theta = 60°$; the area factor $f = 6.9$ (Table 26.3). Such a settler may be about 7

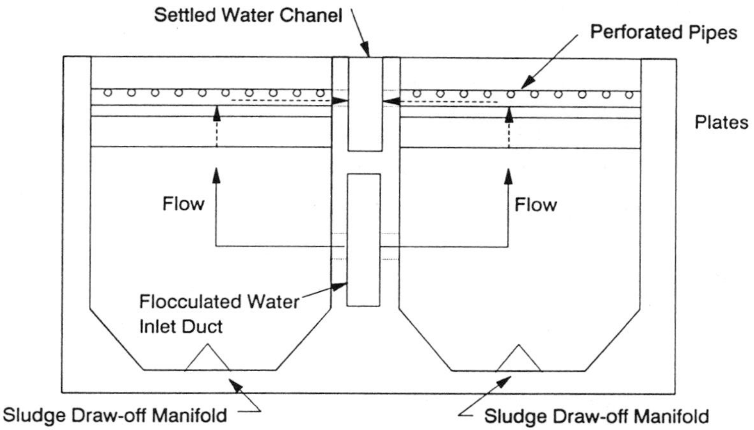

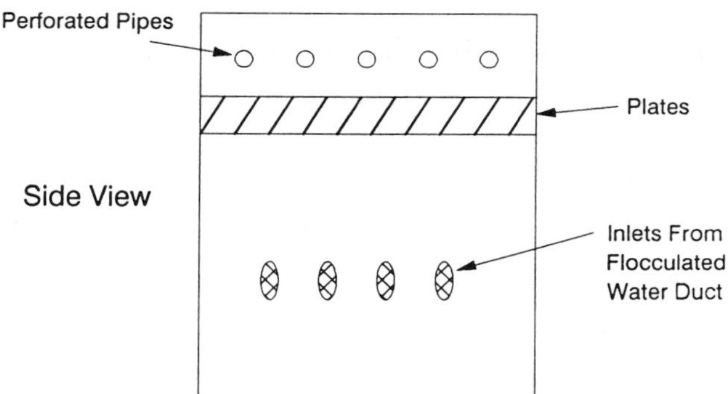

Figure 26.2. Plate settler with longitudinal hoppers at the bottom and an inlet water manifold (*source:* reference 21).

times smaller in area than a conventional settler. The number of plates to cover a horizontal area, A_0 (m^2) will be:

$$n = A_0 \sin \theta / L(e + e_p)$$

where

e = spacing between plates, m
e_p = thickness of plates, m
L = length of plate, m

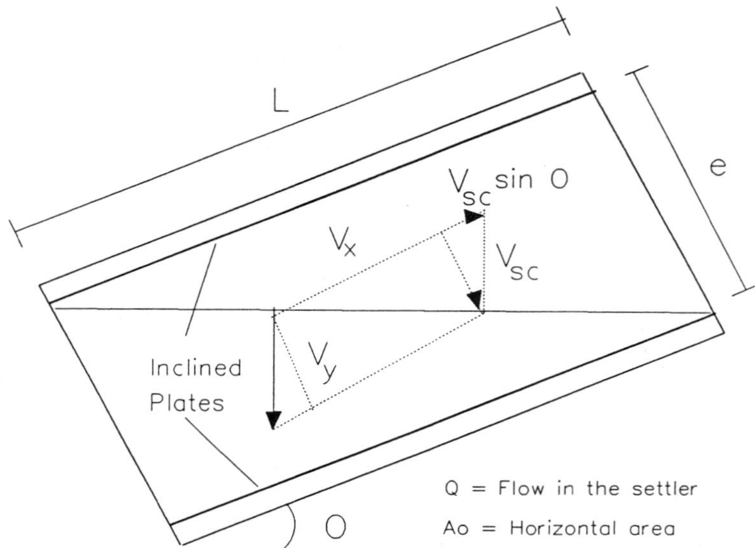

Figure 26.3. Geometrical relationships of an inclined-plate settler (*source:* reference 22).

The total area A_T (m^2) is equal to:

$$A_T = (Q/V \sin \theta)(1 + e_p/e)$$

where

$V = Q/(L \cdot e \cdot n)$
Q = capacity of settler

Recommended surface loading rates for horizontal-flow sedimentation basins equipped with inclined-plate or tube settlers are listed in Table 26.5 for two categories of raw water turbidity; 0 to 100 NTU and 100 to 1000 NTU. These loadings apply specifically to warm water areas (temperature nearly always above 10°C) and apply to most developing countries. For efficient self-cleaning, tubes or inclined plates are usually arranged at an angle of 40° to 60° to the horizontal. The most suitable angle for a particular design depends on the sludge characteristics of the water being treated, usually 55° above the horizontal. The distance between parallel-inclined plates or, similarly, the diameter of settling tubes, is about 5 cm. The passageways formed by the plates, or inside the tubes, are commonly about 1-m long.

Table 26.3. Area Factor (f) (Unitless)

	USEFUL RELATIVE DEPTH $L_u = L - 0.013\ R_N$		
40 ft	53 ft	67 ft	
ANGLE	(12 m)	(16 m)	(20 m)
0	12.00	16.00	20.00
15	11.84	15.71	19.58
30	10.89	14.36	17.82
45	9.19	12.02	14.84
60	6.86	8.85	10.86
75	4.07	5.11	6.14
90	1.00	1.00	1.00

Source: Reference 21.

Table 26.4. Velocities and Surface Loads Used in Pilot Settler

RUN NUMBER	FLOW		SURFACE LOAD		VELOCITY UNDER THE PLATES	
	CFS	m^3/s	GPM/sq ft	m/H	m/MIN	cm/s
1	11,405	323	5	12.25	1.02	1.7
2	8,298	235	3.6	8.9	0.74	1.24
3	6,850	194	3.0	7.3	0.72	1.02
4	5,826	165	2.5	6.0	0.5	0.82

Source: Reference 21.

Table 26.5. Loading for Horizontal-Flow Settling Basins Equipped with Inclined-Plate or Tube Settlers in Warm-Water Areas (Above 10°C)

SETTLING VELOCITY BASED ON TOTAL CLARIFER AREA (m/DAY)	SETTLING VELOCITY BASED ON PORTION COVERED BY PLATES (m/DAY)	PROBABLE EFFLUENT TURBIDITY (NTU)
	(A) Raw Water Turbidity 0–100 NTU	
120	140	1–3
120	170	1–5
120	230	3–7
170	200	1–5
170	230	3–5
	(B) Raw Water Turbidity 100–1000 NTU	
120	140	1–5
120	170	3–7

Source: Reference 48.

26.4 LIMITATIONS

Plate settlers produce large quantities of sludge. The capacity of the sludge hoppers at the bottom of the tanks must be considered so that too frequent withdrawal of the sludge can be avoided. The main factor is the volume of sludge produced. Sludge volume can be estimated with the following expression (21):

$$V = Q\,(K_1 D + K_2 T)/100$$

where

V = volume of sludge per day (appropriate units)
Q = water flow
D = optimal dosage of coagulants in grams per cubic meter (mg/L); range for application of the above equation is 15 to 60 mg/L
T = water turbidity, range of the equation = 100 to 800 NTU
K_1 = coefficient that varies between 0.015 and 0.025
K_2 = coefficient that varies between 0.004 and 0.0001.

For poor raw water conditions hoppers may require very frequent emptying, perhaps hourly.

26.5 AVAILABILITY

The plate settler is widely used in developing countries. The technology is available with various modifications to suit site specific problems. For example, the cost of covering a settler with plates is between one-half to three quarters of the cost of installing plastic tube modules. In addition, the efficiency of asbestos cement or wood plates is higher than tube settlers because of better length-to-width relationships and hydraulic characteristics.

Local materials and labor may be used for the construction of inclined-plate or tube settlers. For plate settlers, the individual plates/trays can be fabricated from polyethylene (or similar type of plastic) or of wood. Plastic may be a problem because of inadequate stiffness. Asbestos cement plates should be coated with plastic or similar type of protective covering because of their susceptibility to corrosion from alum-treated water and subsequent release of asbestos into the treated water. Where wood is used on low slopes, trays are commonly 30 cm apart. It may also be necessary to drain the tank for cleaning occasionally, because sludge does not readily slide down uncoated wooden trays while the basin is in service. Uncoated wood trays have worked successfully with frequent cleaning.

Tube settlers are easily fabricated from PVC pipes (3 to 5 cm internal diameter), which are packed closely together to form a module. In countries with indigenous plastics industries, commercially available tube modules that are prefabricated at the factory are suitable for larger installations. Plastic tube settler inserts are manufactured in Brazil.

26.6 OPERATION AND MAINTENANCE

Plates may be installed at one end of an existing settling tank allowing easy access for removal and cleaning. Regardless of the steepness of the plate angle, it is necessary to remove the plates and clean them on a regular basis. Algae growth on the plates can be a troublesome problem, especially in warm climates.

Constant attention must be paid to sludge removal and subsequent management and disposal.

26.7 CONTROL

It is necessary to obtain equal flow distribution through all the plates or tubes and to maintain laminar flow. Thus inlet and outlet hydraulic design is critical. Care should be exercised in the design of outlet collection systems to assure even distribution of flow through plate or tube modules. This is more easily done with overflow weir outlets than with submerged launders.

26.8 SPECIAL FACTORS

In both the inlet and outlet systems, the spacing between the plates should not be more than 5 ft (1.5 m), and inlet troughs or pipes must be installed over the entire surface of the tank. Lateral troughs or lateral perforated pipes should be used as distributors from a central influent channel.

26.9 RECOMMENDATION

Plate settlers could be used for upgrading and plant expansion. If conventional designs have been used initially, plate settlers could allow plant capacity expansion by 50 to 150% (25).

27. SIMPLIFIED CHEMICAL STORAGE AND DOSING, MIXING, AND FLOCCULATION

27.1 DESCRIPTION AND OPERATIONS

Chemical dosing (see also Chapter 34 on chemical addition) may be done with simple gravity feeders from the floor above. Coagulants may be applied from solution storage tanks. Where liquid alum is not available, the solution is prepared from solid alum (lumps). No dry feeders or dosing pumps are typically needed in small facilities. Trucks may unload materials directly on the floor above by means of ramps if necessary. In larger plants, such as the 24 m^3/sec Los Berros plant in Mexico City, the chemical storage and solution tanks are placed on the treatment floor, and chemicals in solution are centrifugally pumped to a small constant level elevated dosing tank from which the solution flows by gravity. There have been a number of evaluations performed on the development and application of simplified systems (21, 22, 23, 25).

Mixing is usually done hydraulically rather than with mechanical mixers, usually with Parshall flumes and Creagers weirs (i.e., a round-crested overflow spillway). Parshall flumes have the additional advantages of allowing simultaneous mixing and flow measurement.

In very small plants, such as one seen in Uruguay, alum dosing is controlled simply by allowing a faucet to drip at a reasonably steady rate over alum lumps. The experience of the operator determines the drip rate based on effluent clarity. The drip rate may easily be changed by opening or closing the faucet slightly.

Flocculation systems are commonly constructed with hydraulic drives for small treatment plants and mechanical drives for large treatment plants. See Figure 27.1 (21).

Mechanical flocculation is more versatile than hydraulic flocculation and easily allows for an increase or decrease in the velocity gradient and thus the mixing intensity but is more susceptible to short circuiting.

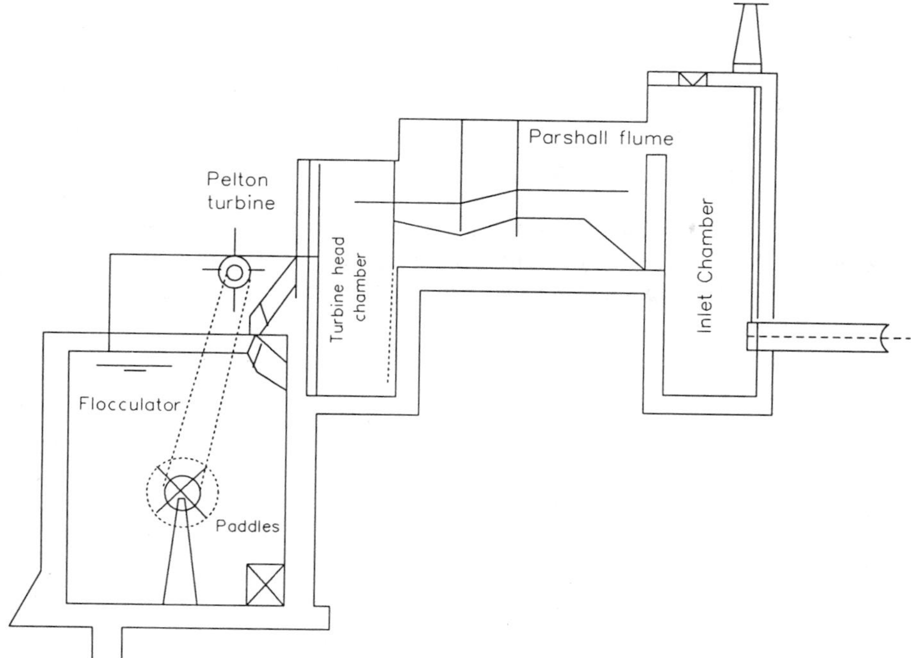

Figure 27.1. Hydromechanical flocculators (*source:* reference 21).

27.2 AVAILABILITY

Simplified chemical addition is widely used in developing countries, especially in South America, and is applicable to small system installations everywhere.

27.3 OPERATION AND MAINTENANCE

A tradeoff for system simplicity is operator awareness and attendance. A sophisticated chemical delivery apparatus which has failed may be replaced with a plastic tube weighted with a rock and inserted into a channel flow. Systems which are operationally simple at the outset are less costly and easier to service, especially where spare parts inventories are difficult to maintain.

27.4 SPECIAL FACTORS

In some cases a hydromechanical flocculator has been used. A Pelton turbine driving a conventional horizontal paddle is introduced into the raw water inflow, inducing rapid mixing at the same time. The head used by the turbine is generally only a few feet and it can produce velocity gradients up to 60 sec^{-1}. This system can be used only in gravity-fed systems, as shown in Figure 27.1.

27.5 RECOMMENDATIONS

System simplicity results in lower O&M costs and such plants may be less costly to construct also (21). Simplicity of construction and operation should be considered for application to all plant designs, but especially in the case of small plants.

28. LAND APPLICATION OF WASTEWATER BY IRRIGATION

28.1 DESCRIPTION

Land application of wastewater is a potential cost effective technique for treatment of municipal wastewater where space is available. The technology may be used where an available irrigation site has suitable soil conditions and ground water hydrology, and where the climate is favorable. Hundreds of efficient systems are currently operating in regions with limited water resources to increase the growth of grass, crops, and forests. In addition, the natural top soil and soil biota provide filtering and stabilization of the organic matter; nutrients in the wastes are used by the plants (Figure 28.1). Some groundwater replenishment potential exists also.

The wastewater is applied by sprinkling to vegetated soil that is moderate to high in permeability (sandy loam to sand gravel mixed loam) and is treated as it travels through the soil matrix by filtration, sorption, ion exchange, precipitation, microbial action, and also by plant uptake. Most chemical, physical, and biological treatment process functions are generally operative to some extent in soil treatment systems.

Sprinklers are categorized as hand moved, mechanically moved, and permanently set. The selection among sprinkler types includes the following considerations: field conditions (shape, slope, vegetation, and soil type), climate, operating conditions, and costs. Vegetation is a vital part of the process and serves to extract nutrients, reduce erosion, and maintain soil permeability. Vegetation harvesting allows reuse of the soil medium for certain treatment functions including waste volume (water content), some organic compounds, nitrogen and nutrients.

The renovated water, after passing through the soil, may be collected by means of a drainage collection system or becomes available for reuse through groundwater recharge. In regions with limited water resources, land application by the irrigation process can improve the growth yield of many crops.

The design of land application of wastewater by irrigation is governed by field area, hydraulic and nutrient application rates, waste application rate, BOD_5 loading, soil depth, crop selection, and other site and waste specific factors. The general design considerations for land treatment approaches are

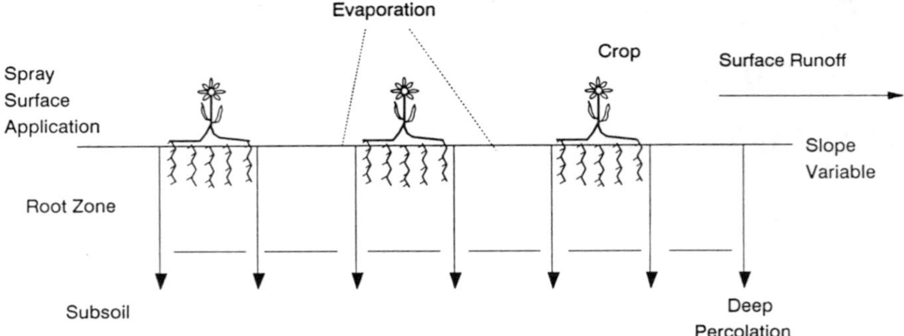

Figure 28.1. Land application of wastewater by irrigation (*source:* reference 23).

given in Tables 28.1 and 28.2 (4, 17). Expected effluent quality is shown on Table 28.3.

28.2 LIMITATIONS

Land application is limited by soil type and depth, topography, underlying geology, climate, crop selection, and land availability. Crop water tolerance, nutrient requirements, and the nitrogen removal capacity of the soil-vegetation complex limit hydraulic loading rate. Land slopes should be less than 15% to minimize runoff and erosion. Pretreatment should be considered for removal of solids and oil and grease, and to maintain reliability of sprinklers and to reduce clogging.

28.3 COSTS

Three types of systems are reported: slow rate irrigation, overland flow, and infiltration percolation. Typical cost ranges are shown in Table 28.4 and cost comparison of capital and operating costs for irrigation at two different sizes is given. Comparative costs for overland flow and infiltration-percolation systems are shown in Table 28.5.

28.4 AVAILABILITY

This technology has been widely and successfully utilized for more than 100 years, beginning long before an understanding of the potential for application of biological treatment was available. The technology is less capital intensive (depending on land costs) and more O&M cost-efficient than conventional treatment processes yielding water of similar quality.

Table 28.1. Comparison of Design Features for Alternative Land-Treatment Processes

FEATURE	IRRIGATION	RAPID INFILTRATION	OVERLAND FLOW	WETLAND APPLICATION	SUBSURFACE APPLICATION
Application techniques	Sprinkler or surface[a]	Usually surface	Sprinkler or surface	Sprinkler or surface	Subsurface piping
Annual application rate, m	0.6–6.0	6–120	3–20	1–30	2–25
Field area required, ha[b]	22–226	1–22	10–44	4–113	5–56
Typical weekly application rate, cm	2.5–10	10–210	6–15[c] 15–40[d]	2.5–60	5–50
Minimum preapplication treatment provided	Primary sedimentation[e]	Primary sedimentation	Screening and grit removal	Primary sedimentation	Primary sedimentation
Disposition of applied wastewater	Evapotranspiration and percolation	Mainly percolation	Surface runoff and evapotranspiration with some percolation	Evapotranspiration, percolation and runoff	Percolation with some evapotranspiration
Need for vegetation	Required	Optional	Required	Required	Optional

[a] Includes ridge and furrow and border strip.
[b] Field area in hectares not including buffer area, roads, or ditches for 0.044 m^3/s (Mgal/d) flow.
[c] Range for application of screened wastewater.
[d] Range for application of lagoon and secondary effluent.
[e] Depends on the use of the effluent and the type of crop.

Note: cm × 0.397 = in
m × 3.2808 = ft
ha × 2.47111 = acre

Source: References 75, 76.

Table 28.2. Comparison of Site Characteristics for Land-Treatment Processes

CHARACTERISTICS	IRRIGATION	RAPID INFILTRATION	OVERLAND FLOW	WETLAND APPLICATION
Climatic restrictions	Storage often needed for cold weather and precipitation	None (possibly modify operation in cold weather)	Storage often need for cold weather	Storage may be needed for cold weather
Depth to groundwater, m	0.6–0.9 (minimum)	3.0 (lesser depths acceptable where underdrainage provided)	Not critical	Not critical
Slope	Less than 20% on cultivated land; less than 40% on noncultivated land	Not critical; excessive slopes require much earthwork	Finish slopes 2–8%	Usually less 5%
Soil permeability	Moderately slow to moderately rapid	Rapid (sands, loamy sands)	Slow (clays, silts, and soils with impermeable barriers)	Slow to moderate

Note: m × 3.2808 = ft
Source: References 75, 76.

Table 28.3. Comparison of Expected Quality of Treated Water from Land Treatment Processes, mg/L

CONSTITUENT	IRRIGATION[a]		RAPID INFILTRATION[b]		OVERLAND FLOW[c]	
	AVERAGE	MAXIMUM	AVERAGE	MAXIMUM	AVERAGE	MAXIMUM
BOD	<2	<5	2	<5	10	<15
Suspended solids	<1	<5	2	<5	10	<20
Ammonia nitrogen as N	<0.5	<2	0.5	<2	0.8	<2
Total nitrogen as N	3	<8	10	<20	3	<5
Total phosphorus as P	<0.1	<0.1	1	<5	4	<6

[a] Percolation of primary or secondary effluent through 1.5 m of soil.
[b] Percolation of primary or secondary effluent through 4.5 m of soil.
[c] Runoff of comminuted municipal wastewater over about 45 m of slope.
Note: m × 3.2808 = ft
Source: References 75, 76.

Table 28.4. Typical Cost Ranges. Land Application of Waste Water by Irrigation[a] (Spray Irrigation Systems)

	0.01 MGD TREATMENT CAPACITY		0.1 MGD TREATMENT CAPACITY	
	CAPITAL COST ($/GPD)	OPERATING COST ($/1000 GAL)	CAPITAL COST ($/GPD)	OPERATING COST ($/1000 GAL)
Spray Irrigation System[b]	5.25–10.50	.88–1.75	1.75–5.25	.18–.50

[a] Cost information based on data presented in the following publications:

1. U.S. EPA (1980) Innovative and Alternative Technology Assessment Manual, MCD-53.
2. U.S. EPA (1980) Construction Costs for Municipal Wastewater Treatment Plants, FRD-11.
3. U.S. HUD (1977) Package Wastewater Treatment Plant Descriptions, Performance, and Cost.

[b] Low rate (0.5–4 inches per week) application rate; not including cost of pretreatment facilities, storage lagoons, and land.

Note: 1 gal = 3.785 L.
2.5 acres = 1 ha.

Source: Based on Reference 78.

28.5 OPERATION AND MAINTENANCE

Proper and regular maintenance of sprinklers is essential. Cleaning of spray nozzles, periodic checks of sprinkler piping, drain outlets and valves is essential. The condition of the soil system, porosity, permeability, organic content, etc., must be checked regularly.

28.6 CONTROL

The application of controls in the management of land application of wastewater should be based on the following factors: land availability, site location, climate, soil and subsurface conditions, wastewater characteristics, hydraulic loading, the capacity and utilization of the plant-soil complex to produce a specific effluent quality, the intended use or reuse of the wastewater, and the ultimate use of the land.

Waste loading rates and pollutant uptakes must be regularly monitored. It may be necessary to modify waste loading rates depending on rainfall frequency and quantity. Thus storage capacity should be provided and maintained.

28.7 SPECIAL FACTORS

Toxic substances, including metals and industrial organic compounds which are not likely to be removed in the renovation process must be controlled in irrigation water or reduced by pretreatment. Extensive soil water

Table 28.5. Comparison of Capital and Operating Costs Costs for 1-MGD Spray Irrigation, Overland Flow, and Infiltration-Percolation Systems

COST ITEM	SPRAY IRRIGATION	OVERLAND FLOW	INFILTRATION/ PERCOLATION
LIQUID LOADING RATE, IN./WK	2.5	4.0	60.0
LAND USED, ACRES	103	64	—
LAND REQUIRED, ACRES	124	77	5
Capital costs ($)			
Land @ $500/acre	109,000	67,000	4,400
Earthwork	18,000	112,000	17,500
Pumping station	88,000	87,500	—
Transmission	230,000	233,000	230,000
Distribution	250,000	112,000	8,800
Collection	—	11,000	52,500
Total Capital Costs	700,000	620,000	314,000
Capital cost per purchased acre	5,600	8,000	63,000
Amortized cost	65,000	61,000	34,000
Capital cost, ¢/1,000 gal	17.7	16.7	9.3
Operating Costs ($/yr)			
Labor	17,500	17,500	13,000
Maintenance	34,000	21,000	6,100
Power	10,000	10,000	3,200
Total Operating Costs	62,000	49,000	22,400
Operating Cost, ¢/1,000 gal.	16.8	13.3	6.1
Total Cost, ¢/1,000 gal	34.4	30	15.4

Note: 1 gal = 3.785 L; 2.5 acres = 1 ha.
Source: Reference 1.

and groundwater pollution are more difficult to correct than waste quality or surface water pollution, and once present in the soil system, refractory pollutants can persist for long periods of time.

28.8 RECOMMENDATIONS

Irrigation should be compared with other land treatment approaches (see Figure 28.2). Knowledge of wastewater characteristics, treatment process mechanisms, and public health requirements are fundamental to the successful design and operation of the land application of wastewater treatment by irrigation.

Land treatment should be considered for application in developing countries. Land is typically available at low cost, even near large population centers. Also, with water shortages in many areas, land treatment of wastewa-

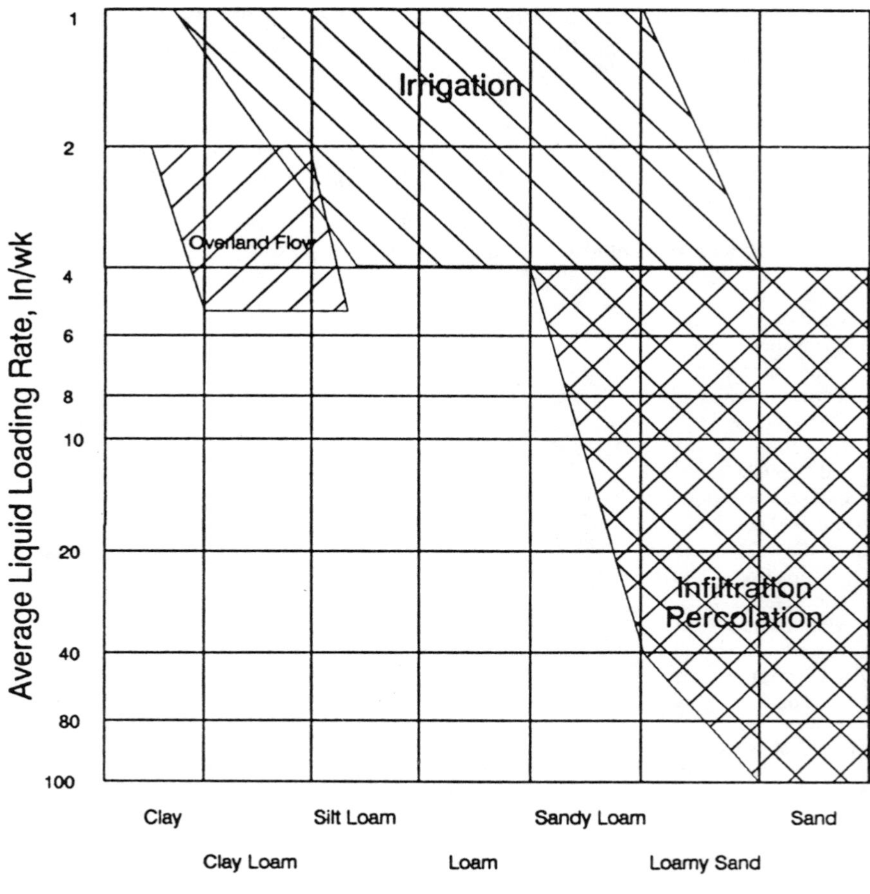

Figure 28.2. Soil type vs. liquid loading rates for different land application approaches (*source:* reference 23).

ter can be a very important water recovery technology. Land treatment can be used in a direct (with follow-on treatment) or indirect (groundwater recharge) water reuse mode. Increased consideration should be given to land application in most waste treatment alternative studies so that a detailed examination of the option can be made.

29. LAND APPLICATION OF WASTEWATER BY OVERLAND FLOW

29.1 DESCRIPTION

Overland flow is a land treatment process in which wastewater is applied at the upper end of sloped vegetated terraces and allowed to flow down the terraces in sheet flow to a series of runoff collection ditches or pipes. Figure 29.1 (2) is a schematic of the overland flow process. The terraces are constructed on tight, nearly impermeable soils and planted with a mixture of grasses.

The wastewater is renovated by a combination of physical, chemical, and biological processes before reaching the toe of the terrace where it is collected in runoff channels and discharged, further treated, or recovered. Overland flow can also be used to treat secondary effluents such as from an oxidation pond/ditch or to provide secondary treatment. Under the right conditions it could be substituted for primary treatment. Table 29.1 provides design parameters of the overland flow process. Additional design considerations are given in Table 29.2.

29.2 LIMITATIONS

The vegetation on the terraces is very important since proper selection will allow the desired smooth sheet flow down the terraces. Vegetation also protects the terraces from erosion. To achieve this two-pronged goal a water-tolerant, tuft grass is preferred. The most commonly used grasses in warm climates have been Bermudagrass, Reed canarygrass, and mixtures of several types of grasses. Such systems will yield good quality effluent, but discharges may require additional treatment.

29.3 COSTS

Typical ranges of construction and operating and maintenance costs are shown in Tables 29.3 and 29.4.

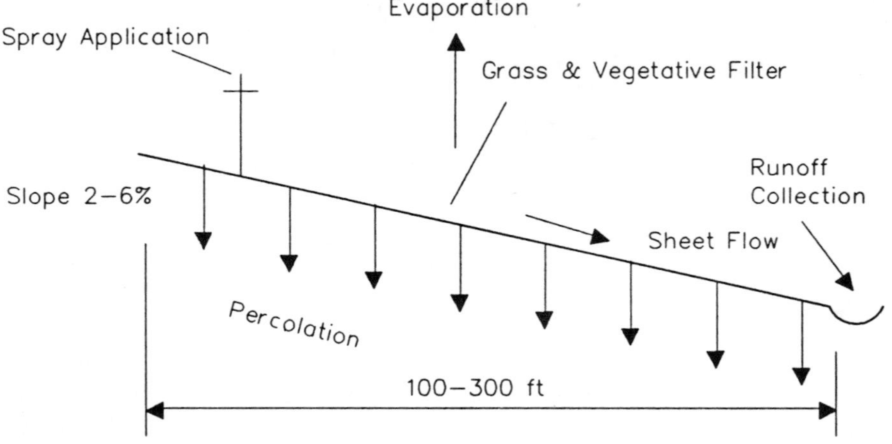

Figure 29.1. Aschematic diagram of overland flow treatment (*source:* reference 2).

29.4 AVAILABILITY

The technology has not been widely applied. There are very few full scale municipal plants for wastewater treatment in operation, and most are in warm, dry climates.

29.5 OPERATION AND MAINTENANCE

See the discussion in Chapter 28. Since overland flow treatment uses basically surface phenomena, soil clogging is typically not a problem. High BOD and suspended solids removal are achieved in raw wastewater applications.

Table 29.1. Suggested Overland Flow Design Ranges

PREAPPLICATION TREATMENT	APPLICATION RATE $M^3/H \cdot M$	HYDRAULIC LOADING RATE CM/D
Screening/Primary	0.07–0.12[a]	2.0–7.0[b]
Aerated Cell (1-day detention)	0.08–0.14	2.0–8.5
Wastewater Treatment Pond[c]	0.09–0.15	2.5–9.0
Secondary[d]	0.11–0.17	3.0–10.0

[a] m^3/hour · m × 80.5 = gal/h · ft
[b] cm/day × 0.394 = in./day
[c] does not include removal of algae
[d] Recommended only for upgrading existing secondary treatment
Source: Reference 27.

Table 29.2. Overland Flow Design Features

• Application Technique	• Sprinkler • Surface
• Process Objectives	• Wastewater Treatment • Crop Production • Augment Surface Streams
• Annual Application Rate	• 10 to 70 ft
• Field Area Required	• 16 to 110 acre/MGD
• Precipitation Treatment (minimum)	• Screening and grit removal
• Slope	• Finish Slope 2 to 8%
• Soil Permeability	• Slow (clay, silt, etc.)
• Depth to Groundwater	• Not critical
• Climatic restrictions	• Storage usually required during extreme cold weather
• Vegetation (secondary objective)	• Grass crop required
• Fate of Applied Wastewater	• Surface runoff • Evaporation • Minimal Percolation

Note: 2.5 acres = 1 ha
acre/MGD = ha/m^3/day
Source: Reference 26.

However, poor design and mismanagement, particularly flow rate overloading with wastewater, can result in health risks, nuisance factors, subsurface water contamination, and potential toxicity from metals buildup in the soils. The success or failure of the process operation primarily depends on careful control of various loading rates, climate, soil type, and depth to groundwater.

A maintenance effort should be based on the type and size of the effluent distribution system, the nature of the vegetative cover, and geographical conditions of the system.

Table 29.3. Typical Cost Ranges for Overland Flow Systems[a]

	10,000 GPD CAPACITY	100,000 GPD CAPACITY
Capital Cost[b] ($/gpd)	6.90–13.80	1.40–2.80
Operating Cost[b] ($/1000 gal)	1.40–2.80	0.35–0.70

[a] Based on cost data from the following sources:
1. U.S. EPA (1980) Innovative and Alterantive Technology Assessment Manual, MCD-53.
2. U.S. EPA (1980) Construction Costs for Municipal Wastewater Treatment Plants, FRD-11.
3. U.S. HUD (1977) Package Wastewater Treatment Plant Descriptions, Performance and Cost.

[b] Complete system including disinfection and discharge; capital costs do not include land costs.
Note: 1 gal/day × 3.785 × 10^{-3} = m^3/day
Source: Based on References 76 and 77.

Table 29.4. Capital and Operating Costs for 1-MGD Overland Flow

COST ITEM	OVERLAND FLOW
Liquid loading rate, in./wk	4.0
Land used, acres	64
Land required, acres	77
Total Capital Costs	$620,000
Capital Cost, ¢/1,000 gal	16.6
Total Operating Costs (/yr)	$49,000
Operating Cost, ¢/1,000 gal	13.3
Total cost, ¢/1,000 gal	30

Note: 3.785 L = 1 gal
2.5 acre = 1 ha
Source: Reference 1.

29.6 CONTROL

See the discussion in Chapter 28. The major concerns and controls in the management of overland flow application should be based on the following factors: site location and geography, the climate especially rainfall frequency and magnitude of storms, soil and subsurface conditions, various loading rates, wastewater characteristics, the capacity and utilization of the plant-soil complex to produce a specific water quality, the intended use or reuse of the wastewater, and the ultimate use of the land.

Operation during extended periods of rain/snow will require storage of untreated wastewater. Storage systems are often the cause of odors and other nuisances.

29.7 SPECIAL FACTORS

See the discussion in Chapter 28. The concentration of potential toxic metals and trace elements which are not likely to be removed in the renovation process must be controlled in overland flow influent or the effluent must be pretreated. Extensive soil, water, and groundwater pollution is difficult to correct. Pollutants, such as heavy metals and some industrial organic compounds which are not easily biodegradable may persist for long periods.

29.8 RECOMMENDATIONS

The overland flow process requires long term commitment of large land areas. Potential odor and vector problems exist, but careful design and operation will keep adverse impacts to a minimum. See the discussion in Chapter 28.

30. LAND APPLICATION OF WASTEWATER BY INFILTRATION—PERCOLATION

31.1 DESCRIPTION

In the infiltration-percolation process, sometimes referred to as rapid infiltration, most of the applied wastewater percolates through the soil, and the treated effluent eventually reaches the groundwater. Thus the process is the most applicable land treatment technology for indirect water reuse. The wastewater is applied to highly permeable soils, such as sandy and loamy soils spreading in basins or by sprinkling, and is treated as it travels through the soil matrix. Vegetation may not be used, but grass cover helps to remove suspended inorganic and organic solids.

A typical cross-section for this process is shown in the schematic view in Figure 30.1 (23). A much greater portion of the applied wastewater percolates to the groundwater than with the irrigation approach and definitely more than with overland flow. There is little or no consumptive use of water or waste constituents by vegetation, and there is less evaporation than with the other land treatment options. Renovated water recovery is accomplished by using underdrains or wells as shown in Figure 30.2 (75). The principal design parameters for infiltration—percolation are shown in Table 30.1.

Spreading basins may be used rather than the trench approach shown in Figure 30.1. These are constructed by removing the fine textured top soil from which shallow banks are constructed. The underlying sandy soil serves as the filtration/treatment medium. Underdrainage is provided by using plastic or clay tile pipes. The distribution system applies wastewater at a rate which constantly floods the basin throughout the application period of several hours to two weeks. The spreading basin water drains uniformly away allowing air movement downward through the soil to fill the voids. A controlled treatment cycle of flooding and drying maintains the infiltration capacity of the soil material.

30.2 LIMITATIONS

The process is limited by soil type, active soil depth, the hydraulic capacity of the soil, the underlying geology, the transmissivity of the superficial aquifer, and the slope of the land. Adverse values for these parameters may result in

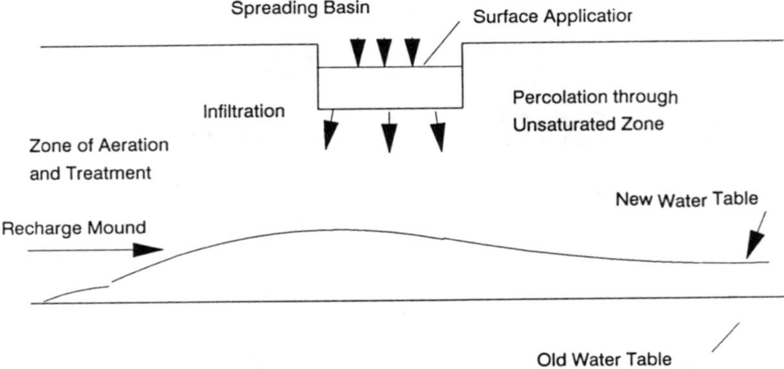

Figure 30.1. Schematic view of land application of wastewater by infiltration-percolation method (*source:* reference 23).

unacceptable treatment results. Nitrate and nitrite removals are generally low but nitrification may occur.

Pretreatment of certain wastes, especially those containing very high TSS, metals, and refractory organic compounds may be necessary to maintain soil treatment capability. See the discussions in Chapters 28 and 29.

30.3 COSTS

The capital and operating costs for infiltration and percolation processes are given in Table 30.2.

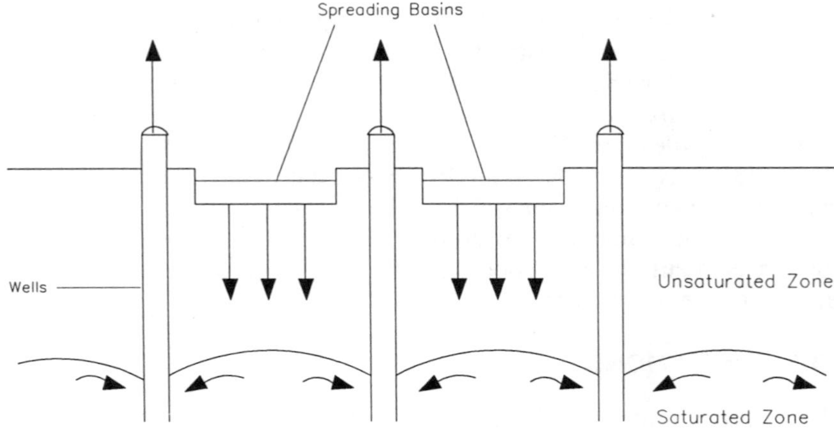

Figure 30.2. Recovery of renovated water by wells (*source:* reference 4).

Table 30.1. Typical Design Parameter for Land Application of Wastewater by Infiltration—Percolation Method

PARAMETER	RANGE VALUES
Field Area	3 to 56 acres/Mgal/day
Application rate	20 to 400 ft/yr
	4 to 92 in./week
BOD_5 loading rate	20 to 100 lb/acre/d
Soil depth	10 to 15 ft or more
Soil permeability	0.6 in./hr or more
hydraulic loading cycle	9 hr to 2 weeks application period
	15 hr to 2 weeks resting period
Soil Texture	Sands, sandy loams
Basin Size	1 to 10 acres, at least 2 basins/site
Height of dikes	4 ft
Underdrains	6 or more ft deep well or drain spacing site specific
Application techniques	Flooding or sprinkling
Preapplication treatment	Primary or secondary

Note: acre/MGD $\times 1 \times 10^{-4}$ = Ha/m^3/day
1 in. = 2.54 cm
1 ft = 0.3048 m
Source: Adapted from References 75 and 76.

30.4 AVAILABILITY

This process has been used for decades, before an understanding of the operative processes evolved. It has been widely used for municipal and certain industrial wastewaters throughout the world.

30.5 OPERATION AND MAINTENANCE

Wastewater spreading basins must be prevented from clogging due to suspended solids in the wastewater, therefore, occasional tillage of the surface layer is necessary. See the discussions in Chapters 28 and 29.

30.6 CONTROL

Preapplication treatment of wastewater is essential in the case of highly contaminated sources. Removal of solids will improve distribution system reliability, reduce nuisance conditions, and may reduce clogging rates. Common preapplication treatment practices include the following: settling for isolated locations with restricted public access; biological and or chemical treatment for urban locations where odors may become a nuisance.

Table 30.2. Capital and Operating Costs for 1-MGD Infiltration-Percolation Systems

COST ITEM	INFILTRATION-PERCOLATION
Liquid loading rate, in./wk	60.0
Land used, acres	—
Land required, acres	5
Total Capital Costs	$314,000
Capital cost, ¢/1,000 gal	9.3
Total Operating Costs (/yr)	$22,400
Operating Cost, ¢/1,000 gal	6.1
Total Cost, ¢/1,000 gal	15.4

Note: 1 gal = 3.785 L
2.5 acre = 1 hectare (ha)
Source: Adapted from Reference 1.

30.7 SPECIFIC FACTORS

This treatment process has potential for contamination of groundwater by nitrates and heavy metals. The heavy metals may be eliminated by pretreatment as necessary. Monitoring for metals and toxic organics is needed especially where not removed by pretreatment.

Application of the technology, as with all land treatment options, requires long term commitment of relatively large land areas (although small by comparison to other land treatment types). Surface water resources are diverted to groundwater. Crops grown and harvested from the system require monitoring for heavy metal content.

30.8 RECOMMENDATIONS

With the continuous usage of spreading basins, the infiltration rate diminishes slowly with time due to clogging. Infiltration capacity of the soil may be partially or wholly restored by occasional tillage of the surface layer and when appropriate, removal (and replacement) of several inches from the surface of the basin.

See the discussion in Chapter 28 related to application options for land treatment.

31. ANAEROBIC PONDS/LAGOONS

31.1 DESCRIPTION

The reader should refer to other sections on pond treatment, Chapters 12 and 21. Anaerobic ponds have steep side walls and are deeper (up to 20 ft) than aerobic and facultative ponds. Anaerobic conditions are maintained by keeping the loading so high that complete deoxygenation is prevalent. Although some oxygenation is possible in shallow surface zones, once greases form an impervious surface layer, complete anaerobic conditions develop. The stabilization/treatment results from thermophilic anaerobic digestion of organic wastes. During the treatment process acid forming bacteria will break down organics in the untreated anaerobic digestion of sludge. The resultant acids are then converted to carbon dioxide, methane, and other end products.

These lagoons are constructed in parallel or series. The typical detention time is 20 to 50 days; depth, 8 to 20 ft (2.4 to 6.1 m); water temperature range, 35°F to 120°F (optimum 86°F); organic loading, 200 to 2200 lb BOD_5/acre/day (kg/Ha/day). In the typical anaerobic pond, wastewater enters near the bottom of the pond (see Figure 31.1) (2) and mixes with active microbial mass in the sludge blanket, which is usually 6 ft (1.8 m) deep. The discharge is located near one of the sides of the pond, submerged below the liquid surface. Excess undigested grease floats to the top, forming a heat retaining and relatively air tight cover.

Anaerobic and facultative ponds are effective in Brazil (64). Tables 31.1, 31.2, and 31.3 show results of studies in Campina Grande in northeast Brazil. So-called maturation ponds (see Table 21.1) which are aerobic during the day from algal activity, may be used as post-treatment for effluent from facultative and anaerobic ponds.

31.2 LIMITATIONS

Anaerobic ponds may generate odors. Relatively large land area is required.

31.3 COSTS

Cost information is given in Figure 31.2 (2, 11).

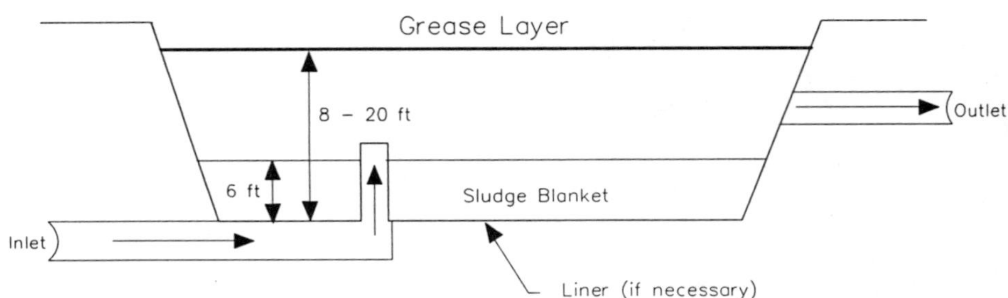

Figure 31.1. Typical anaerobic lagoon (*source:* reference 2).

31.4 AVAILABILITY

The process is used world-wide and is well demonstrated for stabilization of highly concentrated organic wastes, such as cannery and dairy wastes.

31.5 OPERATION AND MAINTENANCE

The principal advantages in developing countries are that high loadings of organics are possible at a much lower cost than comparable forms of treatment achieving similar effluent quality results. There are minimum operating and maintenance requirements. Periodic long-term sludge removal may be required with high organic loadings. Vegetation in peripheral zones around the edges of ponds should be regularly removed since eventually the volumetric capacity of the ponds is reduced.

Table 31.1. Experimental Results from a Series of Five Ponds. Mean Results During May 1979–June 1988 Pond Temperature, 26°C; overall retention time, 28.1 days

	RETENTION TIME (DAYS)	BOD_5 (MG/L)	TSS[1] (MG/L)	FC[2] (NUMBER/100 ML)
Raw sewage	—	240	205	4.6×10^7
Effluent from pond				
1	6.8	63	56	2.9×10^6
2	5.5	45	74	3.2×10^5
3	5.5	25	61	2.4×10^4
4	5.5	19	43	450
5	5.8	17	45	30

[1] Total suspended solids.
[2] Fecal coliform.

Table 31.2. Experimental Results from Four Faculative Ponds Selected mean results obtained in three experiments during June 1977–December 1981; temperature, 26°C

BOD_5 LOADING (KG/HA/D)	RETENTION TIME (DAYS)	BOD REDUCTION (%)
162	18.9	84
255	12.0	79
322	9.5	77
425	6.8	73
529	6.8	74
577	6.3	74

31.6 CONTROL

The major concern in the anaerobic process is to maintain oxygen free conditions. For efficient operation, water temperature should be maintained above 75°F.

31.7 SPECIAL FACTORS

High concentrations of organic matter and decomposition may result in odor problems. Also, there is potential for seepage of partially treated waste-water into ground water unless ponds are lined with clay or impervious material.

Table 31.1. Experimental Results from Anaerobic Ponds Mean Results During June 1977–March 1979 (Pond Temperature, 26°C)

	RETENTION TIME DAYS)	BOD_5 (MG/L)	SS[1] (MG/L)	FC[2] (NUMBER/100 ML)
Raw sewage	—	245	310	4.7×10^7
Effluent from pond[3]				
1	0.8	59	82	8.1×10^6
2	0.4	46	64	5.0×10^6
3	1.9	49	57	4.7×10^6

[1] Total suspended solids
[2] Fecal coliforms
[3] Ponds 1 and 2 in series; ponds 1 and 3 raw sewage

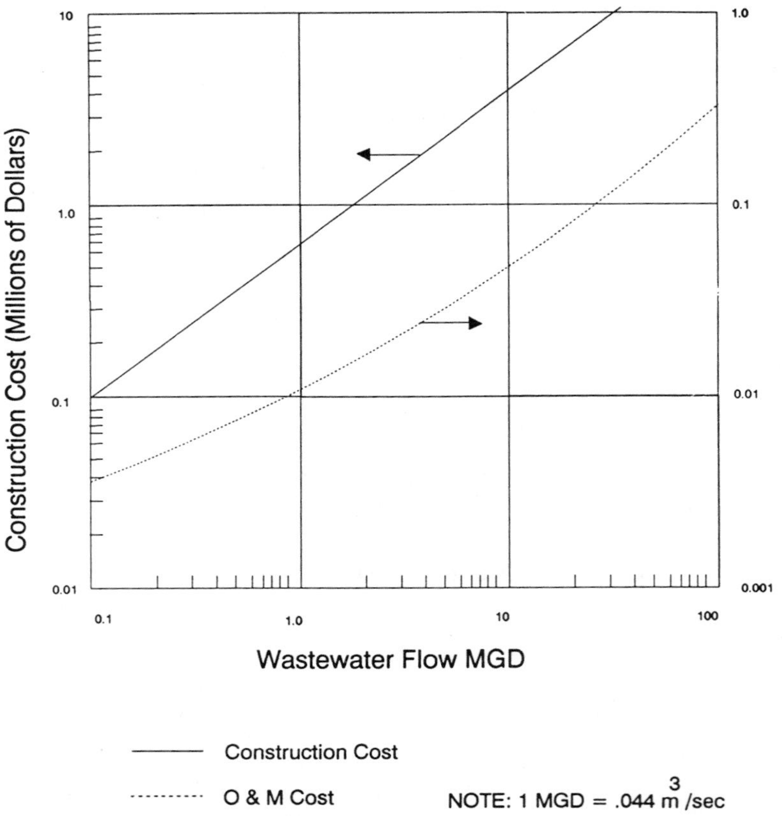

Construction Cost (Millions of Dollars)

10
1.0
0.1
0.01

Annual O & M Cost (Millions of Dollars)

1.0
0.1
0.01
0.001

0.1
1.0
10
100

Wastewater Flow MGD

——— Construction Cost

········· O & M Cost

NOTE: 1 MGD = .044 m^3/sec

Figure 31.2. Construction, operation, and maintenance costs for anaerobic lagoons.

31.8 RECOMMENDATIONS

Ponds and lagoons are simple and efficient waste treatment technologies. (Also see Chapters 12 and 21.) Waste treatment should be considered for application where simplicity is required.

32. OZONE DISINFECTION

32.1 DESCRIPTION

Ozone (O_3) may be used for disinfection and oxidation of organics in water and wastewater treatment plants. As a disinfectant (dosages of 3 to 10 mg/L are common), ozone is an effective agent for deactivating common forms of bacteria, bacterial spores, and vegetative microorganisms, as well as eliminating harmful viruses. Additionally, ozone acts to chemically oxidize materials found in the water and wastewater and can reduce the BOD_5 and COD, forming oxygenated organic intermediates and end products. Ozone treatment also reduces water and wastewater color, odor, and taste (2, 48, 73).

Ozone injection into the wastewater flow is accomplished by mechanical mixing devices, countercurrent or co-current flow columns, porous diffusers or jet injectors. Ozone acts quickly and consequently, requires a relatively short contact time. (See Figure 32.1.) Results of disinfection by ozonation and design criteria are provided in Table 32.1.

Ozone has been found to be a good oxidant for removal of cyanide, phenol, and other dissolved toxic organic materials. Combination of ozonation and activated carbon treatment can achieve 95% chloroform and other trihalomethanes removals.

32.2 LIMITATIONS

Ozonation may not be economically competitive with chlorination under some conditions. Although ozone is effective in disinfecting water, its use is limited by its solubility. The temperature of water being treated determines the amount of ozone that can be dissolved in water. In addition, ozone residuals cannot be maintained in metallic conduits for any period of time because of ozone's highly reactive nature. The inability of ozone to provide a residual in the distribution system is a major drawback to its use.

32.3 COSTS

See Figures 32.2 and 32.3 for construction and operation and maintenance costs, respectively.

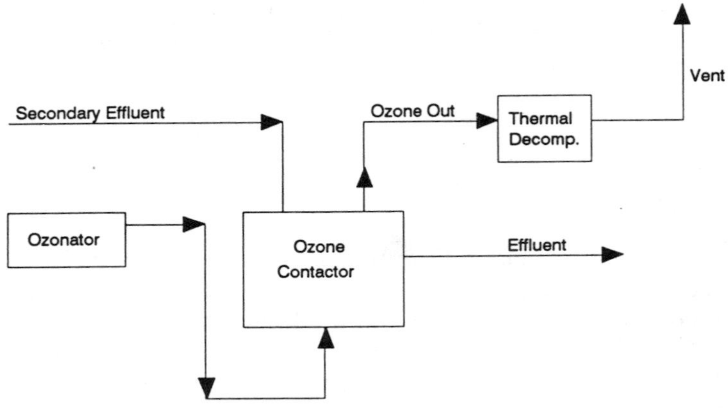

Figure 32.1. Ozone disinfection: flow diagram (*source:* reference 2).

32.4 AVAILABILITY

Ozone has been used in the water industry since the early 1900s particularly in France, and it has been fully demonstrated but not widely used in the U.S. because of its relatively high cost. Recent developments in ozone generation have lowered the cost, thus making it more competitive with other disinfection methods.

32.5 OPERATION AND MAINTENANCE

The ozone disinfection system is a complex series of mechanical and electrical units, requiring substantial maintenance, and is susceptible to a variety of malfunctions as is chlorine gas delivery. In the case of ozonation however, data on consistent, long-term experience are relatively unavailable, especially for large plants. It is not possible to assess long term maintenance

Table 32.1. Design Criteria for Ozonation

INFLUENT	DOSE, mg/l	CONTACT TIME MINUTES	EFFLUENT
Secondary effluent	5.5 to 6.0	≤1	<2 fecal coliforms/100 mL
Secondary effluent	10	3	99% inactivation of fecal coliform
Secondary effluent	1.75 to 3.5	13.5	<200 fecal coliform/100 mL
Drinking water	4	8	Destruction of virus

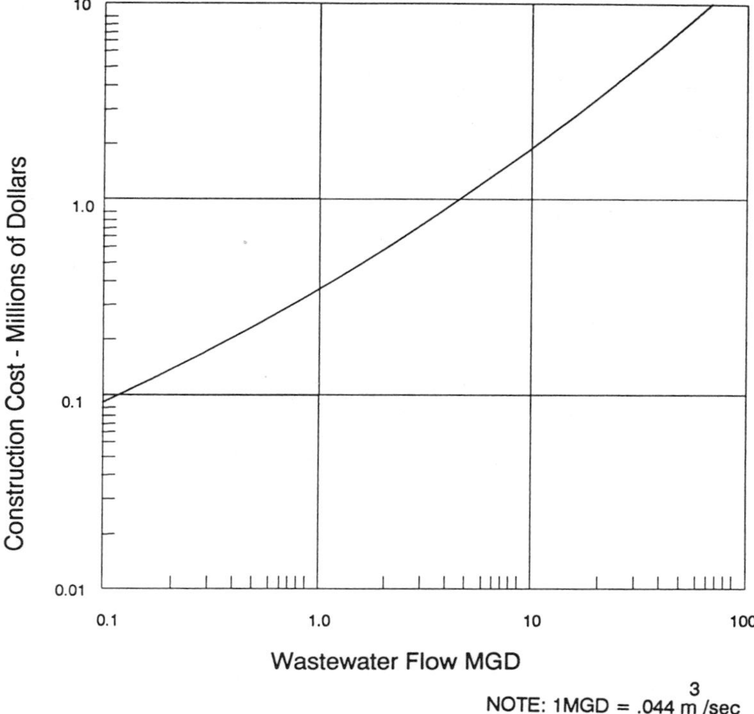

Figure 32.2. Construction cost for ozone disinfection.

requirements for equipment, compressors, cooling and drying equipment, and other appurtenances. It is estimated that power requirements are between 8 and 10 kWh/lb of ozone generated. Monitoring requirements are similar to those for UV disinfection and chlorination, including bacterial analyses and routine ozone monitoring.

32.6 CONTROL

Ozone breaks down to elemental oxygen in a relatively short period of time (half life about 20 minutes), therefore it is difficult to store. Consequently, it is generated on site using air as the oxygen source. The ozone generation process utilizes an electric arc, or corona, through which air or oxygen passes, yielding a certain percentage of ozone. Automatic devices are commonly applied to control voltage, frequency, gas flow, and moisture, all of which influence the ozone generation rate.

Easily oxidizable wastewater organic materials consume ozone at a faster

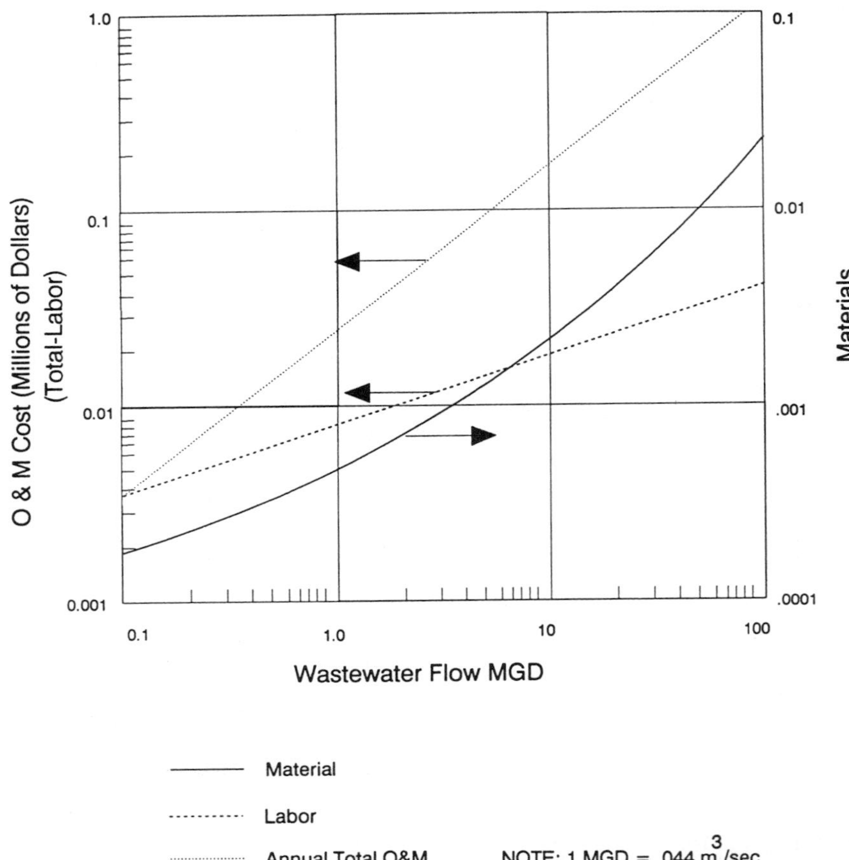

Figure 32.3. Annual operation and maintenance costs for ozone disinfection.

rate than chlorine; effectiveness of disinfection is inversely correlated with effluent quality but directly proportional to ozone dosage. When sufficient ozone is introduced, ozone is a more complete disinfectant than chlorine.

32.7 SPECIAL FACTORS

Ozone is an air pollutant which can discolor or kill vegetation. Residual ozone in off-gas streams must be processed for ozone decomposition prior to release. Ozone is also toxic when inhaled in sufficiently high concentration in air.

Organic removal is improved with combined ozonation and ultraviolet radiation. It is postulated that the UV activates the O_3 molecule and may also activate the substrate. Ozone — UV is effective for the oxidative destruction of pesticides to terminal end products of carbon dioxide and water.

As with any disinfection process, waters containing high levels of suspended solids may require filtration to make ozonation more cost-effective.

32.8 RECOMMENDATIONS

Ozonation is applicable where chlorine is either deficient or where chlorine disinfection produces potentially harmful chlorinated organic compounds (such as trihalomethanes). If oxygen — activated sludge is employed in the system, ozone disinfection is economically attractive, since a source of pure oxygen is available facilitating ozone production.

In many ways, the desirable properties of ozone and chlorine as disinfectants are complementary. Ozone provides fast acting germicidal and viricidal potency, commonly with beneficial results regarding taste, odor, and color. Chlorine provides sustained, flexible, controllable germicidal action and will retain a residual in water supplies. Thus, it would seem that a combination of ozonation and chlorination might provide an almost ideal form of water supply disinfection.

33. WATER COLLECTOR

33.1 DESCRIPTION

The collector may be used where underground water supply is available for development. Figure 33.1 shows an application with a pump facility for direct recharge. The Ranney collector method consists of sinking a reinforced concrete caisson from the bottom of which screens are projected horizontally like the spokes of a wheel; as much as 3000 lineal feet from a single unit is practicable. The caisson is used as a clear well from which the water can be pumped to users. Refer to Figure 33.2 (30).

The screens, which may be 8, 12, 18, or 24 in. (1 in. = 2.54 cm) in diameter, are fabricated from heavy steel and perforated with longitudinal slots. One of the chief advantages of the horizontal collector system is the fact that the entire depth of the aquifer may be utilized. Another major advantage is that the area of screen openings can be varied to control the entrance velocity (and thus flow) of water into the laterals. The design capacity is usually in the range of 2 to 50 MGD (0.09 to 2.2 m^3/sec).

The approach also applies to river water percolated over adjacent sand and gravel formations. The sand and gravel formation is generally very permeable thus allowing for free flow of water to the collector. In the course of percolation, natural filtration occurs thus producing a low suspended solids supply. Use of the system allows development of surface water supplies without construction of dams. Caissons are usually designed to rise a little above the highest flood level of the stream or lake, providing the indirect groundwater source. But sometimes requirements of the design call for limited interruption of the natural character of the shoreline in which case the caisson would stay below the ground surface.

Head induced by the groundwater forces water through the lateral screens into the caisson until the water in the caisson reaches the level of the surface water source. Existing head difference during operation replaces water in the caisson. A pump house (usually on top of the caisson) houses the pumps and controls. Typically a structure on the shore houses the chlorinator, the flow meters, and the electrical controls.

33.2 LIMITATIONS

In the case of development of surface water supplies or indirect groundwater sources, alternate supplies may be required on an intermittent basis as stream or small lake levels vary. As with any water source, contaminated

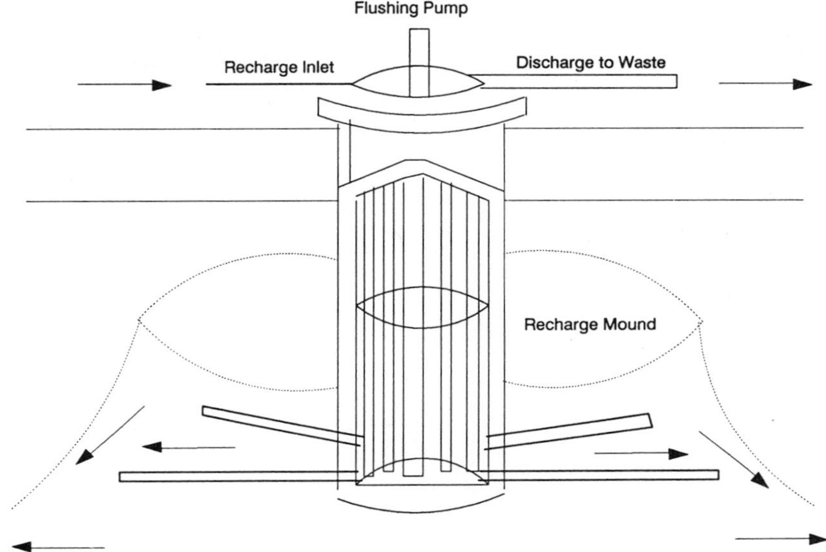

Figure 33.1. Flushing pump showing recharge mound developed by Ranney (*source:* reference 30).

water may require treatment. The collector system itself provides only screening, but the application to groundwater supplies near streams, and small lakes makes use of natural nearby formations for a certain degree of treatment. Local water quality requirements should always be checked against the quality of the developed source for the determination of additional treatment needs.

33.3 COSTS

Construction data are presented in Figure 33.3. It should be kept in mind that the comparison shown to surface water treatment plant costs is only valid where the screening provided in the water collector system is comparable to that provided by an alternate approach. Additional treatment may be required and often is. See section 33.2.

33.4 AVAILABILITY

This manufactured system is available worldwide. The collectors have been installed in many locations in North America and Europe.

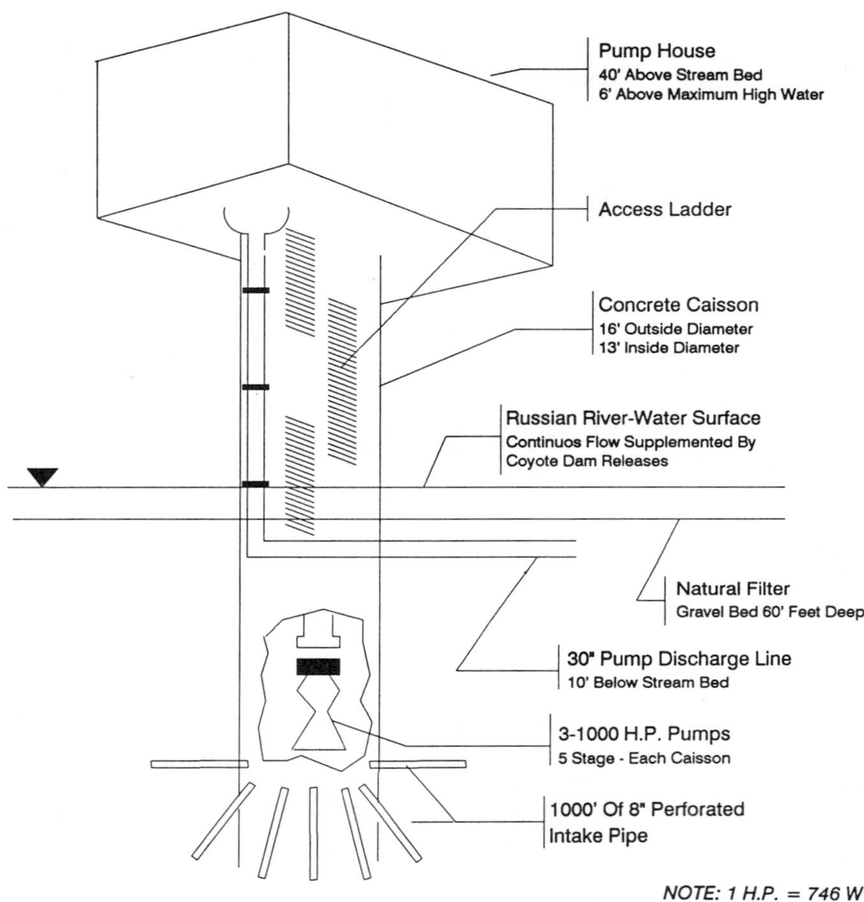

Figure 33.2. Wohler pumping plant—Ranney water collectors (*source:* reference 30).

33.5 OPERATION AND MAINTENANCE

Operational costs for this system may be small when quality and supply of the resource is consistent, and where the cleaning function of the soil system is retained over a long period.

The system may require some specialized labor to monitor system operation and to control the chlorinator. Also, power is necessary to run the pumps and other mechanical equipment.

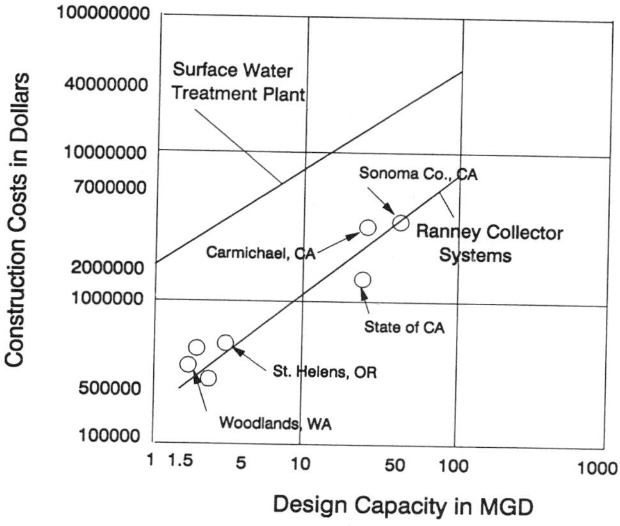

Figure 33.3. Construction costs of surface water treatment plant and Ranney collector systems (*source:* reference 30).

33.6 CONTROL

An automatic control mechanism determines the amount of water that is extracted from the aquifer by controlling the surface area of the laterals available for water transmission. There may also be automatic control of the amount of water that is pumped to the user. A control center located in the onshore structure constantly determines demand and controls necessary pumping. The control system can be expanded to control everything from the chlorinator feed to the area used for water collection in the laterals.

33.7 SPECIAL FACTORS

Artificial recharge of the extraction zone may be utilized independently but designed to be operated along with the system. Recharge laterals accomplish recharge.

33.8 RECOMMENDATIONS

The collector with treatment systems should be explored for water recovery and reuse. This system can produce anywhere from as little as half a million gallons per day to twenty million gallons per day. It may be considered for application directly where groundwater treatment requirements are met by the collector system (screening).

34. CHEMICAL ADDITION FOR WATER AND WASTEWATER TREATMENT

34.1 DESCRIPTION

34.1.1 Introduction

Colloidal solids and very finely divided suspended matter often can not be effectively removed from wastewater by plain sedimentation unless they are rendered settleable by the addition of chemicals. Three theories have been proposed which explain the effects of chemical addition to wastewater:

1. Certain heavy metal salts when treated with alkaline materials form heavy precipitates which enmesh and carry down colloidal suspensions by mechanical entrapment. Salts of iron and aluminum fall into this category.
2. Colloidal particles possess an electrical charge, which because they are alike, repel each other keeping the particles in suspension. If a precipitant with an opposite charge is added, the charges neutralize each other and settling is enhanced. This explains the effectiveness of multivalent ions, such as ferric and aluminum.
3. Insoluble substances which have a large surface area can effectively sorb pollutants and act as the nuclei for the start of flocculation/precipitation. Activated carbon is such a large surface area material.

To be effective, the chemicals must be distributed evenly throughout the wastewater being treated. Rapid mixing with inline devices, turbulence in channels, paddles, propellers, or diffused air will achieve effective mixing with a minimum tank volume. A common rapid mixer consists of a constant speed motor driving a propeller through a speed reducer. Detention periods for mixers vary from 0.5 to 3.0 min. and G values from 200 to 300 sec^{-1}.

Once floc growth has begun, maximum opportunity for contact between floc and colloidal material should be encouraged by controlling the mixing rate. This process of floc growth normally enhanced by slow, non-shear mixing, is called flocculation and takes place in separate flocculation tanks in water treatment plants. Flocculation tanks are usually designed with detention periods of from 10 to 30 minutes. In general, G values are maintained between 10 and 75 sec^{-1}, and Gt values between 10,000 and 100,000. Normally, settling tanks following flocculation are designed with detention periods of from 1 to 4 hours and settling rates of no more than 1,000 gal/ft^2/day (2, 69, 70, 71, 72).

34.1.2 Commonly Used Chemicals

The most common economical chemical additives are alum, ferric chloride, lime, polymers, and powered activated carbon.

Alum. Alum is used for suspended solids/phosphorous removal. It is added directly to the influent which is mixed, flocculated, and settled. Also, alum has been used as a filter aid in tertiary wastewater filtration. Alum is marketed as a solid (lump, ground, rice, or powdered forms) or liquid (~50% solution). The choice between liquid or dry (solid) alum is dependent on cost and application requirements of the process chosen. In general, dry alum is utilized in wastewater treatment. Alum weighs 40 to 75 lb/ft^3 (600 to 1125 kg/m^3) depending on its form, and its solubility in water varies from 50 lb/gal (6 kg/L) at 32°F, to 66 lb/gal (8 kg/L) at 76°F. Alum is best fed dry in the ground rice form. A dry feeder supplies a measured quantity to a dissolver tank and feeding of the alum solution is accomplished by gravity or pumping (see Figure 34.1). For a minimum detention period of 5 minutes, 2 gallons of water (7.6 L) per pound of alum are required. Powdered alum is dusty but slightly hygroscopic and may cake in storage hoppers. Dry alum is stored in mild steel or concrete bins. Required storage space for dry alum is between 30 and 55 ft^3/ton (1 to 1.8 m^3/mt). Because alum solution is corrosive, the dissolving tank, pumps and all piping in contact with it, must be constructed of resistant materials, e.g., rubber, 316 stainless steel, FRP, or plastic.

Ferric Chloride. Ferric chloride is used for suspended solids and/or phosphorous removal. In these applications, solids contact or separate mixing/flocculation tanks are used in treatment of raw water, wastewater or pretreated effluent, often following biological treatment of domestic or mixed industrial/domestic wastes. Ferric chloride is available in either liquid or dry (hydrated crystal or anhydrous powder) form. The commercial solution which dissolves completely in water, is supplied in concentrations from 35 to 45% $FeCl_3$ and weighs between 11.2 and 12.4 lb/gal (1.3 to 1.5 kg/L). Crystalline ferric chloride comes in lumps or sticks with a density range 60 to 64 lb/ft^3 (900 to 960 kg/m^3) and which contain 60% $FeCl_3$. Solubility is 5.4 lb/gal (0.7 kg/L) at 50°F and 7.6 lb/gal (0.9 kg/L) at 68°F. The powder weighs 85 to 90 lb/ft^3 (1275 to 1350 kg/m^3) and contains 96 to 97% $FeCl_3$.

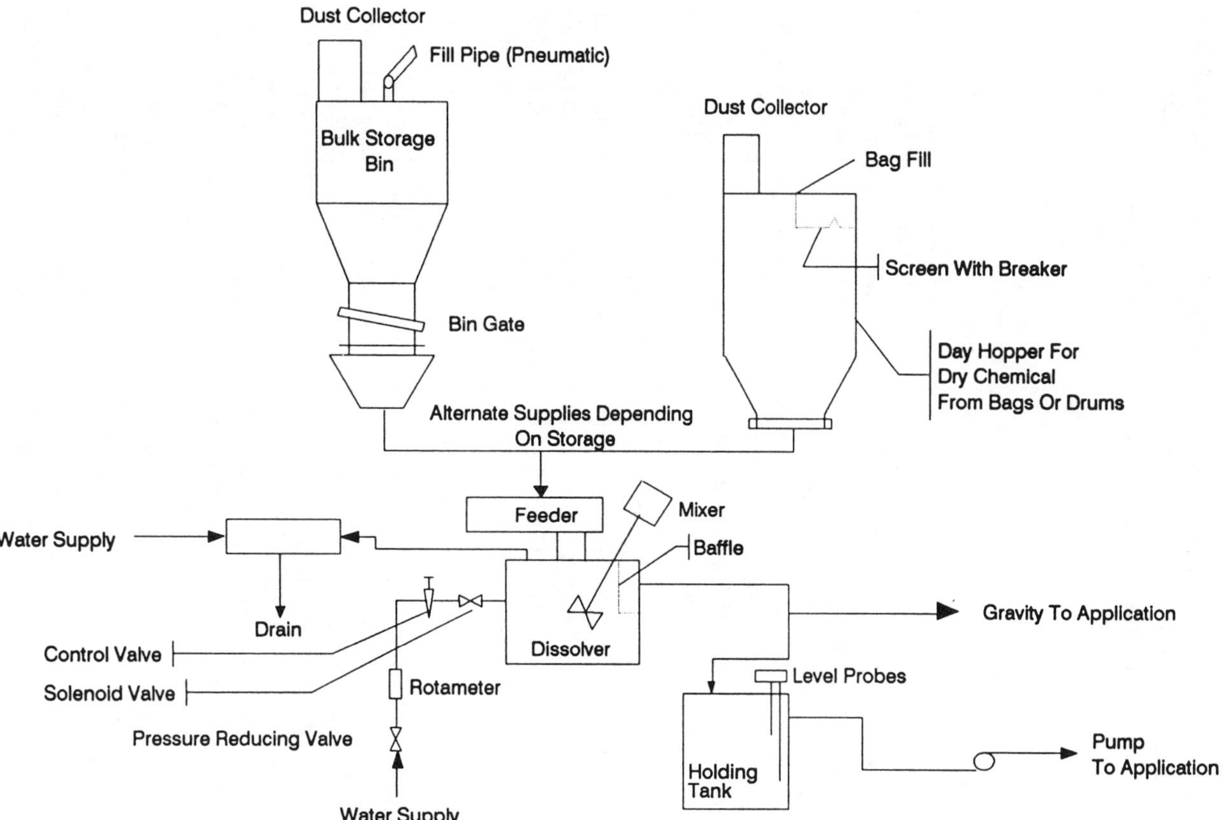

Figure 34.1. Typical chemical mixing and feed system.

Powder solubility in water is 6.2 lb/gal (0.75 kg/L) at 32°F. All forms of $FeCl_3$ are best fed as solutions in concentrations up to 45% $FeCl_3$. Dry $FeCl_3$ is dissolved on site before use in treatment. The stored solution is transferred to a "day tank" and controlled quantities are fed to the mixer by gravity or pumping. Ferric chloride solutions are staining and corrosive, hence must be handled with care.

Lime. Lime addition is used for improved removal of suspended solids, removal of toxic metals, and the removal of phosphates. The primary use of lime has been for phosphorous removal. Lime is available in many forms, however quicklime (CaO) and hydrated lime ($CaOH_2$) are the most common forms marketed. Quicklime or unslaked lime is available in lumps, pebbles, crushed or ground forms. When water is added CaO slakes to hydrated lime with the evolution of heat. Quicklime weighs 55 to 75 lb/ft^3 (825 to 1125 kg/m^3) and contains 70 to 90% CaO. CaO is best fed dry. Applied solutions should be about 10%. Steel and concrete are suitable materials for storage. Quicklime should not be stored for more than 60 days.

Hydrated (slaked) lime is a fine, white powder, density 35 to 40 lb/ft^3 (525 to 600 kg/m^3) and has a commercial strength of 82 to 99% Ca $(OH)_2$. Its solubility in water is low, 1.5 lb/100 gal (0.18 kg/100 L) at 32°F and 1.3 lb/100 gal (0.16 kg/100 L) at 68°F. Hydrated lime may be fed at a maximum rate of 0.5 lb/gal (0.06 kg/L) for continuous dissolving or at a rate of 0.93 lb/gal (0.11 kg/L) as a slurry. Hydrated lime is caustic, irritating, and dusty and should be stored dry. Steel and concrete are suitable materials for lime storage.

Polymer. Polymers (or polyelectrolytes) are high molecular weight compounds (usually synthetic) which can be used as coagulants, coagulant aids, filter aids or sludge conditioners. Polymers are utilized alone or in conjunction with other chemicals such as lime, alum, or ferric chloride to improve performance. Polymers are also used to strengthen flocs to facilitate effluent filtration. Polymers are available in liquid or dry forms. Dry polymers are supplied in relatively small quantities (100 lb bags) and must be dissolved prior to use in treatment. A solution of from 0.2 to 2.0% concentration is usually used. Many competing polymer formulations with differing characteristics are supplied as stock solutions ready for direct application. Manufacturers of these should be consulted as to their use in the treatment of water. Polymer solutions are fed using equipment similar to that commonly used for coagulant (alum-ferric chloride) addition. Stock polymer solutions may be very viscous and special attention must be paid to the diameter of pipes and sizes of orifices used in the feed system. Corrosion resistant materials, such as 316 stainless steel, FRP, or plastic should be used when handling polymer solutions.

Powdered Activated Carbon. Powdered activated carbon is used in water and wastewater treatment to sorb soluble organic materials and as an aid in the settling process. Over the past several years a new application has been developed in which powdered carbon is added to the aeration basins of biological treatment systems. This application achieves high BOD and COD reductions, improved settling in final clarifiers and adsorption of color, toxic organic compounds and detergents. Powdered carbon is marketed in bags or bulk, weighs 15 to 30 lb/ft^3 (225 to 450 kg/m^3) and requires from 72 to 135 ft^3 of storage space per ton. It is insoluble in water and is usually fed dry or as a suspension (slurry). Powdered carbon is fed using chemical feed equipment (either dry or slurry feeders) similar to those used for feeding other chemicals. Spent carbon is removed with the sludge and either discarded or regenerated. Regeneration can be accomplished with a furnace or wet air oxidation process but is generally not applicable to mixed chemical matrices. Because powdered carbon is combustible, storage should be isolated because of the possibility of fire. Suitable materials for handling dry powdered carbon; iron, and mild steel, and for wet carbon; rubber, silicon iron, and 316 stainless steel.

34.1.3 Handling and Storage of Chemicals

Suitable provision must be made for the handling, storage, and feeding of chemicals. Depending on plant size, bags, drums, barrels, or bulk transportation via truck, rail, or barge should be considered. Chemicals must be unloaded, conveyed to and from storage, weighed and added to the treatment process in measured amounts. Dry chemicals are transported by hand, on belt conveyors, bucket elevators, pneumatic tubes, or screw conveyors. Liquids are pumped or flow by gravity. Storage containers and/or hoppers, solution/slurry tanks, conveyors, buckets, tubes, pipes, valves, fittings orifices, and pumps must be constructed of materials resistant to attack by the chemicals being handled. Pipes and pneumatic tubes should be straight and provided with clean-outs to prevent clogging.

The amount of chemical fed is measured by dry (Figure 34.1) or solution feed apparatus (Figures 34.2 and 34.3), depending on the nature of the chemical applied. Dry feed machines control chemical dosage by regulated volumetric or gravimetric displacement of dry chemicals. Gravimetric techniques are more accurate, reliable, and amenable to automatic control but more expensive. For handling certain chemicals, agitators and dust control are provided. Measured dry chemicals are usually dissolved in water before being added to the treatment process.

Solution feed devices control chemical dosage by regulating displacement of raw liquid chemicals or chemical solutions of known strength. Measurement is by orifices, flow tubes, meters or positive displacement, plunger or diaphragm pumps.

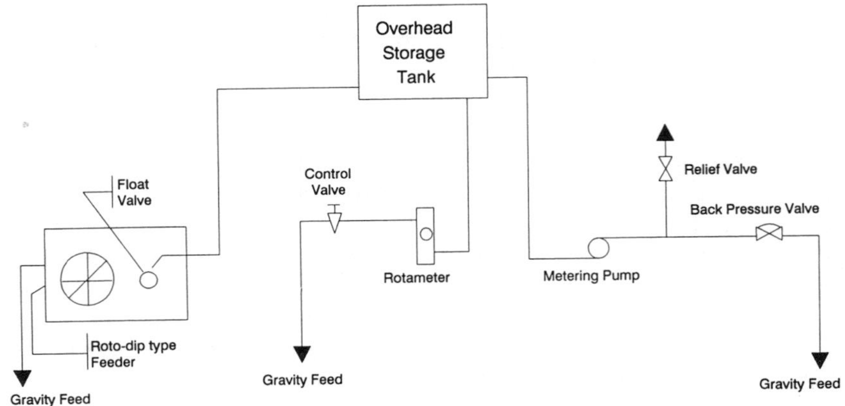

Figure 34.2. Alternative liquid feed systems for overhead storage.

34.1.4 Chemical Dosage

The amount of chemical that must be added to the treatment process varies (1) with the quality of the water, (2) with the degree of treatment (removal) required, (3) with treatment conditions, and (4) with the quantity of water to be treated. Optimal chemical dosage is determined by repeated bench scale testing and laboratory analysis. In most cases, frequent "jar" testing is necessary to determine proper chemical dosages. Results of jar tests should be compared to, or correlated with, actual plant performance. Bench scale

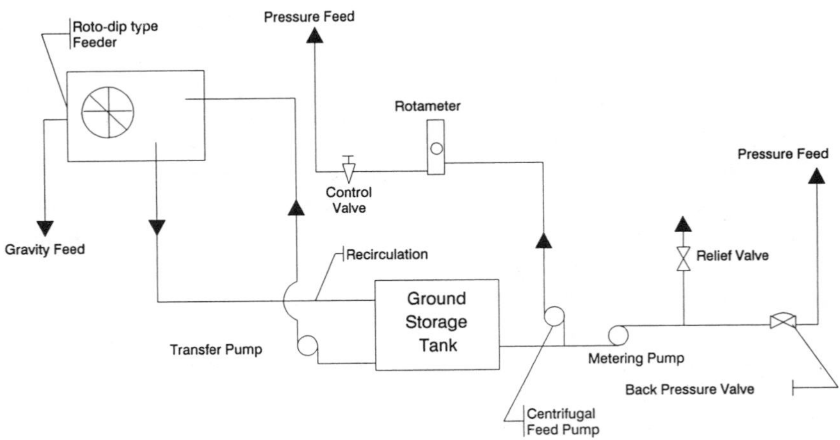

Figure 34.3. Alternative liquid feed systems for ground storage.

analyses should include testing to determine optimum pH, *G*, and *Gt* values for chemical addition. Operating and maintenance costs depend significantly on careful control of chemical doses.

Dosages commonly employed in the design of chemical treatment at municipal wastewater treatment plants are as follows:

- Alum — 5 to 20 mg/L for SS removal; ± 100 mg/L for P removal
- Ferric chloride — 20 to 200 mg/L
- Lime — 100 to 500 mg/L as CaO
- Polymer — 1 to 10 mg/L
- Powdered Carbon — 50 to 300 mg/L

34.1.5 Effectiveness of Chemical Treatment

As a treatment process, addition of chemical coagulants (alum, lime, ferric chloride, polymer) accomplishes removal of BOD and solids about midway between plain sedimentation and secondary biological treatment. Removal of about 80% of the suspended solids and 65% of the BOD can be achieved with typical doses. With higher doses, removals as high as 90% of the suspended solids and 80% of the BOD may be achieved. Addition of chemicals has little or no effect on the concentration of soluble BOD or COD. The addition of lime, alum, or ferric chloride to either effluent from a biological treatment process or untreated wastewater can achieve phosphorous removals in the range of 80 to 95%. Phosphorous concentrations as low as 0.1 to 0.5 mg/L can be achieved utilizing chemical addition.

Chemical wastewater treatment utilizing powdered carbon along with coagulating chemicals can achieve SS, BOD, and P removals of over 90%.

34.2 ADVANTAGES AND DISADVANTAGES

The advantages of chemical addition in wastewater treatment include the following:

- Reliability. With proper operational control, it is capable of producing consistently high effluent quality.
- Flexibility. Treatment efficiencies can be varied as needed, e.g., seasonally to protect streams during periods of low runoff or beaches during summer months.
- Process is suitable for treatment/pretreatment of wastewaters containing toxic or otherwise refractory chemicals.
- Removes phosphorous from untreated wastewater and treated effluents.
- Removes heavy metals from raw wastewater and treated effluents.

The disadvantages of chemical addition are the following:

- Relatively high cost if operated on a full-time basis.
- Increased amount of sludge generated, thereby increasing sludge treatment and disposal costs.
- High concentrations of chemicals and/or heavy metals or other toxic materials in the sludge may render it unsatisfactory for use as a fertilizer and/or soil conditioner.
- Skilled labor is required for operation and maintenance of systems.
- Certain mechanical equipment items and chemicals required in the process may not be readily available in developing countries.

34.3 COSTS

Estimated capital and operation and maintenance costs for chemical treatment including alum, ferric chloride, lime, polymer, and powdered carbon addition are presented in Figures 34.4 and 34.5 respectively. The curves presented include the costs for handling, storage, and feeding of chemicals, but do not include provision for the cost of mixing, flocculating, and/or settling facilities.

34.4 AVAILABILITY

The equipment and appurtenances typically utilized in the simpler chemical application systems; e.g., solution tanks, water level controllers (float controlled valves), piping, valves, and orifices are available in most areas of the world and are easily serviced when applied to small systems. Similarly, alum, lime, and ferric chloride are readily available in many parts of the world.

Gravimetric and volumetric dry feed machines, slakers, mechanical/pneumatic conveyors, piston and diaphragm pumps, and remote/automatic controls may not be readily available in developing countries. Such equipment and chemical additives, such as polyelectrolytes and powdered carbon may not be generally available in developing countries and may be costly to import.

34.5 OPERATION AND MAINTENANCE

The effectiveness of chemical treatment is a function of the concentration of chemical used, the contact time, and conditions during treatment. For best results, dosage, time of contact, and contact conditions must be optimized. In plants that employ mechanical systems, the speed of mixers and flocculator paddlers must be adjusted to optimize dispersion of chemicals and pro-

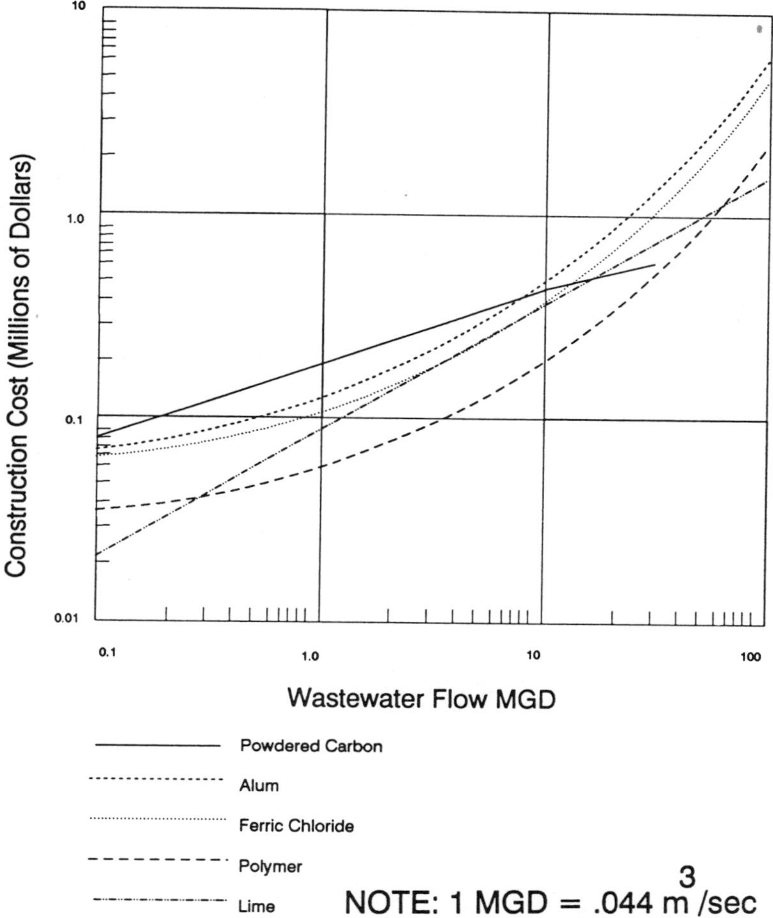

Figure 34.4. Construction costs for chemical addition.

vide opportunity for floc build-up and contact. In tanks that employ turbulence and baffles to accomplish mixing and flocculation, opportunity to control contact time and conditions is limited unless baffles are movable.

In plants which use relatively simple solution feeder systems, the makeup of chemical solutions of known strength and the adjustment of solution flow rates are critical steps in successful operation of the process.

In plants employing more sophisticated chemical feed systems, all mechanical equipment must be periodically and systematically inspected, lubricated, and overhauled. Flow metering devices should be checked at least

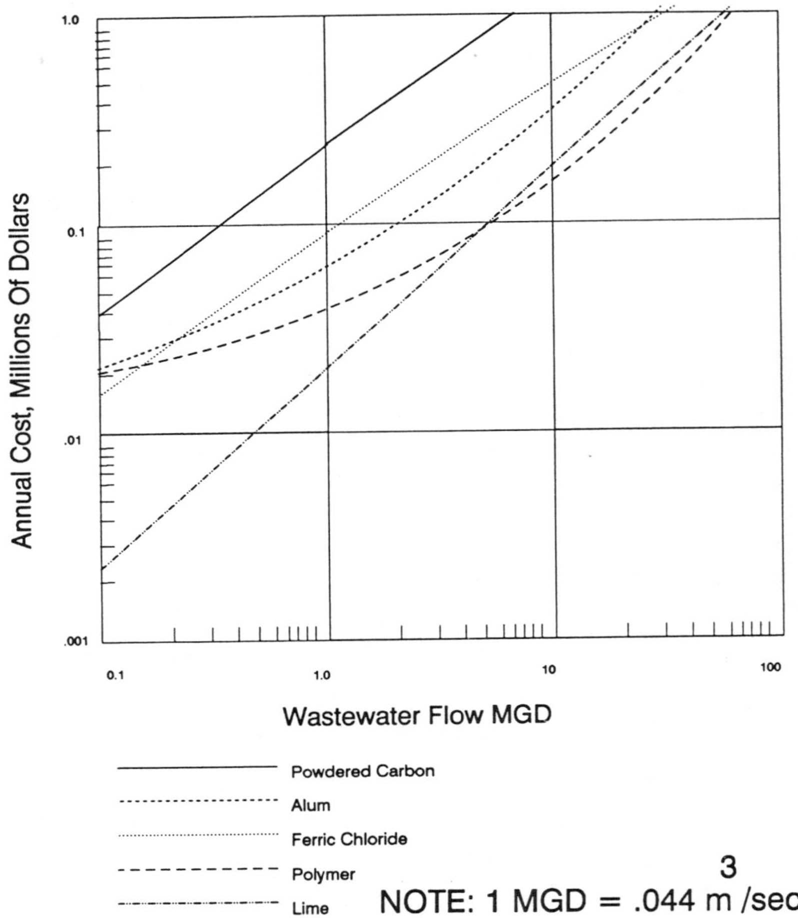

Figure 34.5. Operation and maintenance costs for chemical addition.

yearly. Similarly, feeders must be checked regularly for accuracy. In the case of dry feeders it is desirable to collect the feed for a short period of time, then weigh it on scales. Solution feeders can be checked by comparing their rated output with the time-volume displacement of the solution in a tank of known geometry.

Routine operation and maintenance include unloading and storage of chemicals, and dust control and cleaning of conveyor, pipeline, solution tanks and storage hoppers.

A very important aspect of plant operations is the promulgation of information concerning the safe handling of wastewater treatment chemicals.

Improper handling of most chemicals used can cause eye injury, throat irritation, and other problems.

34.6 CONTROL

The amount of chemical added to the treatment process is a function of the dosage required to achieve the requisite degree of treatment. The dosage required varies with raw water quality and the degree of variation may be determined by repeated bench/laboratory testing of the untreated water. Results of the laboratory "jar" tests are compared to actual plant performance and these results are used to select concentrations used in the treatment process. Also, concentration data concerning critical water characteristics are recorded, which are useful in determining actual dosage amounts. These characteristics include turbidity, BOD, suspended and settleable solids concentrations, and pH. Flow is determined by taking repeated flow meter readings in conjunction with the laboratory tests.

Chemicals are fed in either dry or solution form. In either case, adequate controls must be available to set proper feed rates. Dry feeders are generally equipped with various means to control feed rate, including loss-in-weight type hoppers, adjustable speed belts, screws, and other devices. Solution feeders control chemical dosage rates by regulating the flow of a solution of known strength. In gravity feed systems, flow measurement is achieved using a constant head tank and a primary meter, such as an orifice. In pumped systems, dosage rates are controlled by varying pump speed and/or displacement. Varying solution strength offers another means of control.

It is possible to adjust dosage automatically with varying wastewater flow, varying influent strength, and required effluent quality, but automatic devices are relatively costly and must be maintained frequently.

34.7 SPECIAL FACTORS

The cost of sludge treatment and disposal is a very important consideration in the planning and design of treatment systems in which the use of chemical addition is contemplated. Chemical addition will substantially increase the volume of sludge generated. For example, lime addition will generate from 1.0 to 1.5 lb (0.45 to 0.7 kg) of dry solids per lb (kg) of lime added; powdered carbon will produce 1.0 lb (0.45 kg) per lb (kg) of carbon added; the addition of ferric chloride will produce about 1.0 to 1.3 lb (0.45 to 0.6 kg) of solids per lb (kg) of $FeCl_3$ added.

Certain chemical sludges; e.g., alum sludges, in addition to being voluminous are difficult to dewater. Also, these sludges may contain high concentrations of the chemical used in treatment or of certain metals; e.g., cadmium, chromium, lead, or copper, removed during treatment. This could render the sludges unsatisfactory for use as fertilizers or soil conditioners. Care must be used in the siting of landfills for chemical sludges to ensure that pollution of underground water sources does not occur. Recovery (regeneration) of chemicals used in wastewater treatment such as alum, lime, and powdered carbon is practiced but, appears feasible only at large scale facilities (plant capacities >50 mgd).

34.8 RECOMMENDATIONS

The process of addition of chemical coagulants to wastewater can provide BOD and SS removals of 65 to 80% and 80 to 90%, respectively. Chemical addition will raise the efficiency of existing treatment facilities without appreciable additional construction cost. However, O&M cost may be high if the process is used regularly. The process is attractive where seasonal variations in treatment efficiency are required to meet receiving water quality standards. Also, if nutrient removal is required, the process can produce an effluent with phosphorous concentrations as low as 0.1 mg/L. Removal of heavy metals is possible with lime addition, and significant reductions in toxic organic compounds can be achieved using powdered activated carbon. The process is well suited to the treatment/pretreatment of wastewater containing toxic or otherwise objectionable industrial wastes.

35. GRANULAR ACTIVATED CARBON ADSORPTION

35.1 DESCRIPTION

Granular activated carbon (GAC) adsorption is generally utilized for the removal of suspended and/or colloidal matter in wastewater and the removal of tastes and odors in water supplies. Generally, applications for water supply use powdered activated carbon (PAC). GAC can also be used either as a tertiary treatment process in advanced wastewater treatment plants or as a secondary treatment process. Carbon adsorption is also used in conjunction with biological treatment or less frequently in independent physical/chemical treatment plants. See Figures 35.1 and 35.2 (2, 31).

GAC can be used to upgrade water quality following existing sand filtration systems. Used as a complete replacement for sand or coal, activated carbon functions in a dual way—providing both filtration and adsorption. In this mode frequent backwashing is required and carbon utilization may increase even with regeneration.

The advantages of GAC are (31) the following:

- granular carbon adsorption is a reliable taste and odor removal process,
- keeping a reserve capacity of granular carbon can effectively control sudden water quality fluctuations and unexpected contamination,
- GAC beds are generally more cost effective than powdered carbon; treatment effectiveness is more reliable,
- GAC can be used for polishing organics in industrial wastewaters and for cleaning up accidental spills of toxic and hazardous chemicals to municipal wastewater.

Water treatment with GAC consists of the carbon contact system and the carbon regeneration system. Activated carbon removes soluble, suspended, and colloidal matter from water in three steps:

1. the transport of the dissolved substances to be removed (solute) through a surface film to the exterior surface of the carbon;
2. diffusion of the solute within the pores of the activated granular carbon;

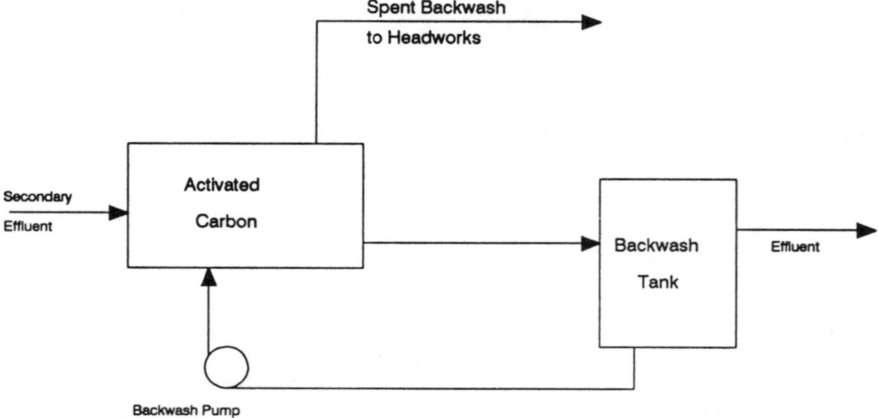

Figure 35.1. Flow diagram of granular activated carbon adsorption (*source:* reference 2).

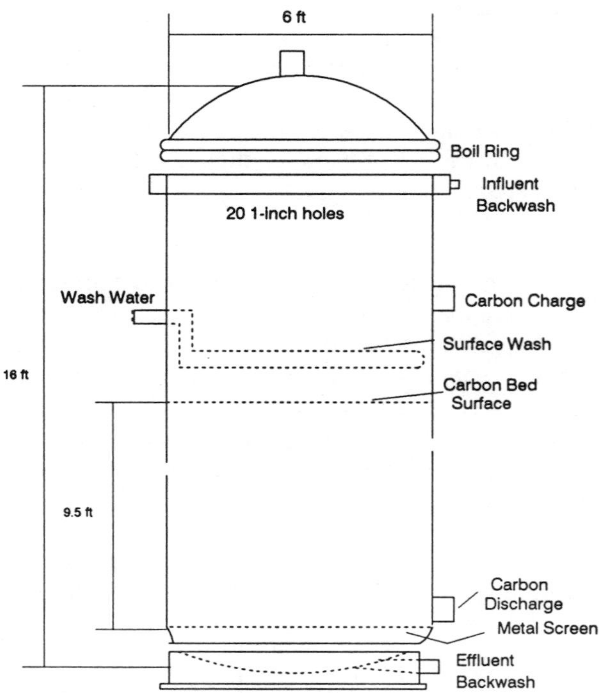

Figure 35.2. Downflow type granular activated carbon system (*source:* reference 45).

3. separation by adsorption of the solute on the interior surfaces, bonding to the pore and capillary spaces of the activated carbon and by filtration.

Several alternative configurations for carbon contacting systems are used for treating wastewater:

- downflow of the wastewater through a carbon bed,
- parallel or series operation of contactors (single or multistage),
- pressure or gravity operation in downflow systems (Figure 35.2).
- packed or expanded bed operation in upflow systems.

35.2 DESIGN CRITERIA

Typical granular activated carbon performance data are as follows:

Performance	Influent	Effluent
BOD mg/L	10 to 50	5 to 20
COD mg/L	20 to 100	10 to 50
TSS mg/L	5 to 10	2 to 10

Typical design parameters for granular activated carbon are as follows (2):

Size = vessels 2 to 12 ft diameter commonly used
Area loading = 2 to 10 gal/min/ft^2
Organic Loading = 0.1 to 0.3 lb BOD_5 or COD/lb carbon
Backwash = 12 to 20 gal/min/ft^2
Bed depth = 5 to 30 ft.
Contact time = 10 to 50 min
Land area = minimal

Note: 1 ft = 0.3048 m
gal/min/ft^2 × 0.04 = m^2/day/m^2
2.2 lb = 1 kg

35.3 LIMITATIONS

Activated carbon adsorption systems may be complicated to operate. Column operation involves cycling different columns into operation as carbon adsorption capacity is exhausted. Carbon regeneration is required for intermediate size systems for economical operation.

Short chain, low molecular weight organic compounds are not effectively removed by adsorption. These compounds may degrade biologically in the columns and in the presence of little or no oxygen, may cause hydrogen sulfide emissions at objectionable levels. Hydrogen sulfide odors are objectionable, but gasification may cause bed instability and channelling.

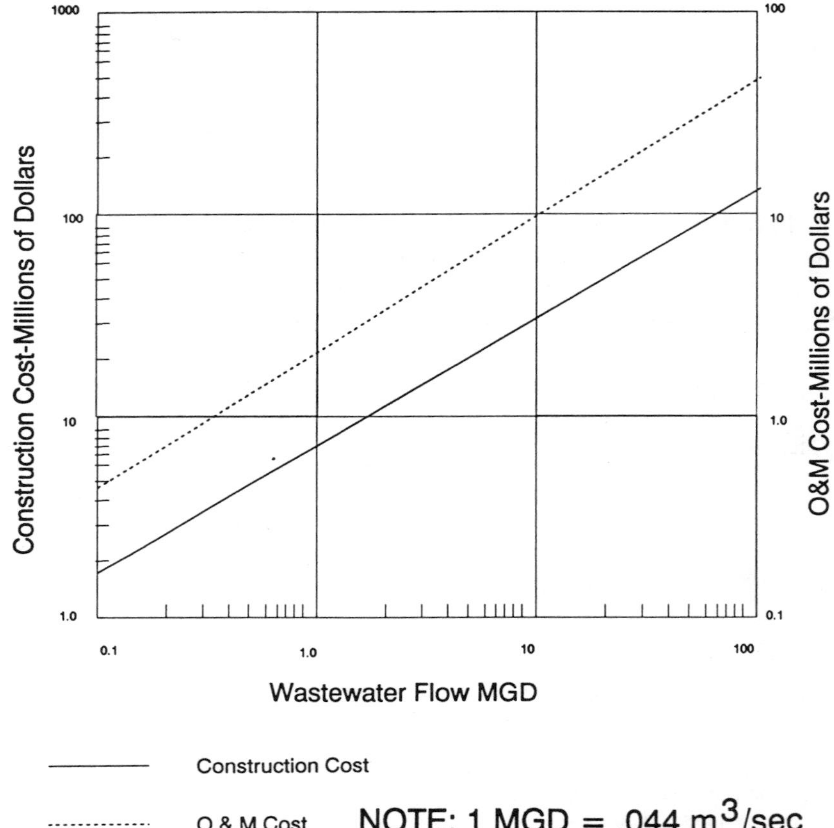

Figure 35.3. Construction, operation, and maintenance costs for granular activated carbon adsorption.

Adsorption is dependent on pH; it is more effective at low pH values. In the presence of recalcitrant compounds pH adjustment may be required.

Adsorption may be required for water reuse systems, where less costly systems such as lagoons have been used beforehand. Also, adsorption may be required where toxic materials such as pesticides are in common use. Carbon adsorption has been used successfully for removal of halomethanes after disinfection.

35.4 COSTS

Figure 35.3 (2, 11) shows construction costs. Values include vessels, media, pumps, carbon storage tanks, controls, and an operation building; loading rate 30 pounds carbon per Mgal (3.6 kg/million L, or 3.6×10^{-3} kg/m^3), contact time = 30 minutes.

35.5 AVAILABILITY

Activated carbon of various grades and types are available worldwide. It is used for a variety of industrial purposes including sugar refining and treatment of distilled liquors. Consistent shipments could be a problem when the material has to be delivered to remote areas.

35.6 RECOMMENDATIONS

Carbon treatment should only be used where trained personnel are available for operation on a continuous basis. Removal of organics, especially toxic compounds is becoming a more important requirement universally.

36. DISSOLVED AIR FLOTATION

36.1 DESCRIPTION

Dissolved air flotation (DAF) is used to remove suspended solids from water by decreasing the apparent density of the solids (flotation). It consists of saturating a portion or all of the water feed with air at a pressure of 40 to 50 psi (250 to 300 kPa) (see Figure 36.1). The water is held at this pressure for 0.5 to 3 min in a detention tank and then released to atmospheric pressure in the flotation tank. The sudden reduction in pressure releases microscopic air bubbles which attach themselves to oil and suspended particles. Agglomerated particles have increased vertical rise rates, in the order of 0.5 to 2 ft/min (0.15 to 0.6 m/min). The floated materials rise to the surface to form a froth layer. The froth layer may be removed continuously by mechanical skimmers. The retention time in the flotation chamber is between 20 to 60 minutes. The process effectiveness depends on the attachment of the air bubbles to the oil and particles to be removed. The attraction between the bubbles and the particles is largely the result of surface charges on the particles.

There are several commercially available dissolved air flotation units on the market. One type is shown in Figure 36.1 (65).

36.2 DESIGN CRITERIA

PARAMETER	RANGE
Pressure, lb/in.^{2} (kPa)	40 to 50 (275 to 300)
Air to solids ratio, lb/lb (kg/kg)	0.01 to 0.1
Float detention, min	20 to 60
Surface hydraulic loading, gal/day/ft^2	500 to 8,000
Recycle, %	$<120\%$
recycling of a portion of the effluent (15 to 120%) is usual in larger units	

Note: gal/day/ft^2 $\times$ 0.04 = m^3/day/m^2

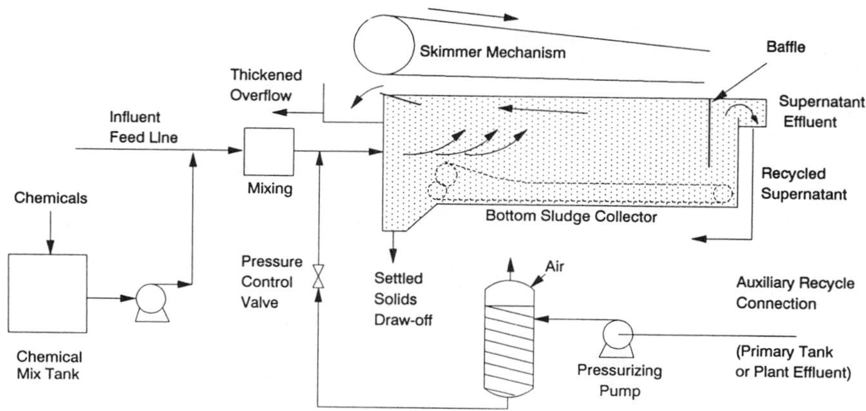

Figure 36.1. Schematic of dissolved air flotation system with recycle.

Performance may be expected as follows:

	% REMOVAL WITHOUT CHEMICALS	% REMOVAL WITH CHEMICALS
Suspended Solids	40 to 65	80 to 93
Oil and Grease	60 to 80	85 to 99

36.3 LIMITATIONS

Additional dissolved air filtration thickener capacity is required for higher input solids and/or output sludge densities. The process is most effective when particle density is near that of water.

Separate sludge thickening and filtration may be required for producing disposable sludge of 15 to 30% solids content.

36.4 COSTS

See Figures 36.2 and 36.3 for construction and operation and maintenance costs, respectively.

36.5 AVAILABILITY

DAF has been used for many years to treat industrial wastewater and has been used to some degree to treat municipal wastewater. The technology is well understood and relatively easy to operate. Operating costs are higher than for simple sedimentation.

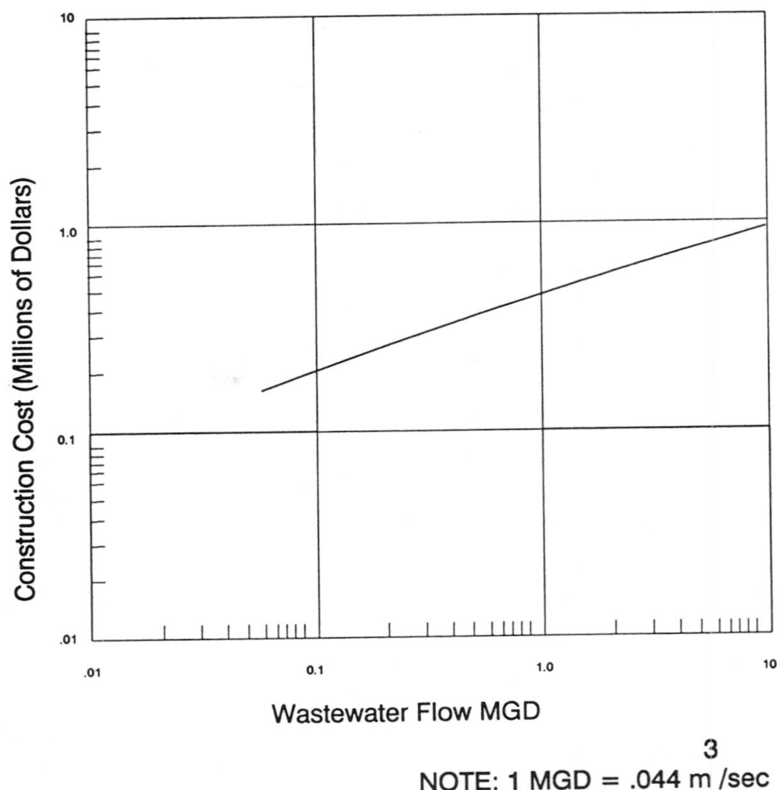

Figure 36.2. Construction cost for dissolved air flotation (*source:* reference 2).

36.6 OPERATION AND MAINTENANCE

A froth layer is usually generated and must be skimmed off the top of the unit. It is usually denser than the clarifier sludge. The thickened froth can be stable over a short period of time but undue delay in removal may result in resuspension of the particulate matter in the froth into the liquid phase.

The froth sludge may be disposed along with the clarifier sludge. If the oil content is high, consideration may be given to recovery of oils. This option is often applicable when the industrial waste fraction in the wastewater is high and when a pretreatment program is not in place in the case of municipal wastewater treatment.

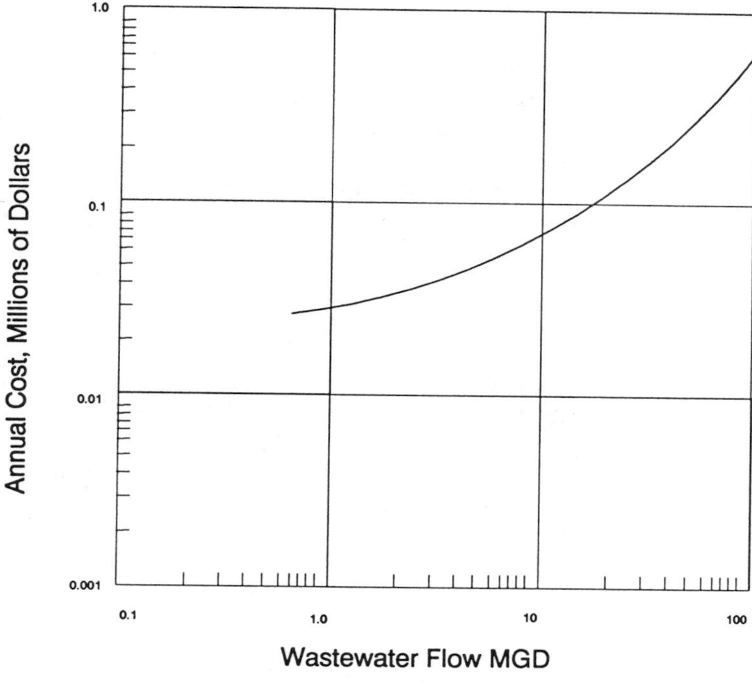

Figure 36.3. Operation and maintenance costs for dissolved air flotation (*source:* reference 2).

36.7 CONTROL

The operation is completely automatic. The outlet is connected to the distribution piping system, and varying demand is automatically adjusted by the inlet flow regulating valve which maintains a constant level in the flotation tank. The backwashing system is automatically operated by a clear well level control and a timer. The installation may be equipped with an alarm system that transmits a beeper signal via telephone and radio to the operator.

The more uniform the distribution of microbubbles, the shallower the flotation tank can be. The depth of effective flotation units is 4 to 9 ft (1.2 to 2.7 m).

36.8 SPECIAL FACTORS

Units can be circular, square, or rectangular. Gases other than air may be used. The petroleum industry has used nitrogen and closed vessels to reduce the possibilty of fire. Other considerations include: the air is likely to strip volatile organic compounds into the air; the air compressors may need silencers to control noise but air may be supplied from plant air in activated sludge plants; if chemicals are used the froth and sludges are likely to contain high concentrations of coagulant chemicals.

36.9 RECOMMENDATIONS

Dissolved air flotation is used to remove lighter suspended materials whose specific gravity is slightly in excess of 1.0. It is used to remove oil and grease, and sometimes used when existing clarifiers are overloaded hydraulically. In this case, clarifiers may be converted to DAF which requires less surface area. DAF should be considered when wastewaters contain industrial wastes high in oil and grease fractions.

37. IMHOFF TANKS

37.1 DESCRIPTION

Imhoff tanks are two-storied treatment units in which the lower story provides for sludge storage and the upper for sludge settling. Thus the two technologies of settling and digestion are combined in one unit process. The tanks may be either round or rectangular. Settling particles slide from the upper into the lower compartment (see Figure 36.1). The two stories are constructed so that rising gas bubbles and sludge particles cannot enter the settling compartment. Separation of sewage and sludge is required in Imhoff tanks. Thus, putrescible solids are removed from the flowing sewage and stay removed. Mechanical cleaning mechanisms are not needed.

Although the sludge chamber can be designed to hold sludge for only a few days, it is generally designed for much greater capacity, to serve as a digestion chamber. It is given sufficient capacity to hold the sludge until it is well digested, dried or otherwise treated and disposed of. Sedimentation and digester compartments of Imhoff tanks are sized on the basis of conventional design criteria for each. Multiple Imhoff tanks are normally provided.

37.1.1 Sedimentation Chamber

Generally the sedimentation compartment of Imhoff tanks is designed with a maximum surface rating based on average dry-weather flow of 600 gpd/ft^2 (25 m^3/m^2/day); the sloping bottom of the sedimentation compartment has 1.4 vertical to 1.0 horizontal slope, and the slot between upper and lower compartments has a minimum opening and a minimum overlap of 6 in. (15 cm). The sludge slots *must* overlap or otherwise the passage of rising sludge gases or sludge particles into the upper settling chamber is possible, potentially interfering with effective settling.

Where gas is not collected, gas vents are given a minimum width of 18 in. (46 cm), an area equal to about 20% of the total superficial area of the tank, and a freeboard of 18 to 24 in. (46 to 61 cm) above the flow line of the tank. At least one vent should be 24 in. (61 cm) in diameter to facilitate entrance for tank repairs.

Baffles across the ends of the sedimentation chambers of Imhoff tanks are used to distribute the flow more uniformly and to serve as a trap to collect floating solids. Where the tanks are long, an extra beam is placed across the center of the tank to help support the sedimentation chamber and act as an intermediate baffle, collecting scum and preventing surface currents.

Inlet and outlet arrangements are particularly important because of the desirability of reversing the direction of flow through the tanks to give an even distribution of solids in the digestion chamber. A weir running the width of the sedimentation chamber is the most common arrangement as it serves effectively to distribute the flow and control outlet velocity. It is desirable to keep a relatively constant water level in the sedimentation chamber; therefore, the weir crest is set above the maximum flow line of the effluent channel.

A walkway along the tank to facilitate cleaning the sides of the sedimentation chamber and slot is usually provided for the operator.

Submerged drawoff pipes have been installed in the flowing-through chambers and gas vents of some recently designed Imhoff tanks for the removal of skimmings and scum and have resulted in reduced manual attention. Sludge may be transferred to sludge beds or lagoons. Settled, digested sludge requires additional treatment before disposal.

37.1.2 Digestion Chamber

Required capacity of anaerobic digestion chambers is the same as for unheated separate tanks. The criteria for digestion tank loading vary between 1 to 4 ft^3 (0.11 m^3) per capita depending on digestion temperature, climate, and operating conditions. Where data on the solids are not available, the foregoing unit per capita capacities are given for plants treating domestic sewage.

The digestion chamber (lower story) is generally subdivided by cross walls into several compartments. These walls provide structural economy and keep sewage from following a path of low resistance through the digestion chamber and fouling the effluent. Openings in the cross walls for equalization of sludge storage in successive compartments must lie below the normal sludge level. Each settling chamber should have its own sludge chamber otherwise there will be cross currents in the lower story owing to unequal distribution of flow between parallel sedimentation chambers. These cross currents will drive shocks of septic digestion tank liquor into the effluent. Longitudinal dividing walls should therefore, be built into tanks that have more than one settling compartment.

The bottom slope of sludge compartments is generally 1 vertical to 1 or 2 horizontal.

Sludge is withdrawn through a central riser pipe controlled by its own valve. The pipe should have a free outlet at which the sludge can be seen and sampled. An irrigating or flushing ring at the mouth of the riser will improve sludge withdrawal capability.

Sludge drawoff from the digestion chamber is usually accomplished by utilizing the hydrostatic head; a differential of at least 6 ft (1.8 m) is necessary.

Digestion compartments need no outlets to the surface other than the gas stacks. The entire tank surface is effective for sedimentation. If sludge slots are kept high, tank depth is utilized to the fullest extent for digestion and tank surface for settling.

As noted above, an area equal to 20% of the total surface area of the tank is normally provided for venting gas from the digestion compartment. Large Imhoff tanks are sometimes equipped for gas collection. This adds little to the cost of construction because the slabs that separate the upper and lower stories form the necessary gas collecting covers (Figure 37.1). Vertical stacks penetrate to the digestion compartment and house gas domes. These end about 1 ft (0.3 m) below the water surface. Scum is kept out of them by porous concrete or wooded slabs. Scum can then be broken up mechanically or by water jets. Heavy layers of scum can be drawn off through gated, lateral openings or skimmed off.

37.1.3 Variations in Design

A variation of the conventional type is a two story mechanized clarifier superimposed over a mechanized digester with a concrete tray between the two compartments. A one way seal permits sludge to pass downward from the settling to the digestion compartment but prevents gas, scum, or warm liquor from passing upward. Gas from the digestion compartment is led off through an opening at one side to a gas dome and take-off located above the water level. Heating of the digestion compartment is practiced in colder climates utilizing auxiliary fuel or generated gas and circulating hot water

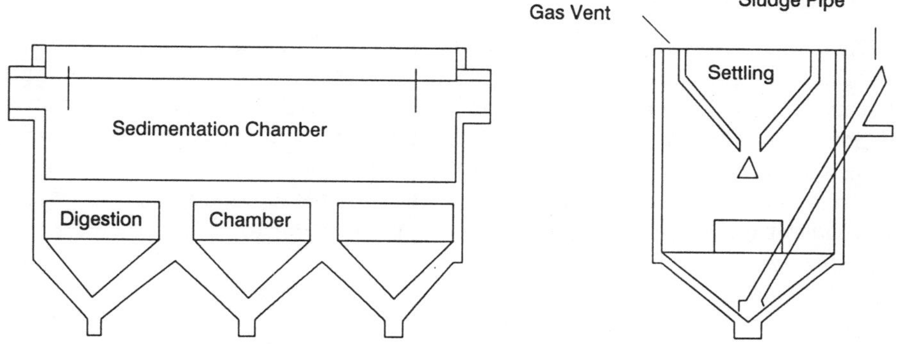

Figure 37.1. Imhoff tank.

through internal pipe coils or by means of an external heat exchanger. Radiation losses through the concrete tray are moderate because during operation a layer of scum packs against the underside and acts as insulation. Digestion temperatures of 85°F to 90°F are typical.

37.2 ADVANTAGES AND DISADVANTAGES

Advantages for developing countries and small systems are: construction cost is relatively low (manual labor is largely required); design and operation is simple; unskilled maintenance labor is normally sufficient for effective operation; no chemicals are required; power requirements are low; large quantities of wash water are not required; sludge disposal is simpler. Disadvantages are: heating the digestion compartment in colder climates may be impractical because heat is lost through slot and gas vents; tanks must be isolated from populated areas because digestion gases are vented to the atmosphere and may result in objectionable odors; deeper tanks are required for conventional designs as opposed to the more modern design variants.

37.3 COSTS

Estimated capital and O&M costs for Imhoff Tanks, exclusive of land and special foundations are presented in Figure 37.2.

37.4 AVAILABILITY

The technology and construction components are readily available everywhere.

37.5 OPERATION AND MAINTENANCE

Digesting sludge or scum must not be permitted to reach the slots. There should be a neutral zone of at least 18 in. (46 cm) both below and above the slots. The upper neutral zone offers protection against excessive scum formation. To maintain the neutral zone, sludge and scum must be withdrawn even if not fully digested. Appearance of a row of gas bubbles on the sewage surface above the slots is a sign that sludge has encroached on the slots. Sufficient sludge must be left in the tank as a seeding and buffering reserve. Repeat breaking-in periods should be avoided.

Sludge digestion tanks will foam in the presence of foam causing substances in the sludge liquor. Rising gas bubbles then drive the foam into the gas stacks, sometimes causing overtopping. As a rule, foaming is confined to breaking-in periods. It can be controlled by flushing the stacks with clean water or sewage thereby diluting the foaming sludge liquor. Sludge with-

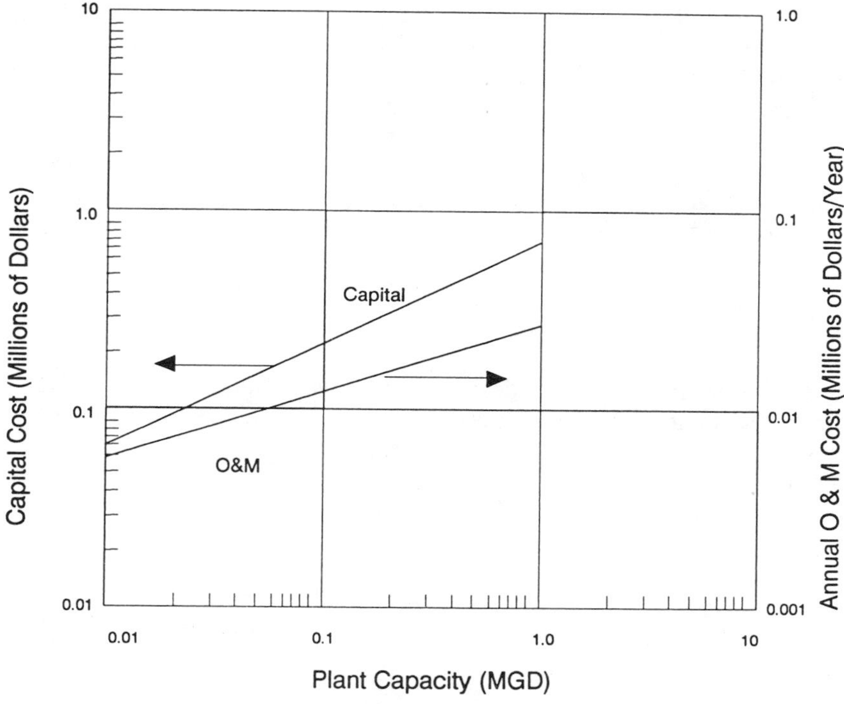

Figure 37.2. Imhoff tank costs.*

drawal has much the same effect because it lowers the sludge level and pulls in fresh sewage from the settling chamber. The use of anti-foaming agents may be considered.

Because the system is simple to operate and maintain an "out of sight, out of mind" attitude often develops at Imhoff tank facilities, and the process is ignored for long periods of time. Effective operation demands regular attention to the sedimentation-digestion interface to prevent interfering interactions between the two as described herein. Sludge compaction in the digestion chamber can result in virtually unremovable accumulations over a period of months.

37.6 CONTROL

Opportunities for process control in a conventionally designed Imhoff tank are somewhat limited. Maintenance of a clear zone above and below the sedimentation compartment slots is essential for good clarifier performance. Thus devices and means for locating the top of the sludge and bottom of the scum layers will improve the sedimentation process performance. One approach to locating the limits of the clear zone is by lowering a suction pump into the digestion compartment.

Controls to govern the withdrawal rate and amount of sludge removed from the digestion compartment improve that process performance. A means of tracking the location of the top of sludge layer will ensure that enough sludge is left for seeding thereby improving system performance.

Simple controls to govern the rate of sludge removal help to maximize the solids content of the sludge withdrawn from the tank.

37.7 SPECIAL FACTORS

Imhoff tanks should be preceded by grit removal to reduce the frequency of digester compartment cleaning as the result of excessive accumulation of grit in these units. Also, measures for sludge treatment and disposal must be provided as part of the treatment system.

37.8 RECOMMENDATIONS

Imhoff tanks with ample design capacity can provide TSS and BOD removal efficiencies of from 50 to 75% and 25 to 30%, respectively. Imhoff tanks are suitable for use as a primary treatment step preceding oxidation ponds, intermittent sand filters, subsurface filters, leaching systems, and trickling filters. They may be suitable for pretreatment before ocean discharge and be used for treatment of some industrial wastes.

38. ROUGHING FILTERS

38.1 DESCRIPTION

Roughing filters are used predominantly for wastewater treatment but can also be used in water treatment (see Figure 38.1). They allow deep penetration of suspended materials into a filter bed and have a large solids/silt storage capacity. The solid materials retained by the filters are removed by either flushing or excavating the filter media, washing it and replacing it. Roughing filtration uses much larger media (more than 2.0 mm diameter) than either slow or rapid filter media (0.15 to 0.35 and 0.4 to 0.7 mm diameter, respectively). The roughing filter operates much like a trickling filter. Indeed, the biological activity has been observed to be much the same, including a potential for nitrification of wastewater (4). The rate of filtration can be as low as those used for slow sand filters or higher than those used for rapid filters, depending on the type of filter, the nature of the turbidity and the desired degree of turbidity removal. There are basically two types of roughing filters which are differentiated by their direction of flow: vertical flow (Figure 38.1) and horizontal flow (Figures 38.2 and 38.3). Vertical flow roughing filters may be either upflow or downflow (4, 25, 34).

38.1.1 Vertical Flow

Design Criteria and performance expectations for two types of wastewater roughing filters are provided in Tables 38.1, 38.2, and 38.3.

Gravel media upflow units consist usually of several gravel layers tapering from a coarse gravel layer located directly above the underdrain system, to successively fine gravel layers to permit deep penetration of suspended solids into the filter bed.

Where coconut is abundant, a low cost filter medium of shredded coconut fibers has been employed and proven successful (25). Small villages in Thailand and Southeast Asia have found the following very useful: Shredded coconut fiber may be prepared manually by soaking the husk for 2 to 3 days in water (until the fiber does not impart any more color to the water), then shredding the husk by pulling off binding particles. Shredded coconut fibers may also be purchased directly from upholstery stores or coconut fiber factories. The depth of the coconut fiber in the filter box is usually 60 to 80 cm.

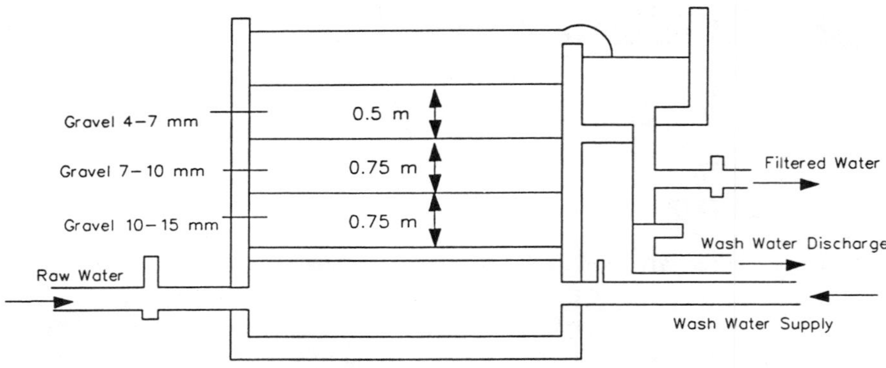

Figure 38.1. Gravel upflow roughing filter.

Several small filter plants ranging in capacity from 24 to 360 m^3/day were constructed from 1972 to 1976 in the Lower Mekong River Basin countries (Thailand, Vietnam, Cambodia) and in the Philippines. Two-stage filtration, using shredded coconut fibers and burnt rice husks for the roughing and polishing filter, respectively, was typical for all filter plants. (25)

38.1.2 Horizontal Flow

Horizontal flow roughing filters (HRF) have a large solids storage capacity because of their coarse filter media and long filter length. For overall efficiency it is best to use a graded gravel scheme for the filter medium. The horizontal flow filter is usually divided into several zones, each with its own uniform grain size, tapering from large sizes in the initial zone to small sizes

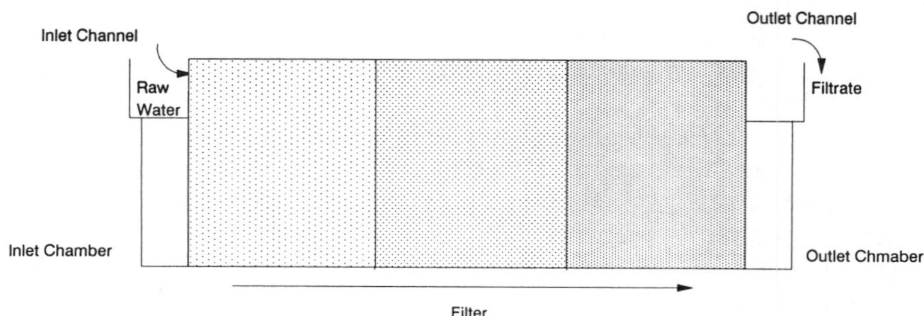

Figure 38.2. Basic features of a horizontal-flow roughing filter (*source:* reference 25).

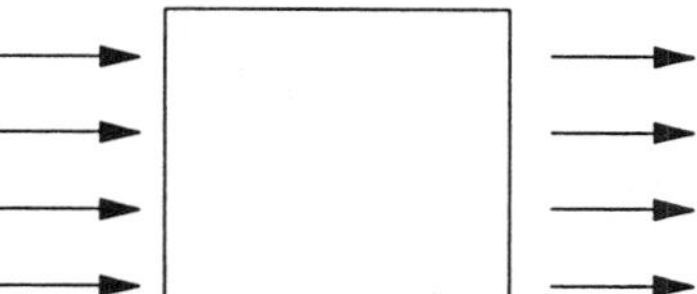

HRF acts as a multistore sedimentation tank

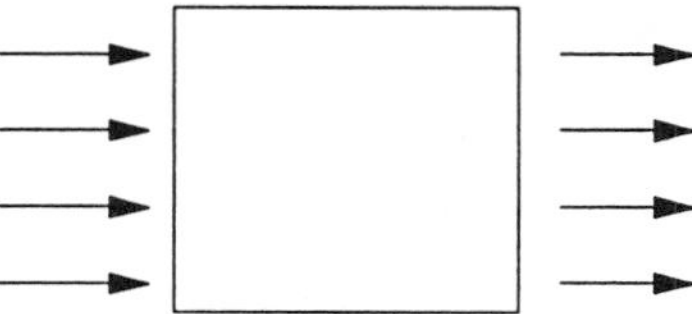

Accumulation of solids on the upper collector surface

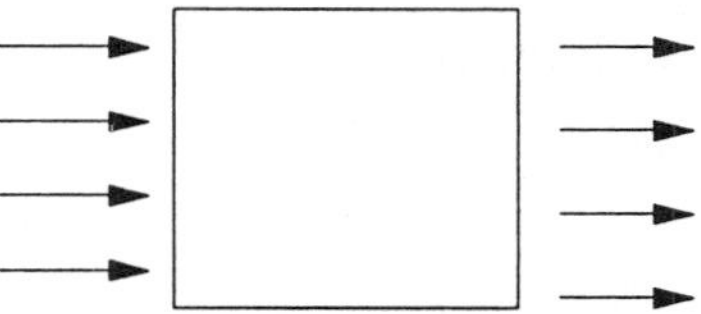

Drift of separated solids to the filter bottom

Figure 38.3. Mechanism of horizontal-flow roughing filter (*source:* reference 34).

in the final zone. In this way, penetration of suspended solids will more easily take place over the entire filter bed and result in longer filter runs.

Horizontal flow roughing filters may also be constructed adjacent to a stream bed so as to allow raw water to flow through a porous stone wall and into a gravel bed. The drain system is made of perforated PVC pipe that leads to a junction box. To avoid the infiltration of surface runoff, an impermeable layer of clay or a polyethylene liner can be placed over the gravel bed. This particular design has a capacity ranging from 85 to 860 m^3/day, and is intended to operate at a filtration rate of 0.5 m/hr. It can treat waters of turbidities less than 150 NTU prior to slow sand filtration. The length of the filter run is variable.

38.2 LIMITATIONS

Roughing filters are limited, because higher turbidities (greater than 150 NTU) will result in frequent clogging of filter media and cause discontinuous operation. Higher solids loadings may be used if cleaning is practiced regularly.

Table 38.1. Typical Design Parameters for Roughing Filters

PARAMETER	ROUGHING FILTER TYPE	
	VERTICAL	HORIZONTAL
Filtration rate (m/hr)	<20	0.5 to 4.0
Filter thickness (mm)	60 to 80	4 to 40
Sequence of filter arrangement	vertical	longitudinal
Total length of filter (m)	—	9 to 12
Filter side wall height (m)	—	1 to 1.5
Coarse gravel layer thickness (m)	10 to 15	10 to 20
Fine gravel layer	11 to 17	—
Filter Slope	—	1% in direction of flow
Discharge measurement	—	V-notch weir
Filter medium	various	—
Removal Efficiency (%)	60 to 90	60 to 70
Influent water turbidity (NTU)	<150	30 to 100

Source: Reference 19.

In the case of wastewater, treatment performance is not considered equivalent to filtration. Roughing filters may be used for pretreatment of relatively high solids content wastewaters or for treatment of bypassed combined sewer overflows.

38.3 COSTS

The construction costs, and operation and maintenance costs of vertical flow and horizontal flow roughing filters are about the same as trickling filters, depending on the type of construction.

Table 38.2. Typical Wastewater Design Parameters for Coconut Fiber Media Roughing Filters

PARAMETERS	VERTICAL ONLY
Filtration rate (m/hr)	1.25–1.5
Filter thickness (mm)	60–80
Sequence of filter arrangement	vertical
Removal efficiency (%)	60–90
Influent water turbidity (NTU)	<150

Source: Reference 25.

Table 38.3. Performance of Horizontal-Flow Settling Basins in Colombia

LOCATION OF WATER TREATMENT PLAN	DETENTION PERIOD (HR)		SETTLING VELOCITY (m/DAY)		TEMPERATURE (°C)		INFLUENT TURBIDITY (NTU)		EFFLUENT TURBIDITY (NTU)	
	MAX.	MIN.	MAX.	MIN.	MAX.	MIN.	MAX.	MIN.	MAX.	MIN.
Cali	5.05	3.92	22.2	17.3	18	2	600	5	8	2
Pasto	3.96	3.1	27.2	21.3	14	14	120	5	4	0.9
Pereira	2.66	2.13	35.4	28.2	18.5	18.5	130	7	9	4
Santa Marta	3.31	1.83	50.2	27.6	28	28	4690	59	6	2

Source: Reference 8.

38.4 AVAILABILITY

Roughing filters are a very economical method of sewage treatment for smaller communities. Materials of construction are available on a wide scale basis. Several water treatment plants exist in South and Central America (see Table 38.3). Roughing filters may be used for high rate and intermittent treatment applications of any kind, including stormwater treatment, sewage and industrial waste pretreatment, and other applications.

For the coconut flush system, the availability of the raw coconut husks at low cost and ease of construction and operation combine to make this manual filter bed regeneration process economical in areas where coconut trees are common. The use of such indigenous materials for filter media is also a practical alternative to conventional filter design.

38.5 OPERATION AND MAINTENANCE

Filtration rates in gravel upflow filters are relatively high because of the large pore spaces in the filter bed which are not likely to clog rapidly. Low backwashing rates are used because no attempt is made to expand the bed; but longer time periods for adequate cleaning of the gravel are usually necessary (about 20 to 30 minutes).

There are no backwashing arrangements for cleaning the coconut fibers because the fibers do not readily relinquish entrapped particles due to their fibrous nature. Instead, water is drained from the filter box and the dirty fibers are removed and discarded. Coconut fiber stock which has been properly cleaned is then packed into the filter. The filter medium generally must be replaced every three to four months.

Frequent cleanings are required for horizontal roughing filters when the filter lengths are short. Short filter lengths could be used, especially in areas of low labor costs. On the other hand, if filter beds are long enough, 2 to 5 years can go by before cleaning.

Structural constraints limit the depth of the filter bed in vertical flow filters but higher filtration rates and backwashing of the filter media are possible. On the other hand, horizontal flow filters enjoy practically unlimited filter run length but normally are subject to lower filtration rates, and they generally require manual cleaning of the filter media.

38.6 SPECIAL FACTORS

To use the roughing filters effectively, the raw wastewater characteristics and treatment objectives should be clearly defined.

The filter length is the most critical dimension in the design of horizontal flow roughing filters and should be selected after considering an appropriate balance between construction costs and the frequent cleanings required when filter lengths are short.

Upflow roughing filters are used predominantly to replace the unit processes of flocculation and sedimentation used in conventional rapid filtration plants. They are similar in design and construction to gravel bed flocculators. They are used in an alternating upflow-downflow (UDF) operating mode.

Roughing filters are often used before slow sand filters because of their effectiveness in managing high concentrations of suspended solids.

38.7 RECOMMENDATIONS

Roughing filters are very effective for treatment of sewage wastewater when turbidity is in the range of 20 to 150 NTU to eliminate too frequent clogging and to ensure continuous operation for an extended period of time. Roughing filters can also be applied to water treatment.

The UDF is a simple and economical treatment process for smaller water treatment plants in developing countries and for small systems. In this application, the UDF system replaces conventional arrangements for mixing, flocculation, and sedimentation used in rapid filtration plants. This results in reduced construction and operation and maintenance costs.

39. ROTATING BIOLOGICAL CONTACTORS (RBC)

39.1 DESCRIPTION

The RBC process is a fixed film biological reactor using plastic media mounted on a horizontal shaft and placed in a tank. A general layout is shown in Figure 39.1 (2). The process functions in a similar fashion to other fixed film reactors, such as trickling filters. Common media forms are discs made of styrofoam and a denser lattice type made of polyethylene. The discs are slowly rotated while wastewater flows through the tank. The discs are about 40% immersed in contact with the wastewater. Organic matter is removed by the biological film that develops on the media. Rotation results in exposure of the film to the atmosphere as a means of aeration. Excess biomass on the media strips off (sloughs) from time to time, often during seasonal changes when the character of the organisms change. Solids are maintained in suspension by the mixing action of the rotating media. Multiple staging of RBCs increases treatment efficiency and could aid in achieving nitrification year round. Winter nitrification rates will depend on the ambient temperature range; greater efficiency is achieved in moderate temperature zones. A complete system generally consists of two or more parallel trains with each train consisting of multiple stages in series (see Figure 39.1).

The typical design criteria for RBC technology are shown in Table 39.1 (2, 35, 36, 37, 38, 39).

39.2 LIMITATIONS

Performance of this process may fall off significantly at temperatures below 55°F (13°C) if the biologically coated media is exposed directly to the ambient air. High organic loadings may result in first stage septicity and supplemental aeration may be required. Use of dense media for early stages can result in media clogging.

Alkalinity deficiency may result when nitrification rates are high; a supplemental alkalinity source may be required.

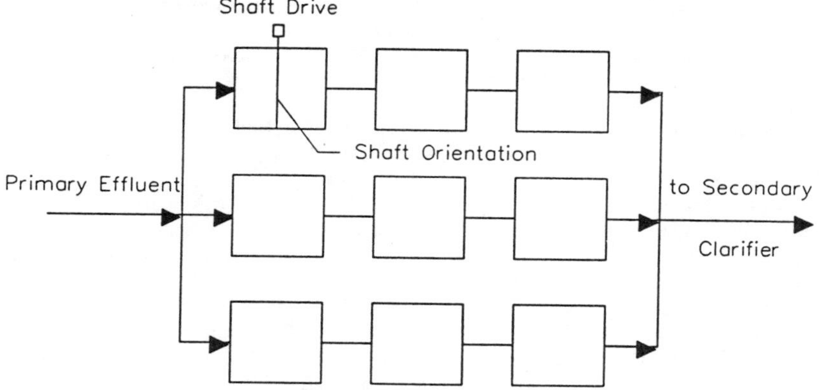

Figure 39.1. Typical staged RBC configuration (*source:* reference 2).

39.3 COSTS

Costs are based on including RBC shafts (standard media 100,000 ft^2/shaft, 9290 m^2/shaft), motor drives (5 hp/shaft), molded fiberglass covers and reinforced concrete basins. Operation and maintenance costs include power, labor and materials (see Figure 39.2).

39.4 AVAILABILITY

The process has been in use in the United States since 1969 and may be considered ready for widespread application. It should work well in warm climates. Its use is growing because of its characteristic modular construction, low hydraulic head loss and shallow depth which make it adaptable to either new or existing treatment facilities (38).

39.5 OPERATION AND MAINTENANCE

Ease of maintenance should be provided by considering the following:

1. Ease of access to shafts, media, and other mechanical equipment needing regular inspection and maintenance and possible periodic removal or replacement.
2. Use of self-aligning, moisture resistant bearings.

Rotating biological contactors have few moving parts and require minor preventive maintenance. Chain drives, belt drives, sprockets, rotating shafts

Table 39.1. Typical Design Parameters for RBCs

PARAMETER		TYPICAL VALUE RANGE
Organic loading		
(without nitrification)	=	30 to 60 lb BOD_5/d/1000 ft^2 media
(with nitrification)	=	15 to 20 lb BOD_5/d/1000 ft^2 media
Hydraulic loading		
(without nitrification)	=	0.75 to 1.5 gal/d/ft^2 media
(with nitrification)	=	0.3 to 0.6 gal/d/ft^2 media
Number of stages/train	=	1 to 4 depending upon treatment objectives
Number of parallel trains	=	At least 2
Rotational velocity	=	60 ft/min for mechanically drivers
Peripheral velocity		30 to 60 ft/min for air drivers
Typical media specific surface		
(Disc type)	=	20 to 25 ft^2/ft^3
(Standard Lattice type)	=	30 to 40 ft^2/ft^3
(High density Lattice type)	=	50 to 60 ft^2/ft^3
Percent media submerged	=	40%
Tank volume	=	0.12 gal/ft of disc area
Detention time based on 0.12 gal/ft^2/d		
(without nitrification)	=	40 to 120 minutes
(with nitrification)	=	90 to 250 minutes
Secondary clarifiers overflow rate	=	500 to 800 gal/d/ft^2
Horse power	=	3 to 5/25 ft shaft
		5 to 7.5/25 ft shaft

Note: ft × 0.3048 = m
BOD_5/day/1000 ft^3 × 33 = BOD_5/day/1000 m^3
gal/day/ft^2 × 0.04 = m^3/day/m^2
Source: References 2, 35, 36, 37, 38, 39.

and any other moving parts should be inspected and maintained on a regular basis. All exposed parts, bearing housing shaft ends and bolts should be painted or covered with a layer of grease to prevent rust damage. Motors, speed reducers and all other metal parts should be painted for protection.

As in the case of any mechanical treatment system, an active preventive maintenance program that keeps equipment properly lubricated and adjusted to help reduce wear and breakage is more effective in maintaining system operability than maintenance on an as-needed basis.

Properly designed systems have sufficient turbulence so that solids or sloughed slime growths will not settle out on the bottom of the bays. If grease balls appear on the water surface in the bays, they should be removed with a dip net or screen device.

39.6 CONTROL

RBC units perform most effectively under conditions of low hydraulic loadings and organic loadings, compared to other biological treatment processes.

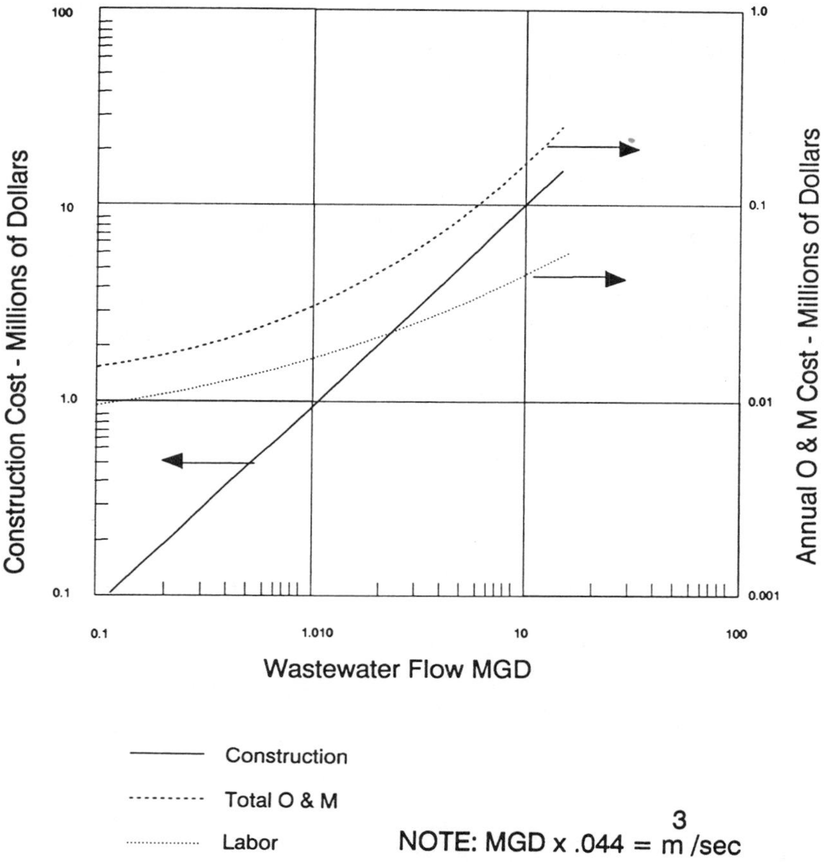

Figure 39.2. Construction, operation, and maintenance costs for rotating biological contactors (*source:* references 2 and 5).

For instance, with a four stage system with final clarifier preceded by primary treatment, the expected percent removals are: BOD_5 = 80 to 90%, suspended solids = 80 to 90%, phosphorus = 10 to 30%, and NH_3-N up to 95%.

39.7 SPECIAL FACTORS

There are special factors to be considered when considering the RBC system. These include: longer retention times (8 to 10 times longer than trickling filters); less susceptibility to upset is likely from changes in hydraulic or organic loading than conventional activated sludge. While loadings must be lower, unit costs are significantly lower as well.

For high strength wastewaters, an enlarged first stage may be employed to maintain aerobic conditions. An intermediate clarifier may also be employed when high solids are generated in order to avoid anaerobic conditions in the contact basin.

39.8 RECOMMENDATIONS

RBC technology may be considered for applications in warm climates. Construction costs are lower per unit of treatment capacity purchased. Operational efficiency is more consistent than other biological processes and maintenance is simpler.

40. ACTIVATED SLUDGE TREATMENT

40.1 DESCRIPTION

Activated sludge treatment is used to remove dissolved and colloidal biodegradable organics. It has been applied in many design variations, and a complete discussion of the many forms is beyond the scope of this book. The technology is a continuous flow, stirred, biological treatment process with recycling of the biomass. The completely mixed variant is shown on Figure 40.1. The process is characterized by a suspension of aerobic microorganisms maintained in a relatively homogeneous state by mixing- and turbulence-induced or diffused aeration. The microorganisms are used to oxidize soluble and colloidal organics to carbon dioxide (CO_2) and water (H_2O) in the presence of molecular oxygen. The process may or may not be preceded by sedimentation. The mixture of microorganisms and wastewater (called mixed liquor) formed in the aeration basins is transferred, following secondary clarifiers for liquid and solids separation. The major portion of the microorganisms settling out in the secondary clarifiers is recycled to the aeration basins to be mixed with incoming water, while the excess which constitutes the waste sludge, is sent to sludge handling facilities. The rate of return and concentration of activated sludge recycled to the aeration basins determines the mixed liquor suspended solids (MLSS) level developed and maintained in the basins. During the oxidation process, a certain amount of the organic material is synthesized into new cells, some of which then undergoes auto-oxidation (endogenous respiration) in the aeration basins, with the remainder forming net growth or excess sludge. Oxygen is required in the process to support the oxidation and synthesis reactions. Volatile organic compounds may be removed to a certain extent during the aeration process. Metals will also be partially removed, with accumulation in the sludge.

Activated sludge systems are classified generally as high rate, conventional, or extended aeration (low rate), based on organic loading. In the conventional activated sludge plant, the wastewater is commonly aerated for a period of 4 to 8 hours (based on the average daily flow) in a plug flow hydraulic mode. Either surface or submerged aerators may be used. Compressors are used to supply air to submerged systems. Common modifications to the process are: step aeration, contact stabilization, and completely mixed flow regimes. Alum or ferric chloride may be added for phosphorus

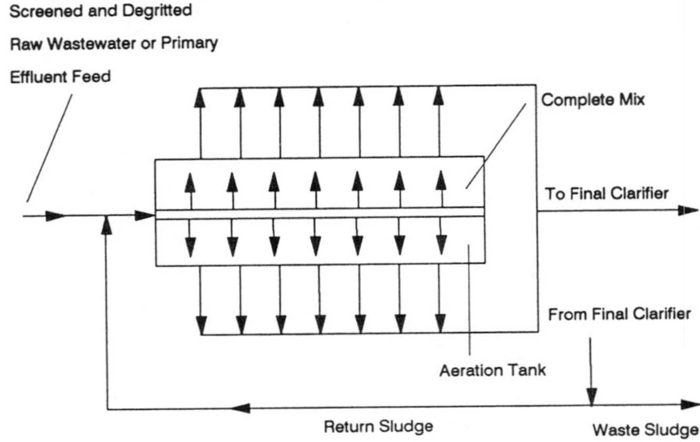

Figure 40.1. Typical completely mixed activated sludge treatment (*source:* reference 2).

removal. Design parameters for a wide range of process modifications are shown in Table 40.1 (4). The typical design parameters for high rate activated sludge treatment are given in Table 40.2 (2).

40.2 LIMITATIONS

This process alone may not provide effluent quality sufficient to meet stringent discharge requirements. The concentration of BOD_5 and suspended solids in final clarified effluent are in the range of 15 to 30 mg/L.

40.3 COSTS

The construction costs include aeration basins, air supply equipment and piping. Clarifiers and recycle pumps are not included. See Figure 40.2 (2, 11). Costs shown are for completely mixed conventional loading rates. High rate systems construction costs may be as much as 20% lower. There is however usually a trade off in that O&M costs are higher for high rate systems. The decision about which option to select should be made based on local conditions and waste treatment requirements.

40.4 AVAILABILITY

This technology is the most versatile and widely used biological municipal wastewater treatment process in the world. On the other hand, the process is difficult to operate for consistent effluent quality and therefore is rarely used

Table 40.1. Design Parameters for Activated-Sludge Processes

PROCESS MODIFICATION	SOLIDS RETENTION TIME DAYS	F/M, kg BOD_5 APPLIED/ kg MLVSS/D	VOLUMETRIC LOADING kg BOD_5 APPLIED/m^3/D	MLSS, mg/L	HYDRAULIC RETENTION TIME, HR	RECIRCULATION RATIO
Conventional	5–15	0.2–0.4	0.3–0.6	1,500–3,000	4–8	0.25–0.5
Tapered aeration	5–15	0.2–0.4	0.3–0.6	1,500–3,000	4–8	0.25–0.5
Continuous-flow stirred-tank reactor	5–15	0.2–0.6	0.8–2.0	3,000–6,000	3–5	0.25–1.0
Step aeration	5–15	0.2–0.4	0.6–1.0	2,000–3,500	3–5	0.25–0.75
Modified aeration	0.2–0.5	1.5–5.0	1.2–2.4	200–500	1.5–3	0.05–0.15
Contact stabilization	5–15	0.2–0.6	1.0–1.2	(1,000–3,000)[a]	(0.5–1.0)[a]	0.25–1.0
				(4,000–10,000)[b]	(3–6)[b]	
Extended aeration	20–30	0.05–0.15	0.1–0.4	3,000–6,000	18–36	0.75–1.50
Kraus process	5–15	0.03–0.8	0.6–1.6	2,000–3,000	4–8	0.5–1.0
High-rate aeration	5–10	0.4–1.5	1.6–1.6	4,000–10,000	0.5–2	1.0–5.0
Pure-oxygen systems	8–20	0.25–1.0	1.6–3.3	6,000–8,000	1–3	0.25–0.5

[a] Contact unit
[b] Solids stabilization unit
Note: kg/m^3/d × 62.4280 = lb/10^3 ft^3/d
kg/kg/d × 1.0 = lb/lb/d
mg/L = g/m^3
MLSS = mixed liquor suspended solids
F/M = food to microorganism loading
Source: Reference 4.

Table 40.2. Design Parameters for Activated Sludge Treatment (modified aeration and completely mixed aeration)

	MODIFIED AERATION	COMPLETELY MIXED AERATION
Volumetric loading, lb bod_5/d/1000 ft^3	50–100	50–125
MLSS, mg/L	800–2000	3000–5000
Aeration detention time, hours	2–3	2–4
F/M, lb BOD_5/d/lb MLVSS	0.75–2	0.4–0.8
Std ft^3 air/lb BOD_5 removed	400–800	800–1200
lb O_2/lb BOD_5 removed	0.4–0.7	0.9–1.2
Sludge retention time, days	0.75–2	2–5
Recycle ratio (R)	0.25–1.0	0.25–0.5
Volatile fraction of MLSS	0.7–0.85	0.7–0.8

Source: Reference 2.

in developing countries. Air compressor equipment and materials of construction are readily available. Air diffuser systems, mechanical sludge removal and collection equipment and other appurtenances are less widely available and require regular replacement and repair.

40.5 OPERATION AND MAINTENANCE

In conventional activated sludge treatment the aeration basins are typically designed to operate in either completely mixed or plug flow hydraulic configurations. Either surface or submerged aeration can be employed to transfer oxygen from air to wastewater. Compressors are used to supply air to submerged aeration systems through a network of diffusers. Diffusers used in activated sludge treatment processes are porous ceramic plates, porous ceramic domes, ceramic or plastic tubes connected to the pipe header, and lateral systems for fine or coarse air bubbles. The air delivery system or mixers and all mechanical parts require regular and complete maintenance.

To operate an activated sludge treatment process efficiently, it is necessary to understand that the microorganisms in the system form the operational basis of the removal and conversion system for organic compounds. It is especially important to operate the system on the basis of solids retention time (SRT).

40.6 CONTROL

Problems of instability in activated sludge treatment will significantly reduce process efficiency. Under uniform conditions of wastewater flow and sludge recirculation, the biological process will operate at a constant food to micro-

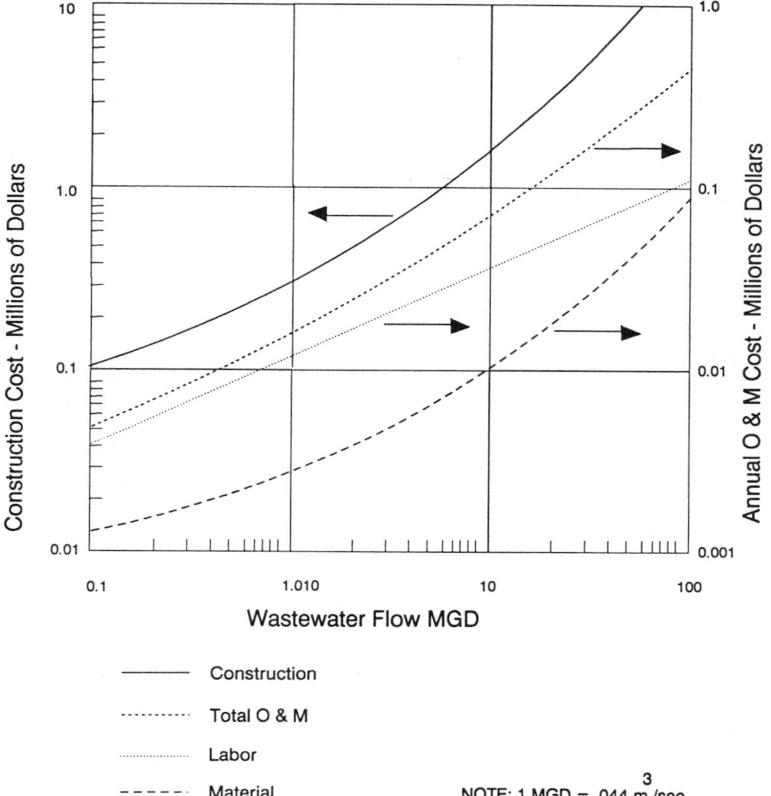

Figure 40.2. Construction, operation, and maintenance costs for activated sludge treatment (completely mixed).

organism (F/M) ratio. The best control option is the management of solids retention time (SRT). Sludge wasting should be based on predetermined SRT values based on the process design and wastewater characteristics.

40.7 SPECIAL FACTORS

High rate activated sludge and extended aeration processes are completely mixed systems. Complete mixing in high rate systems allows increased BOD loadings and shortened aeration periods. Thorough mixing in combination with long retention periods used in extended aeration plants provides the assimilative capacity necessary to accept intermittent loading without loss of efficiency. For example, extended aeration plants have been used successfully for schools in the United States where the load enters during a 10 to 12 hour period each day for only 5 days per week. Activated sludge organisms can be conditioned (acclimatized) to accept reasonably high concentrations of toxic organic compounds and metals. Thus, industrial waste may be treated. These conditions, however require special operational attention by specially trained operators.

40.8 RECOMMENDATIONS

Proper and regular maintenance of the activated sludge treatment process is required to achieve high percentage removal of BOD and suspended solids from the influent wastewater. This can be achieved only with knowledgeable, trained, and experienced plant operators. Therefore, a highly trained plant operator or engineer is required for process effectiveness. Many plants in the U.S. and elsewhere do not achieve the design efficiencies because of lack of trained operators continuously on duty at the plant. The plant requires 24-hour operator attention and regular maintenance.

41. STEEP SLOPE SEWERS

41.1 DESCRIPTION

Traditionally, sewers are constructed on slopes sufficient to allow a minimum 2 ft/sec (0.6 m/sec) velocity of wastewater. This is to prevent accumulation of settled material in the sewer which might eventually lead to blockage. Also, traditional design practice requires that maximum velocities (i.e., maximum installed slopes) not be exceeded in order to keep erosion damage to the sewers to a minimum. There are also requirements for minimum spacing of manholes in sewer runs to allow for cleaning and access. Manholes may be a significant source of infiltration of water and particulate matter which exacerbate erosion and treatment problems.

Traditional practices are costly, especially in geographic areas where steep slopes are common. Steep slopes are common along the eastern and western portions of the North American continent, throughout Central America, and the northern and western sectors of the South American continent. Most of these areas would benefit significantly from altered sewer design practices.

Within the past twenty years the nature of construction materials has changed drastically. Plastic sewer pipe products which have formerly been questionable for "high stress" applications have improved. The erosion characteristics have now been demonstrated to be comparable, and even better than cement-based sewer products (40).

In addition, new cementaceous sewer products, reinforced concrete pipe, asbestos cement[1] pipe and others have improved erosion resistance properties. Testing should be conducted to determine the nature of new products for steep slope applications.

Steep slope sewers may be used to reduce costs relative to gravity and pressure sewer systems where populations live in rough and hilly terrain. Installation practice for sewer lines depends on the terrain slope, soil type and bedding type. Steep slope sewers may be constructed in a number of segments moving down hill for example. Diagrams of some variations are given on Figure 41.1. Access to a steep slope sewer is by drop manholes or

[1] Asbestos cement pipe manufacture has been banned by regulations of the U.S. EPA in July 1989, pursuant to the Toxic Substances Control Act. New applications of this material are thus unlikely in the future in the U.S. There are major manufacturers of asbestos cement pipe in Latin America (e.g., in Peru) and currently applications are widespread.

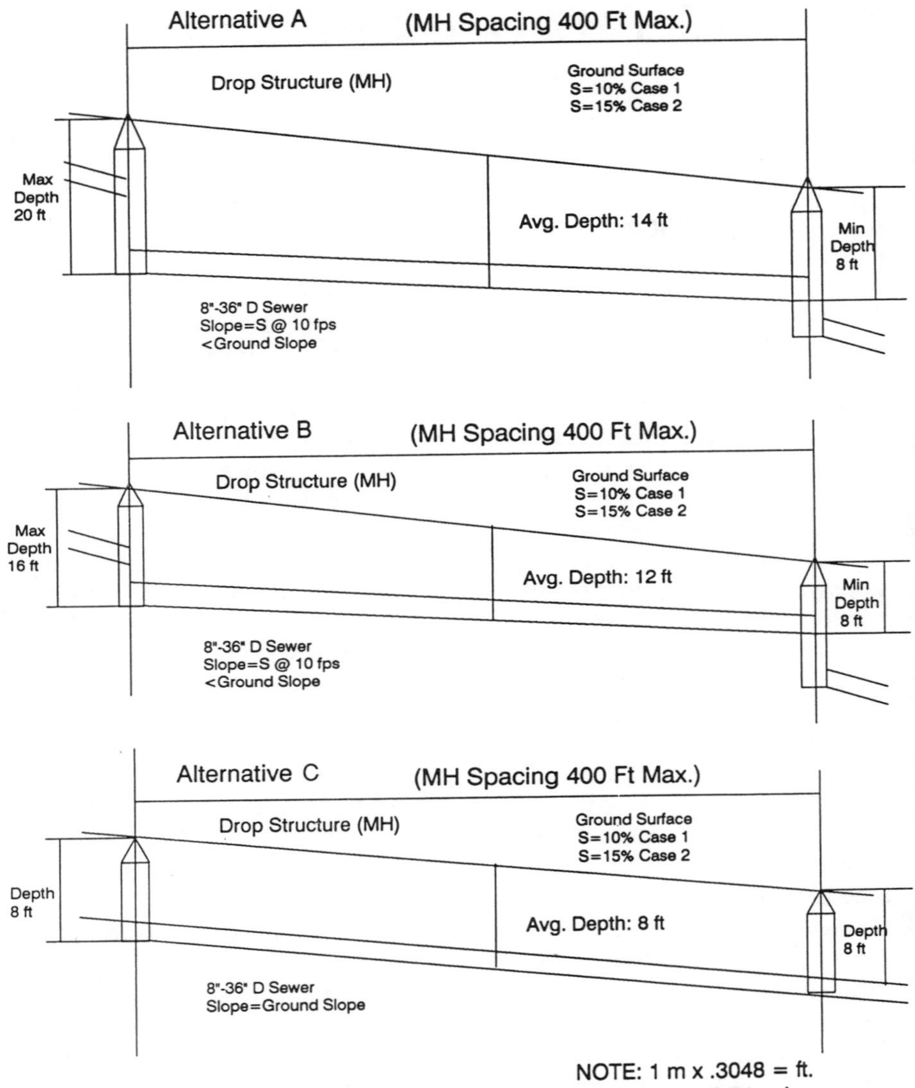

Figure 41.1. Alternatives A, B, and C.

clean outs located conveniently at intervals or at changes in terrain slope. Manhole installation need not be at every major change in direction which is common in traditional practice; since slopes are steep, manholes are not necessary at such frequent intervals to provide access for construction purposes.

41.2 LIMITATIONS

There may be increased erosion of sewer materials with the higher velocities likely in steep slope designs. Use of the approach requires changes from traditional design and construction practice, training of laborers and construction workers, and different maintenance practices.

41.3 COSTS

Considerable sewer installation cost savings can be realized from application of the steep slope approach. Examples are included herein since it is not possible to include a wide range of applicability. The costs were prepared using a Lotus program spreadsheet. A similar approach may be used by designing engineers for specific locations. Costs have been calculated for three alternatives sewer design scenarios (see Figure 41.1):

Alternative A: traditional design factors with a maximum depth to sewer of 20 feet.

Alternative B: traditional design factors with a maximum depth of 16 feet.

Alternative C: a design approach wherein the depth to sewer remains constant (at 8 ft, 2.7 m). This approach is only possible where sewer construction material is erosion resistant, since if depth to sewer is constant the sewer installation will basically follow the slope of the land surface.

The depth in Alternatives A and B is variable, as seen in Figure 41.1, while the depth is constant at 8 feet for Alternative C.

The calculations for comparative costs have been done for several different conditions as given in Summary Table 41.1. Note that for Alternative C, the manhole depths and spacing (and thus the costs) are fixed for various pipe sizes, since it is not necessary to control maximum slope. For Alternates A and B the manhole costs increase for increasing pipe size. Alternates A and B costs are higher because the manhole spacing must be reduced (decreased length between manholes) in order to maintain proper traditional design slopes (allowing velocities between 2 ft/sec, and about 10 ft/sec, 0.6 to 3 m/sec).

The cost per lineal foot (Cost per LF, on Table 41.1) for the three alternatives may be compared. Manholes may not be eliminated completely (cleaning, direction changes, and other requirements). With alternative C, manholes may represent as little as 3% of the system costs. For Alternates A and B, manhole costs do not get below about 20% of system costs.

The total costs should also be compared. This can be done in various ways.

Table 41.1. Summary—Sewer Costs Plus Manhole Costs

ALTERNATIVE A
AVERAGE SEWER DEPTH 14 FEET
MAXIMUM SEWER DEPTH 20 FEET
MINIMUM SEWER DEPTH 8 FEET

Condition 1—Rock Case 1: Ground Slope = 10%

	Cost per LF						
Dia.	8″	10″	12″	15″	18″	24″	36″
Item							
Sewer	60.28	62.07	63.30	95.80	105.79	146.11	173.49
Manhole	12.85	17.13	21.42	27.34	30.41	34.73	38.07
Total	73.13	70.20	84.72	123.14	136.20	180.84	211.56

Condition 1—Rock Case 2: Ground Slope = 15%

	Cost per LF						
Dia.	8″	10″	12″	15″	18″	24″	36″
Item							
Sewer	60.28	62.07	63.30	95.80	105.79	146.11	173.49
Manhole	27.34	38.65	42.83	48.95	51.92	55.87	59.77
Total	87.62	100.72	106.13	144.75	157.71	201.98	233.26

Condition 2—Shoring Case 1: Ground Slope = 10%

	Cost per LF						
Dia.	8″	10″	12″	15″	18″	24″	36″
Item							
Sewer	35.55	37.46	38.80	47.37	57.59	74.39	102.86
Manhole	8.57	11.42	14.28	18.23	20.28	23.16	25.39
Total	44.12	48.88	53.08	65.60	77.87	97.55	128.25

Condition 2—Shoring Case 2: Ground Slope = 15%

	Cost per LF						
Dia.	8″	10″	12″	15″	18″	24″	36″
Item							
Sewer	35.55	37.46	38.80	47.37	57.59	74.39	102.86
Manhole	18.23	25.77	28.56	32.64	34.62	37.25	39.85
Total	53.78	63.23	67.36	80.01	92.21	111.64	142.71

Condition 3—Sloped Sides Case 1: Ground Slope = 15%

	Cost per LF						
Dia.	8″	10″	12″	15″	18″	24″	36″
Item							
Sewer	36.60	38.45	39.74	48.22	58.36	74.99	103.12
Manhole	10.38	13.84	17.30	22.09	24.57	28.06	30.76
Total	46.98	52.29	57.04	70.31	82.93	103.05	133.88

Condition 3—Sloped Sides Case 2: Ground Slope = 15%

	Cost per LF						
Dia.	8″	10″	12″	15″	18″	24″	36″
Item							
Sewer	36.60	38.45	39.74	48.22	58.36	74.99	103.12
Manhole	22.09	31.23	34.61	39.55	41.95	45.14	48.29
Total	58.69	69.68	74.35	87.77	100.31	120.13	151.41

Table 41.1. *Continued*

Condition 4—Vertical Sides Case 1: Ground Slope = 10%

				Cost per LF			
Dia.	8″	10″	12″	15″	18″	24″	36″
Item							
Sewer	11.85	13.70	14.99	23.47	33.60	50.24	78.36
Manhole	6.37	8.49	10.62	13.55	15.08	17.22	18.87
Total	18.22	22.19	25.61	37.02	48.68	67.46	97.23

Condition 4—Vertical Sides Case 1: Ground Slope = 15%

				Cost per LF			
Dia.	8″	10″	12″	15″	18″	24″	36″
Item							
Sewer	11.85	13.70	14.99	23.47	33.60	50.24	78.36
Manhole	13.55	19.16	21,23	24,27	25.74	27.70	29.63
Total	25.40	32.86	36.22	47.74	59.34	77.94	107.99

ALTERNATIVE B
AVERAGE SEWER DEPTH 14 FEET
MAXIMUM SEWER DEPTH 20 FEET
MINIMUM SEWER DEPTH 8 FEET

Condition 1—Rock Case 1: Ground Slope = 10%

				Cost per LF			
Dia.	8″	10″	12″	15″	18″	24″	36″
Item							
Sewer	52.26	54.05	55.2	83.77	93.76	130.07	157.45
Manhole	10.50	21.00	26.24	33.59	37.16	42.41	46.66
Total	62.76	75.05	81.52	117.36	130.92	172.48	204.11

Condition 1—Rock Case 2: Ground Slope = 15%

				Cost per LF			
Dia.	8″	10″	12″	15″	18″	24″	36″
Item							
Sewer	52.26	54.05	55.28	83.77	93.76	130.07	157.45
Manhole	33.59	47.18	52.49	59.99	63.62	68.84	72.40
Total	85.85	101.23	107.77	143.76	157.38	198.91	229.85

Condition 2—Shoring Case 1: Ground Slope = 10%

				Cost per LF			
Dia.	8″	10″	12″	15″	18″	24″	36″
Item							
Sewer	31.23	33.14	34.49	42.55	52.77	69.07	97.53
Manhole	7.07	14.15	17.68	22.63	25.04	28.58	31.43
Total	38.30	47.29	52.17	65.18	77.81	97.65	128.96

Condition 2—Shoring Case 2: Ground Slope = 15%

				Cost per LF			
Dia.	8″	10″	12″	15″	18″	24″	36″
Item							
Sewer	31.23	33.14	34.49	42.55	52.77	69.07	97.53
Manhole	22.63	31.79	35.36	40.41	42.86	46.38	48.78
Total	53.86	64.93	69.85	82.96	95.63	115.45	146.31

Table 41.1. *Continued*

Condition 3—Shoring Sides Case 12: Ground Slope = 105%

Dia.	8″	10″	12″	15″	18″	24″	36″
Item				Cost per LF			
Sewer	29.02	30.87	32.16	40.14	50.27	66.40	94.53
Manhole	7.71	15.42	19.28	24.67	27.29	31.15	34.27
Total	36.73	46.29	51.44	64.81	77.56	97.55	128.80

Condition 3—Sloped Sides Case 2: Ground Slope 15%

Dia.	8″	10″	12″	15″	18″	24″	36″
Item				Cost per LF			
Sewer	29.02	30.87	32.16	40.14	50.27	66.40	94.53
Manhole	24.67	34.65	38.65	44.06	46.73	50.56	53.17
Total	53.69	65.52	70.71	84.20	97.00	116.96	147.70

Condition 4—Vertical Sides Case 1: Ground Slope = 10%

Dia.	8″	10″	12″	15″	18″	24″	36″
Item				Cost per LF			
Sewer	10.84	12.69	13.98	21.95	32.09	48.22	76.34
Manhole	5.32	10.63	13.29	17.01	18.81	21.47	23.62
Total	16.16	23.32	27.27	38.96	50.90	69.69	99.96

Condition 4—Vertical Sides Case 1: Ground Slope = 10%

Dia.	8″	10″	12″	15″	18″	24″	36″
Item				Cost per LF			
Sewer	10.84	12.69	13.98	21.95	32.09	48.22	76.34
Manhole	17.01	23.89	26.58	30.37	32.21	34.85	36.66
Total	27.85	36.58	40.56	52.32	64.30	83.07	113.00

ALTERNATIVE C
AVERAGE SEWER DEPTH 14 FEET
MAXIMUM SEWER DEPTH 20 FEET
MINIMUM SEWER DEPTH 8 FEET

Condition 1—Rock

Dia.	8″	10″	12″	15″	18″	24″	36″
Item				Cost per LF			
Sewer	36.23	38.01	39.24	59.71	69.70	98.00	125.37
Manhole	5.80	5.80	5.80	5.80	5.80	5.80	5.80
Total	42.03	43.81	45.04	65.51	75.50	103.80	131.17

Condition 2—Shoring

Dia.	8″	10″	12″	15″	18″	24″	36″
Item				Cost per LF			
Sewer	22.60	24.51	25.86	32.91	43.13	58.42	86.88
Manhole	4.08	4.08	4.08	4.08	4.08	4.08	4.08
Total	26.68	28.59	29.94	36.99	47.21	62.50	90.96

Table 41.1. *Continued*

Condition 3—Sloped Sides

Dia.	8″	10″	12″	Cost per LF 15″	18″	24″	36″
Item							
Sewer	16.90	18.75	20.04	27.00	37.14	52.26	80.38
Manhole	3.72	3.72	3.72	3.72	3.72	3.72	3.72
Total	20.62	22.47	23.76	30.76	40.86	55.98	84.10

Condition 4—Vertical Sides

Dia.	8″	10″	12″	Cost per LF 15″	18″	24″	36″
Item							
Sewer	8.81	10.67	11.95	18.92	29.06	44.17	72.30
Manhole	3.21	3.21	3.21	3.21	3.21	3.21	3.21
Total	12.02	13.88	15.16	22.13	32.27	47.38	75.51

Note: 1 ft = .3048 m

Note for example, that for Condition #1 (rock), a cost saving of about 40% is possible using alternative C compared to A. Differences are both larger and smaller when comparing other modes of construction. Note also that the number of manholes significantly impact overall costs. The cost savings may be considered conservative since it is probably possible to eliminate even more manholes from the higher cost alternatives.

Based on steep slope sewer case studies in different terrains, the construction cost is lower than for traditional gravity systems and probably lower than for pressure sewer systems. The cost comparison is based on PVC pipe material. The operation and maintenance cost is probably the same or higher than for pressure sewer systems. These determinations may be made on a case by case basis.

There is another important consideration. When depths are fixed, for example at the suggested depth of 8 ft (2.4 m), it may be possible to use vertical side construction generally throughout the project. It is not possible to use vertical side techniques at depths of 14 to 20 ft (4.3 to 6 m), unless in solid rock. Using this consideration it is possible to compare different conditions on Table 41.1. For example, if the cost for Alternate C (vertical sides) is compared to Alternate A (with shoring), the cost for the shoring technique (probably required for deep systems), is about 4-½ times higher.

Cost is related significantly to depth of installation. It is probably not possible however, to construct systems at depths less than about 6 to 8 ft (2 to 2.7 m) because of house connection considerations and consideration of friction losses.

41.4 AVAILABILITY

The materials of construction for the steep slope approach are probably plastic pipes as opposed to traditional cementaceous materials of construction. The availability of pipes and cleanout devices will be limited in developing countries. The construction approaches are different also. It is always more difficult to use different techniques from conventional practice; forming joints, installing cleanouts and generally insuring system integrity will be challenges to be met in changeover from traditional practice. The design approach is different and engineers will be challenged also.

41.5 OPERATION AND MAINTENANCE

Steep slope sewers have been studied (40). Solids deposition is possible where slope changes occur, especially where steep slopes change back to flatter slopes. Cleanouts may be added at these points; it is possible to locate manholes at these locations during design. Variations in velocity may cause higher erosion than flatter slope systems. Early maintenance programs will identify such areas.

41.6 CONTROL

The steep slope sewers design should be based on the following factors: terrain slope, bedding type, soil type, piping material type and diameter of pipe, operation and maintenance programs available for sewers by the local utilities.

41.7 SPECIAL FACTORS

Definite advantages are low construction cost and minimal (albeit different from traditional practice) system operation skills required. The technology is best suited for rough and hilly terrains where extensive construction cost is involved.

41.8 RECOMMENDATIONS

Studies indicate that for construction of steep slope sewers in rough terrain, PVC pipe is highly regarded because of its light weight and ease of handling and installation. Applications of the steep slope approach should be considered for appropriate locations.

42. SEQUENCING BATCH REACTORS (SBR)

42.1 DESCRIPTION

A sequencing batch reactor (SBR) is a batch operation very similar to the fill-and-draw activated sludge treatment system. A typical SBR may be composed of one or more tanks. In an SBR system each tank has five basic operating modes or periods based on the primary functions: FILL, REACT, SETTLE, DRAW, and IDLE, all in a timed sequence. FILL (the receiving of raw wastewater) and DRAW (the discharge of treated effluent) must occur in each complete cycle for a given tank. REACT (the time to complete desired reaction), SETTLE (the time to separate the microorganisms from treated effluent), and IDLE (the time after discharging the tank and before refilling) can be modified or eliminated depending on effluent requirements and the pre- and post-treatment facilities available. The time for a complete cycle in a single tank system is the total time between the beginning of FILL and the end of IDLE; in multiple tank systems, the cycle time is the time between the beginning of FILL for the first reactor and the end of IDLE for the last reactor. In multiple tank systems the reactors fill in sequence, see Figure 42.1 (41). The operating sequence is shown in Figure 42.2 (42).

The single tank system is applicable for noncontinuous flow situations, especially for small rural towns where smaller wastewater flows occurs. Minimum operator input is required for satisfactory performance of the single tank system. The operation of multiple tank systems can be either simple with a minimum of operator input or complex, depending on flow and load variations and degree of treatment required.

In contrast to other technology discussions, the design approach for SBRs is presented herein because it is less available generally to the design community than conventional systems. There are no fixed rules for the design approach and wide variations in design criteria, such as F/M ratio, are possible. A somewhat different design approach is given by Arora, et al. (81). The typical design parameters of SBR systems are described in sequence as follows (41, 42, 43, 44, 45, 46, 47, 81):

1. Decide if primary treatment is needed. Primary treatment may be unnecessary in many SBR system applications, especially if the design sludge age or sludge retention time (SRT) is high (more than 20 days).

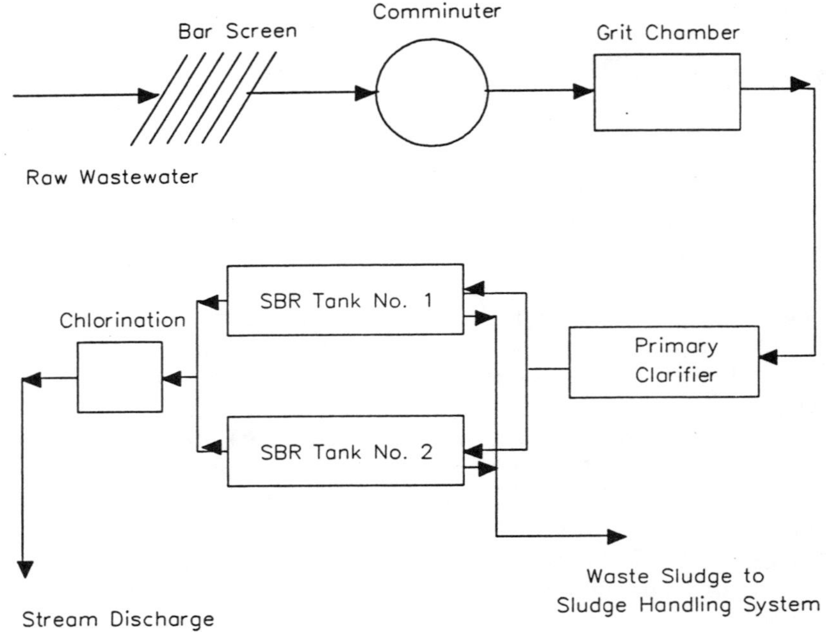

Figure 42.1. Typical sequence batch reactor wastewater treatment process (*source:* reference 41).

A high SRT system will also accomplish some sludge digestion aerobically in the reactor.

2. Select the desired food to microorganism (F/M) ratio. The selection of the design F/M ratio should be based on considerations such as nitrification requirements and desired SRT. From a given influent BOD, F can be calculated in pounds of BOD/day, and application of the selected F/M ratio yields the design sludge mass (M).
   Note: F = BOD (mg/L) × (8.34 lb/gal) × flow (MGD)
   M = F divided by F/M ratio
3. Select a value of mixed liquor suspended solids (MLSS) concentration in the reactor at the end of DRAW. This is slightly different from designing a conventional continuous flow system. The MLSS concentration in an SBR design corresponds to a particular period in the SBR operating cycle since the concentration changes throughout the cycle. In an SBR, the MLSS concentration is lowest at the end of FILL and highest at the end of DRAW. With most SBR systems, the MLSS concentration at the end of DRAW should be higher than the corresponding value used in the design of a conventional continuous flow

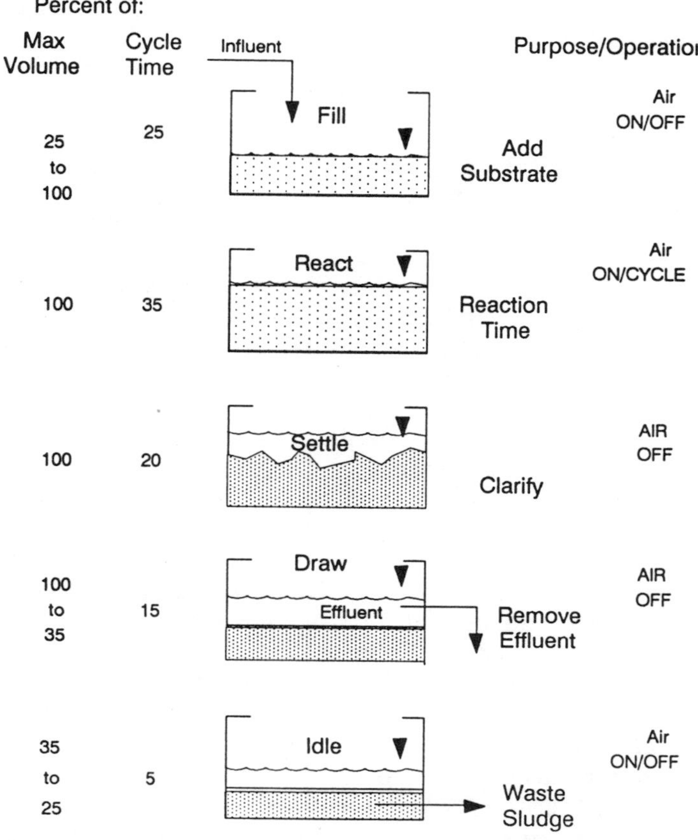

Figure 42.2. Typical SBR operation for one cycle (*source:* reference 42).

system, because the MLSS concentration in the SBR system at the end of DRAW represents a completely settled mixed liquor similar to that in a conventional clarifier underflow. The design mixed liquor volume can then be calculated from the selected MLSS concentration.

4. Select the number of SBR tanks. The number selected will depend on the mixed liquor volume determined in step 3, as well as on considerations of area, unit availability during operation, projected maintenance, and operational flexibility desired. There are no basic rules of judgment in this regard, except that in most cases it is desirable to provide at least two tanks.
5. Select a cycle length, comprised of FILL, REACT, SETTLE, DRAW, and IDLE, for each "batch" treatment. The total time for a cycle will be

the sum (T) of the times allowed for the cycle phases, calculated as follows:

$$T = t_{\text{FILL}} + t_{\text{REACT}} + t_{\text{SETTLE}} + t_{\text{DRAW}} + t_{\text{IDLE}}$$

The time for FILL, t_{FILL}, can be calculated from the peak daily flow rate, considering the number of tanks. The combined times for SETTLE, t_{SETTLE}, and for DRAW, t_{DRAW}, can be estimated to be less than 3 hours. The time for REACT, t_{REACT}, should be determined from kinetic studies but for domestic wastewater the range of t_{REACT} will generally be between 0.5 and 2 hours. The factor t_{IDLE} is selected to provide the operating characteristics needed so that the active part of the cycle will achieve required performance levels.

6. Calculate the volume of liquid per tank per decant. Daily volume per decant V_d = average flow/number of cycles. Volume per tank per decant = V_d/number of tanks.
7. Calculate the tank size. The total volume required per tank is the sum of the volume of mixed liquor per tank at the end of DRAW and the volume of liquid decanted per tank per cycle. The final dimensions of the tanks can be developed by selecting a reasonable tank depth. In most cases a depth of 15 ft (4.6 m) or less is practical from the standpoint of oxygen transfer efficiency. Also, allowance must be made for appropriate freeboard, usually 3 to 4 ft (0.9 to 1.2 m).
8. Size the aeration equipment. This is done in the same manner as in a conventional continuous flow system, except that since the aeration equipment runs for only a portion of the operating cycle in an SBR system (REACT, or REACT and a part of FILL), the calculated daily oxygen requirement must be met in a shorter time frame. The size of the aeration equipment is therefore larger than that of a conventional continuous flow system of the same capacity.
9. Size the decanter and associated piping. The decant rate is calculated from the maximum volume of liquid decanted per tank per cycle. This volume is then divided by the desired decant time, t_{DRAW}. The DRAW period is typically chosen to be approximately 45 minutes.
   Note: Volume of tank = Volume mixed liquor + Volume decant; Area of tank = Volume of tank divided by tank depth.

Factors to be considered that can place constraints on the design process are: the ability to maintain treatment quality in a single tank system, the optimum or maximum sizes for an individual reactor unit in a multi-tank system, desired sludge storage volume.

The design steps outlined above illustrate a simplified approach. In a real situation, many iterative calculations may be necessary to accommodate several conditions simultaneously, such as different MLSS concentrations, different number of operating cycles to achieve flexibility during actual plant operation, diurnal flow variations, and different decant heights to correspond to different conditions of sludge settleability, respectively.

42.2 LIMITATIONS

The SBR is fundamentally a simple treatment system, but since the basis for design is different from traditional continuous flow technology, operation may seem more complicated. In a periodic process such as this, time of aeration and energy input for aeration is very important. The "feast and famine" nature of substrate availability causes varying amounts of biomass to be produced. Higher sludge volume indices (SVI) are possible than for continuous system. The decanter and control systems must be designed to take full advantage of the periodic nature of the process (45).

42.3 COSTS

Table 42.1 (42) presents estimated costs for constructing SBR systems to handle flow rates of 379, 1893, 3785, and 18,925 m³/d (or 1, 5, 10, and 50 MGD, respectively). Table 42.2 (42) defines the operation and maintenance costs. A two tank system is used for the 379 m³/d plant, and a three tank system for the other three daily flow rates. The design criteria for cost purposes are summarized as follows:

FLOW				
(m³/d)	(MGD)	SETS OF TANKS	TANKS PER SET	TOTAL VOLUME (m³)
379	1	2	1	252
1,893	5	3	1	947
3,785	10	3	1	1,893
18,925	50	3	4	9,465

Capital and operation and maintenance costs are lower than the costs for conventional activated sludge.

42.4 AVAILABILITY

The SBR wastewater treatment process is at least a century old. It has to be designed and constructed by experienced engineers based on requirements. SBR wastewater treatment plants are currently operating at several sites in Australia, Canada, and the United States (39). The designs of these plants

Table 42.1. Cost Estimates for SBR for Four Average Daily Flow Rates[a]

PROCESS UNIT	FLOW RATES (m^3/D; MGD IN PARENTHESES) 379 (1)	1893 (2)	3785 (10)	18,925 (50)
Inlet Control System	$ 2,700	$ 4,000	$ 5,300	$ 27,000
Contact Chamber Baffle Walls	2,700	5,300	6,600	32,000
Aerators	33,000	66,000	80,000	342,000
Excavation, Concrete and/ Handrail	93,000	200,000	330,000	1,130,000
Microprocessors	13,000	13,000	13,000	13,000
Level Control/Monitoring	2,700	5,300	3,300	21,000
Dacant System	12,000	21,000	24,000	120,000
Subtotal (1)	$160,000	$316,000	$ 470,000	$1,650,000
Noncomponent Costs[a]	40,000	79,000	118,000	420,000
Subtotal (2)	$200,000	$395,000	$ 585,000	$2,100,000
Engineering, Construction on Supervision and Contingencies[b]	60,000	118,000	177,000	625,000
Total Installed Capital Costs	$260,000	$513,000	$ 760,000	$2,700,000
Annual Operation and Maintenance	18,000	32,000	53,000	200,000
Present Worth Costs[c]	$494,000	$900,000	$1,450,000	$5,400,000
Costs/(m^3/d)	$ 1,300	$ 475	$ 383	$ 285

[a] At 25% of subtotal (1), includes piping, electrical, instrumentation, and site preparation.
[b] At 30% of subtotal (2).
[c] Present worth computed at 7-3/8% interest rate and 20-year life (PWF = 10.29213). Add present worth O&M costs to Total Installed Capital Costs.
Source: Reference 42.

differ in several aspects including inlet design and aeration/mixing system design but they all operate on sequencing batch principles.

42.5 OPERATION AND MAINTENANCE

SBR technology is the same as activated sludge, but all steps of the process take place sequentially in one tank rather than superimposed, as is the case in a continuous flow system. Typical SBR operation (Figure 42.2) involves filling a tank with raw wastewater or primary effluent, aerating the wastewater to convert the organics into microbial mass, proving a period for settling, discharging the treated effluent, and a period identified as IDLE that represents the time after discharging the tank and before refilling. This configuration allows incoming flow to be switched to one tank while the other is going through the aeration, clarification, and discharge functions.

A key element in the SBR process is that a tank is never completely emptied; a portion of settled solids is left in the tank for the next cycle. The

Table 42.2. Operation and Maintenance Cost Estimates for the SBR for Four Average Daily Flow Rates[b]

	FLOW RATES (m^3/D; MGD IN PARENTHESES)			
COST (DOLLAR/YR)	379 (1)	1983 (5)	3785 (10)	18,925 (50)
Operation Labor	$10,600	$13,600	$20,000	$ 45,000
Maintenance Labor	1,800	2,600	3,200	8,000
Power[a]	3,000	13,000	25,000	130,000
Material	2,600	3,500	5,100	17,000
Total O&M (rounded)	$18,000	$32,000	$53,000	$200,000

[a] Includes mixing, aeration, and decanting at a power rate of $0.06/kWh.
[b] Costs based on January 1990.
Source: Reference 42.

remaining portion of this residue (sludge) is transferred to the sludge handling facility. The fraction of wasted sludge will depend upon the desired sludge age.

The retention of sludge within the tank establishes a population of microorganisms uniquely suited to treating the waste. During the process, the microorganisms are subject to periods of high and low oxygen and high and low food availability. This condition develops a population of organisms which is very efficient at treating the particular wastewater.

The maintenance requirements for SBR treatment process systems are similar to activated sludge systems. In addition there are timing devices for control of the aeration systems; electronic control systems include automatic switches, valves, and mechanical devices to implement the process components for each phase of the cycle.

Operators must be trained for SBR operation. Conventional treatment plant operators will require retraining. Overall, the requirements would be similar to those for conventional activated sludge treatment plant operators.

42.6 CONTROL

The SBR treatment process is designed for BOD, suspended solids, nitrogen, and phosphorous removal. As mentioned earlier, in a single tank system the desired treatment phases can be achieved through simple operations. The more complex (and more common) SBR system will involve multiple tanks with capability for adjustable times and liquid levels; dissolved oxygen and turbidity control; flow regulating valves; and compressors, mixers, and pump controls.

Process control is simple but not straightforward. Extensive use is made of motorized control valves, level sensors, automatic timers, and other devices.

Because of the increasing availability of control devices, the system can be over-controlled from the point of view of automatic timing devices, for example. Fixed time sequencing limits treatment flexibility in each of the phases and can adversely affect process performance. On the other hand, the use of fixed treatment sequence timing cycles results in simpler operation. Timing cycles may be interfaced using a microcomputer, allowing the operator to superimpose sequencing modifications based on current wastewater influent and effluent quality and other factors.

42.7 SPECIAL FACTORS

Factors to be considered that can place constraints on the design of the process include the ability to maintain treatment quality in a single tank system, the optimum or maximum sizes for an individual reactor unit in a multiple tank system, and desired sludge storage volume.

There are many factors arguing for use of SBRs. An SBR tank serves as an equalization basin, smoothing load and flow variations. Because effluent discharge is periodic, it is often possible to hold the processed wastes until desired quality is achieved, thus improving ability to meet discharge quality requirements of permits. Design life — system utilization characteristics are improved; early in the design life when flows are low, level sensors may be set lower and the system capacity adjusted to incoming flow rates. Aeration rates and energy utilization may be accordingly lowered. Hydraulic overloading (and thus possible washout of mixed liquor solids for example) is virtually eliminated (in multiple tank systems only) since the treated volume may be held in a tank as long as necessary. There is no need for return sludge recycle or the pumping and valving required. Solid and liquid separation occurs under nearly ideal quiescent conditions and is thus found to be very effective. The fill-and-draw nature of the process possesses inherent potential for the sequential anoxic/oxic phasing necessary for biological phosphorus and nitrogen removal. Effective nutrient removal has been demonstrated in several full scale plant demonstrations.

42.8 RECOMMENDATIONS

The capital cost consideration is playing an increasingly important role for new plant and upgrading programs. The SBR wastewater treatment technology is worthy of consideration because of high treatment performance, likelihood of more trouble free operation; it may be more economical from an operation and maintenance point of view.

43. DRAFT TUBE SUBMERGED TURBINE AERATION (DTSTA)

43.1 DESCRIPTION

Draft tube submerged turbine aeration (DTSTA) consists of a down-pumping, airfoil type axial flow impeller (Figure 43.1) and a compressed air supply (blowers, valves, and piping). Coarse air bubbles are sheared into smaller bubbles by energy provided by the high pumping rate. The manufacturers claim that the system provides superior transfer efficiencies due to high oxygen dissolution from the air phase. DTSTA is employed in conventional completely mixed configurations and in the "barrier oxidation ditch" process which is discussed herein (82).

In conventional completely mixed applications DTSTA is normally supported from a deck inside the tank. Deep tanks, 25 to 30 ft (7.6 to 9.1 m) water depth are used. Horsepower requirements range from 0.15 to 0.20 horsepower (111 to 150 W) per 1000 gallon tank volume. Pumping rates are such that the entire tank volume is turned over about 12 to 15 times per hour.

The two principle features of the barrier oxidation ditch using DTSTA are the use of a U-Tube conduit and a barrier wall (Figure 43.2). Pumping rates are set to cycle basin contents 6 to 8 times per hour.

The DTSTA advantages include:

- highly efficient O_2 transfer:
  Standard Oxygen Transfer Efficiency = 50 to 60%
  Actual Oxygen Transfer Efficiency = 37 to 43%.
- highly efficient mass transfer per horsepower,
- independent control of aeration and mixing,
- minimal basin heat loss,
- stable operation over a wide range of liquid levels.

The U.S. EPA conducted a technological assessment of facilities using DTSTA for oxygenation of activated sludge, to evaluate process performance, mechanical reliability, system operation, and design features. Six facilities were evaluated including those located at Cranston RI, Fairfield IA, Santa Fe NM, Atmore AL, Foley AL, and Presque Isle ME. The Cranston RI

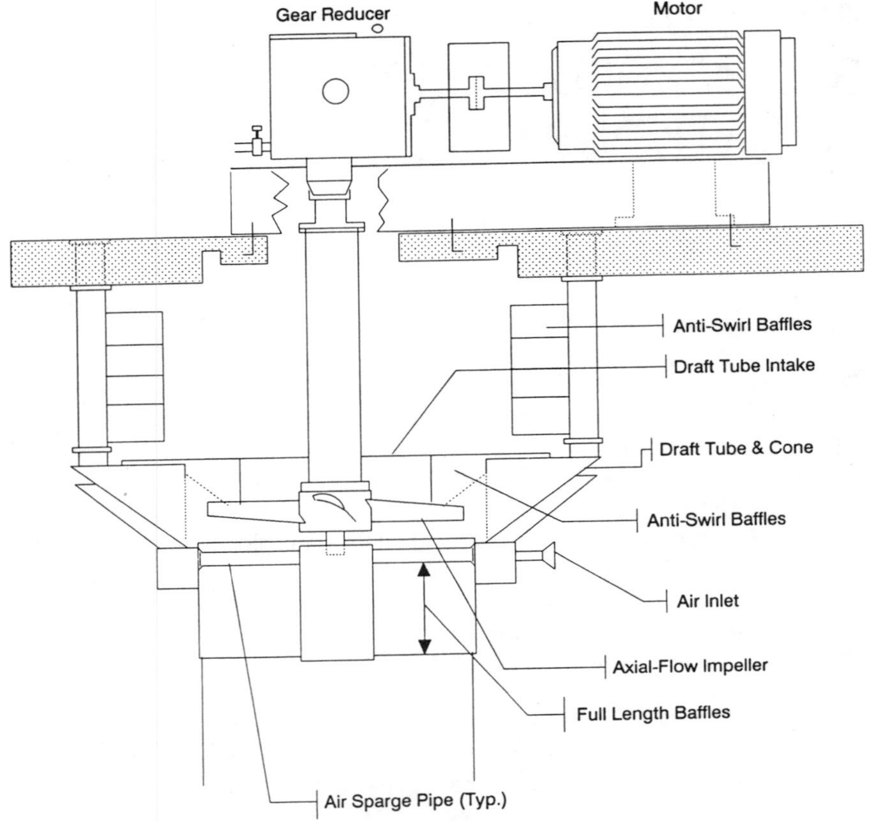

Figure 43.1. Draft tube submerged turbine aerator.

plant employed DTSTA in the completely mixed mode, all others were barrier oxidation ditch applications.

43.2 FINDINGS

Average Standard Aeration Efficiency (SAE) values for DTSTA equipment tested in the factory (3.81 lb O_2/hr/bhp) were over double the average SAEs for field tested equipment (1.87 lb O_2/hr/bhp). It appears that the energy loss of the U-Tube configuration is a very important factor in the design of DTSTA systems.

All facilities studied exhibited a decrease in performance efficiency over time under process conditions. Average reductions for the six plants evaluated are as follows:

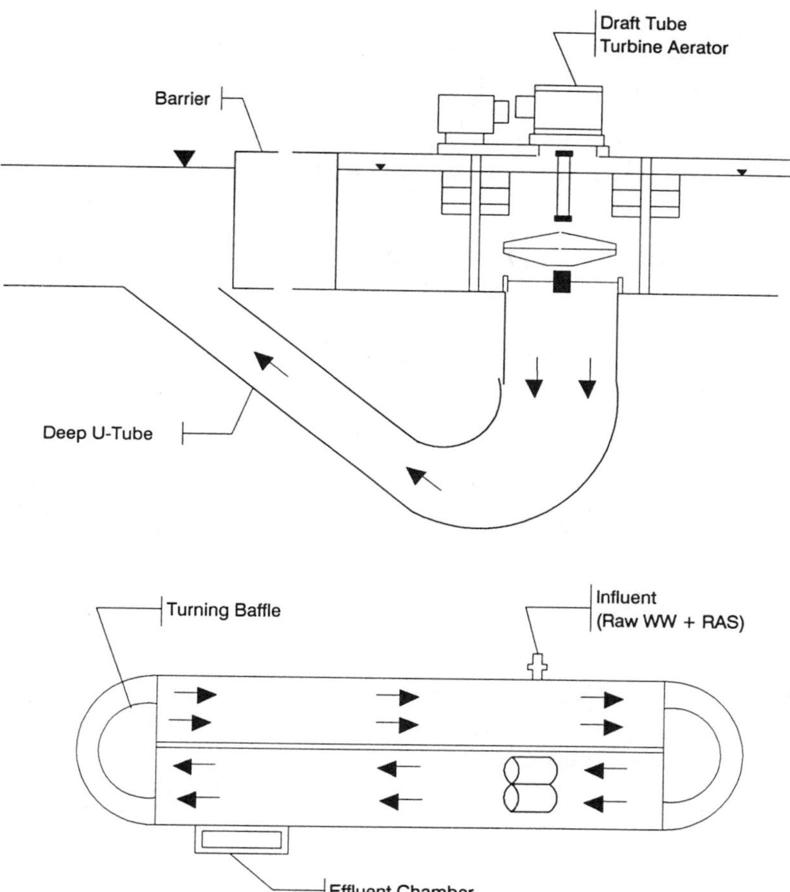

Figure 43.2. Barrier oxidation ditch.

turbine power increase—35 to 40%,
pumping rate decrease—12 to 18%,
air handling capacity decrease—23 to 33%,
SAE decrease—38 to 45%.

Also, the carbon steel impellers at the Cranston RI facility showed significant blade wear attributable to cavitation and erosion and/or corrosion. The DTSTA units at this facility are comparatively high speed, small diameter units. This may have accounted for increased impeller wear.

The decrease in DTSTA performance over time is attributable to the accumulation of minute debris on the leading edge of the impeller which adversely changes its hydrodynamic characteristics. A nonfouling impeller design has been introduced which eliminates the time related deterioration of DTSTA units. The new impeller has a swept back blade and is made of 304 stainless steel. Also, some of the newer designs use a Teflon® material on the leading edge to reduce friction and help with debris shedding.

Manufacturers of DTSTA equipment continue to address performance problems with emphasis being placed on side by side, clean and dirty (wastewater) testing. Recent process O_2 transfer tests reveal that the alpha factor for DTSTA equipment may be in the range of from 0.40 to 0.50 which is considerably less than values of 0.70 to 0.90 cited in the specifications for the six facilities evaluated and those normally found in the literature.

43.3 POTENTIAL BENEFITS

DTSTA is well suited to aeration of deep tanks; deep basins conserve land area. The air sparger is placed at mid-depth which allows deep tank aeration at conventional blower pressures (10 to 15 ft of water). DTSTA eliminates ice and mist formation and conserves basin heat energy. When employed in the barrier oxidation ditch configuration DTSTA offers the following benefits:

- point source aeration of all mixed liquor,
- adjustable channel velocities (0.5 to 2.0 ft/sec, 0.15 to 0.9 m/sec) independent of O_2 transfer rate,
- water level adjustment (1 to 6 ft, 0.3 to 1.8 m) for flow equalization.

43.4 RECOMMENDATIONS

DTSTA equipment is used in conventional completely mixed activated sludge configurations and applied to the barrier oxidation ditch process. The equipment is applicable to aerobic digestion and for post aeration of treated wastewater.

44. INTERMITTENT SAND FILTERS (ISFs)

44.1 DESCRIPTION

Intermittent sand filters (ISFs) are beds of medium to coarse sands, from 2 to 4 ft (0.6 to 1.2 m) deep and underlain with gravel containing underdrains. Effluent is intermittently applied to the surface, and purification of the effluent occurs as it infiltrates and percolates through the sand bed. A resting period is provided and thus multiple beds are required. Underdrains collect the filtrate and convey it to additional treatment processes and/or discharge. The full scale use of (ISFs) as a secondary wastewater treatment process is not a new technology. They were frequently used by sewered communities in the USA around the turn of the century. ISF design concepts include buried filters, open single pass filters, and open recirculating sand filters (19,66).

Buried sand filters are constructed below grade and covered with backfill material. Buried filter designs are most commonly used for small flows, such as those from single homes and small commercial establishments. These filters are designed to perform for very long periods of time, up to 20 years and are similar to buried sand filters except that the surface of the filter is left exposed and higher hydraulic and organic loadings are generally applied (see Figure 44.1). These filters are used for individual homes as well as larger flows from small communities or industries (up to 0.2 MGD, 0.01 m^3/sec). Recirculating sand filters are open filters which utilize somewhat coarser media and employ filtrate recirculation. A portion of the filtrate is diverted for further treatment or disposal during each dose. Recirculating ratios of 3 : 1 to 5 : 1 are typical.

ISFs can produce high quality effluents. Concentrations of effluent BOD_5 and TSS are typically less than 10 mg/L with ammonia nitrogen less than 5 mg-N/L. Only limited removal of phosphorous and fecal coliform bacteria are achieved. Design considerations important for achieving this level of treatment include pretreatment, sand characteristics, hydraulic and organic loading rates, temperature and filter dosing techniques.

44.1.1 Sand Characteristics

Sizes. Sand with an effective size of 0.4 to 1.5 mm and a uniformity coefficient not greater than 4.0 is satisfactory.

Depth. Media depths used in intermittent sand filters were initially 4 to 10 ft (1.2 to 3 m). However, studies revealed that most of the purification of wastewater occurred within the top 9 to 12 in. (23 to 30 cm) of the bed. It is critical to maintain sufficient sand depth so that the capillary zone does not infringe on the zone required for treatment. For these reasons, most media depths used today range from 24 to 42 in. (62 to 107 cm). The use of shallower filter beds helps to keep the cost of installation low. Deeper beds tend to produce a more consistent effluent quality.

44.1.2 Pretreatment

The operation and performance of ISFs are directly related to the degree of pretreatment of the applied wastewater. There appears to be a direct relationship between degree of pretreatment, the wastewater hydraulic performance over the long term and effluent quality.

44.1.3 Rates

The allowable loading of ISFs varies with the nature of the sand, the strength of the sewage, and the method of pretreatment. Average values are presented below:

Maximum Allowable Loading of ISFs

APPLIED SEWAGE	GPD	PERSONS/ ACRE[1]	lb BOD DAILY
Untreated	20,000 to 80,000	250 to 1,000	Up to 75
Settled	50,000 to 125,000	50 to 1,500	—
Biologically treated	Up to 50,000[2]	Up to 5,000	—

[1] Sewage flow = 80 to 100 gpcd.
[2] Up to 125,000 gpd and 5,000 persons for small plants with unskilled supervision.
Note: 2.4 acres = 1 ha

44.1.4 Temperature

Temperature directly affects the rate of microbial growth, chemical reactions, absorption mechanisms, and other factors that contribute to the stabilization of wastewater within an intermittent sand filter. Somewhat better operation and performance, therefore, can be expected from filters in warmer locales.

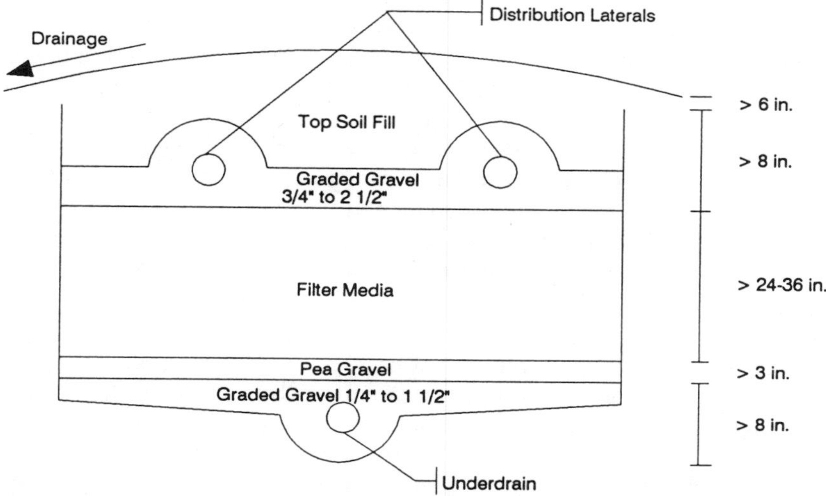

Figure 44.1. Typical buried sand filter.

44.1.5 Dosage

A sufficient amount of sewage is run onto the bed to cover it to a depth of 2 to 4 in. (5 to 20 cm). The higher end of the range tends to load large beds more uniformly. Dosage is regulated so that flooding is completed in 10 to 20 min. The necessary rate of discharge is about 0.2 cfs/1,000 ft^2 (6.1×10^{-5} $m^3/m^2/sec$) under the average available head. Each dose carries from 50,000 to 100,000 gal (190 to 380 m^3) of sewage onto an acre of bed. Two or more doses are applied daily, depending upon the rate at which the sewage is absorbed. Sufficient time must be allowed for recovery of the bed by reaeration (resting period). Dosing is controlled by hand operation of gates or by automatic dosing tanks. Dosing tanks store the full dose of a bed. Siphons, or other regulating devices, are sized large enough to discharge under minimum head, at about twice the maximum expected rate of inflow. Intermittent operation is thereby assured.

Construction of intermittent sand filters involves: 1) clearing and grubbing the area and stripping the top soil; 2) leveling the area and subdividing it into beds of suitable size by building up the soil into partition embankments and access embankments; 3) underdraining the beds; 4) laying distribution mains and constructing distribution manholes and inlets in the access embankments; 5) building sewage carriers or troughs on the beds.

Shape and arrangement of beds depend upon local topographic conditions. Bed size and number should be kept consistent with the storage needed for the intermittent operation of the beds. Also, it must be possible to take at least one bed out of service for cleaning or repairs without overloading the remaining area. Beds more than an acre (0.4 ha) in size are uncommon.

Sewage is usually conveyed to the beds in vitrified tile or concrete pipe, sometimes in a cast iron force main. The conduits are laid in the access embankments which are made wide enough (8 to 10 ft; 2.4 to 3 m) for necessary construction and maintenance equipment. Ramps lead onto each bed. The height of the access embankment is determined by the necessary hydraulic gradient of the influent sewer and depth of cover. A depth of 2 ft (0.6 m) will protect the pipe against breakage and freezing in all but very cold climates. The partition embankments are often low (12 to 18 in.; 30 to 59 cm) but wide enough for a foot path (2 ft; 0.6 m). Embankment dimensions are normally governed by the amount of soil to be disposed of.

Sewage is discharged onto individual beds through branch pipes, generally connected to the main distributor at manholes and controlled by shear or sluice gates inside the manhole. One or more outlets are so placed that lateral travel of the applied sewage is not more than 20 ft (6.1 m). To prevent erosion of the sand surface, the sewage discharges onto a concrete or stone apron covering the nearby surface of the bed or into a trough or carrier made of wood or concrete running almost across the bed and equipped with side outlets, deflectors, and splash slabs.

44.1.6 Perforated Underdrains

Pipes of vitrified tile or concrete or pipe laid with open joints (about ⅜ in.; 1 cm) separation, are placed at least 2 ft (0.6 m) below the sand surface. Underdrain minimum diameter is 4 in. (10 cm), and spacing preferably not more than 10 ft (3 m) apart. The greater the depth and the coarser the sand, the wider can be the spacing. In order to ensure full aeration of the bed after dosing, underdrains are sloped to the outlet to discharge sewage at the maximum rate of percolation. The values given below are ranges of percolation rates:

Maximum Rates of Percolation of Water Through ISFs
(MGD × 0.044 = m^3/sec)

Effective size, mm	0.2	0.3	0.4	0.5	0.6
Maximum rate of percolation, mgd/acre					
at 50°F	20 to 50	50 to 100	100 to 200	150 to 300	200 to 400
at 70°F	30 to 60	70 to 150	120 to 150	200 to 400	300 to 600

As a rule, the lower values are encountered. Underdrains are carefully laid to line and grade in open trenches in the sand bed. The pipe is surrounded by successive layers of coarse stone or gravel on coarse sand.

44.2 ADVANTAGES AND DISADVANTAGES

Advantages are: reasonably low construction cost requiring largely manual labor; design and operation are simple; unskilled maintenance labor is required; no chemicals are required and sand can usually be found locally; low power requirements; high quality effluent.

Disadvantages are: pretreatment (at least sedimentation) is probably required for most applications; there may be odor problems from open filter configurations; systems require substantial land area; colder temperatures can dramatically affect system cost and performance adversely; application is probably not feasible in the absence of natural sand deposits; i.e., construction of sand filters in the absence of natural, sandy areas is unusual.

44.3 COSTS

Costs (both Capital and O&M) of ISFs, exclusive of land cost, are presented in Figure 44.2 (19, 66).

44.4 AVAILABILITY

The technology, equipment and hardware typically utilized in ISFs should be available locally in most areas of the world. The critical component is the sand which often is available locally.

44.5 OPERATION AND MAINTENANCE

ISFs are not backwashed and thus solids accumulate. A mat of solids is deposited on the surface of the bed. On drying, this mat cracks and curls and may be stripped from the bed. The solids remaining after presettling may penetrate further into the sand. This necessitates occasional removal of the surface sand and resanding the bed. Sludge mat and waste sand may be used locally for fill. Raking, harrowing, and plowing will open up a clogged bed but may ultimately intensify clogging by forcing solids deeper into the sand. Light harrowing is useful in releveling beds and loosening sand that has been packed down by heavy rains (for open filters only). Buried filters should be designed to perform without maintenance for up to 20 years.

Pooling should not be allowed to develop on the beds as this tends to produce septic action, obnoxious odors, and an effluent of poor quality.

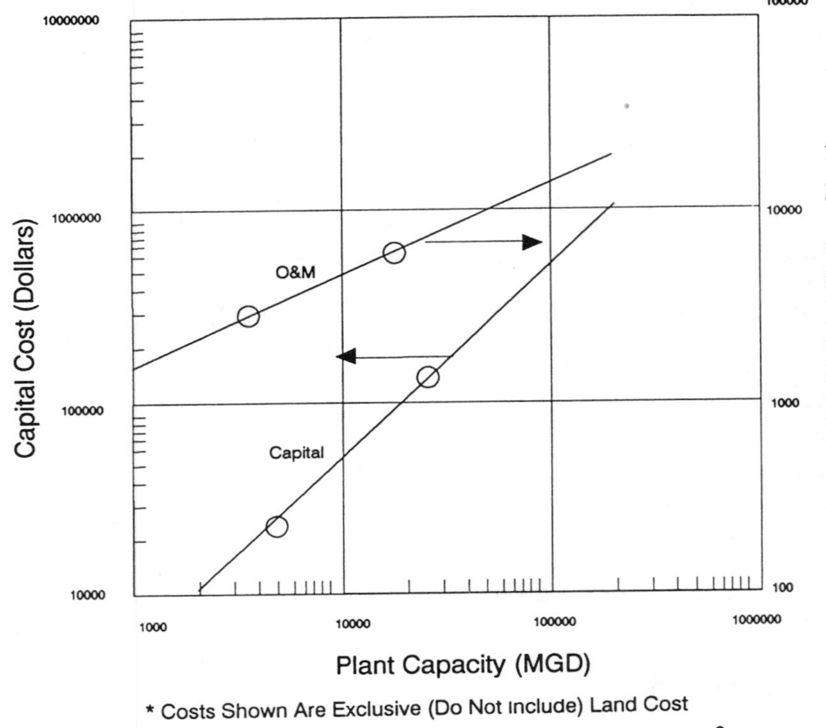

Figure 44.2. Intermittent sand filter costs.*

Pooling indicates that cleaning is necessary. The surface of the beds should be kept level to afford uniform distribution of the sewage; weeds, grass, etc., should not be allowed to grow on the beds. Slight odors are unavoidable in plants filtering raw sewage.

In cold climates, winter operation often requires furrowing the surface and dosing the bed deeply on a cold night so that a sheet of ice will form and span the furrows between the ridges. The bed itself is thus kept from freezing, and the sewage underruns the ice sheet. Beds should be thoroughly cleaned before and after they have been furrowed.

44.6 CONTROL

ISFs require relatively little operational control. Once wastewater is applied to the filter, the sand "matures" in a few days to several weeks. BOD and SS concentrations in the effluent will normally drop rapidly after maturation.

Depending upon media size, rate of application, and ambient temperature, nitrification may take from 2 weeks to 6 months to develop. Cold weather start-up should be avoided since the biological growth on the filter media may not develop properly.

A small measure of process control can be achieved through variations in loading rate, dosing cycle, resting period, and frequency of cleaning.

44.7 SPECIAL FACTORS

Intermittent sand filters should be preceded by some form of pretreatment. Septic tanks, Imhoff tanks, primary sedimentation, trickling filters, and activated sludge can be used for pretreatment depending on the level of effluent quality desired. Filter effluent may be disinfected.

Care must be exercised when siting intermittent sand filtration systems in natural sand deposits to ensure that percolation into the groundwater table and subsequent pollution of groundwater aquifers does not occur. In areas where percolation into an aquifer could pose a threat to groundwater quality, the filter system may be provided with an impermeable base or lining. Construction costs are higher in these cases.

44.8 RECOMMENDATIONS

Intermittent sand filtration is a well proven and a relatively low cost method of wastewater treatment which produces a high quality effluent. The process is highly reliable and requires minimal operator attention. Energy requirements are low, less than 0.3 hp-hr/100 gallons (0.55 KWhr per m^3), and operation and maintenance can be performed by unskilled labor. Removal efficiencies for BOD and SS can exceed 95% and concentrations of ammonia nitrogen in the filter effluent are normally less than 5 mg-N/L with appropriate pretreatment. The process is best suited for treatment of municipal wastes for flows generally less than 0.2 MGD (0.01 m^3/sec).

REFERENCES FOR PART I

1. Reid, G. and Coffey, K., *Appropriate Methods of Treating Water and Wastewater in Developing Countries,* Bureau of Water and Environmental Resources Research, The University of Oklahoma, Norman, OK, 1978, for U.S. Agency for International Development, Wash. D.C.
2. U.S. Environmental Protection Agency, Office of Research and Development, "Innovative and Alternative Technology Assessment Manual," February, 1980.
3. Montgomery, J. M., *Water Treatment Principles and Design,* Consulting Engineers, Inc., John Wiley & Sons, Inc., New York, 1985.
4. *Wastewater Engineering: Treatment, Disposal, Reuse,* Metcalf & Eddy, Inc., McGraw-Hill, New York, 1979.

5. U.S. Environmental Protection Agency, "Estimating Water Treatment Costs, Volume 2, Cost Curves Applicable to 1 to 200 MGD Treatment Plants," August 1979.
6. White, G., *Handbook of Chlorination,* Second Edition, Van Nostrand Reinhold, New York, 1986.
7. Liguori F., "Small Water Systems Serving the Public," U.S. Environmental Protection Agency, July 1978.
8. Banco Interamericano De Desarrollo (Interamerican Development Bank, Washington, D.C.), *Organizacion Panamericana De La Salud, Tecnologias Apropiadas en America Latina Y El Caribe, Columbia, (Vols. I and II),* Bogota, Marzo 28 de 1987.
9. Personal Communication with Jose M. De Azevedo Netto, Sao Paolo, Brazil, June 1987.
10. Martin, E. J. and Martin, E. T., *Examination of the Water Supply and Sewerage Rehabilitation Needs for Selected Cities in Ecuador,* October 1983, for the Interamerican Development Bank, Wash. D.C., 1983.
11. Martin, E. J. and Martin, E. T., Water and Wastewater Cost Analysis Handbook for Latin America and the Caribbean, 1985, Pan American Health Organization (WHO, Washington, D.C.) and the Interamerican Development Bank, Wash. D.C.
12. Martin, E. J. and Kondracki, M. G., Technical Analysis for the Expansion of Potable Water Supply of Metropolitan San Salvador, El Salvador, May 1984, for the Interamerican Development Bank, Wash. D.C.
13. Martin, E. J., Oppelt, E. T., and Smith, B. P., "Chemical, Physical, Biological (CPB) Treatment of Hazardous Wastes, Fifth United States-Japan Governmental Conference on Solid Waste Management," U.S. Environmental Protection Agency, September 1982.
14. U.S. Environmental Protection Agency, "An Emerging Technology Aquaculture. An Alternative Wastewater Treatment Approach," July 1983.
15. U.S. Environmental Protection Agency, "Aquaculture Systems for Wastewater Treatment," Seminar Proceedings and Engineering Assessment, September 1979.
16. U.S. Environmental Protection Agency, "An Emerging Technology, Wetlands Treatment," A Practical Approach, September 1983.
17. Clark, J. W., Viessman, W., and Hammer M., *Water Supply and Pollution Control,* Third Edition, Donnelley, New York, 1977.
18. U.S. Environmental Protection Agency, "Disinfection with Ultraviolet Light, Design, Construct, and Operate for Success," September 1986.
19. U.S. Environmental Protection Agency, Office of Water Program, "Design Manual, Onsite Wastewater Treatment and Disposal Systems," October 1980.
20. Carlson, D. and Seabloom, R., et al., "Ultraviolet Disinfection of Water for Small Water Supplies," U.S. Environmental Protection Agency, Project Summary, September 1985.
21. *American Water Works Association Journal,* Several papers, Vol. 78, July 1986.
22. Arboleda, J., "Latin American Experience in the Design and Construction of Water Treatment Plants, A New Approach," American Water Works Association, July 1983.
23. Reid, G. W., "Appropriate Methods of Treating Water and Wastewater in Developing Countries," Ann Arbor Science, Ann Arbor, Michigan, 1982.
24. U.S. Environmental Protection Agency, "Design Manual, Municipal Wastewater Stabilization Ponds," October 1983.
25. Schulz, C. R. and Okun, D. A., *Surface Water Treatment for Communities in Developing Countries,* John Wiley & Sons, New York, 1984.
26. U.S. Environmental Protection Agency, "An Emerging Technology, Overland Flow, A Viable Land Treatment Option," May 1982.
27. U.S. Environmental Protection Agency, "Overland Flow Update, New Information Improves Reliability," October 1984.
28. Brazilian Stabilization-Pond Research Suggests Low-Cost Urban Applications, World Water, July 1983.
29. Pan American Health Organization (PAHO), "Evaluation of the Utilization of New Tech-

nology in Water Treatment in Latin America," Seventeenth Meeting of the PAHO Advisory Committee on Medical Research, Lima, Peru, May 1978.
30. Ranney Method, Developed by Ranney Western Corporation, Kennewick, Washington, 1985.
31. U.S. Environmental Protection Agency, "Granular Activated Carbon Systems, Problems and Remedies," August 1984.
32. U.S. Environmental Protection Agency, "Tertiary Granular Filtration, Problems and Remedies," August 1984.
33. Boyle, W. C. and Wallace, A., "Status of Porous Biomass Support Systems for Wastewater Treatment: An Innovative/Alternative Technology Assessment," U.S. Environmental Protection Agency, April 1986.
34. World Health Organization, "Horizontal-flow Roughing Filtration: An Appropriate Pretreatment for Slow Sand Filters in Developing Countries," WHO International Reference Centre for Wastes Disposal, *IRCWD NEWS,* August 1984.
35. Lin, S. D., Schnepper, D., and Evans, R., "A Close Look at Changes of BOD_5 in an RBC System," *Journal of Water Pollution Control Federation,* Vol. 58, No. 7, July 1986.
36. Brenner, R., Heidman, J., Opatken, E., and Petraser, A., "Design Information on Rotating Biological Contractors," Project Summary, U.S. Environmental Protection Agency, July 1984.
37. Echelberger, W. and Zogorsri, J. et al., "Rotating Biological Contactors Hydraulic Versus Organic Loading," Project Summary, U.S. Environmental Protection Agency, April 1986.
38. U.S. Environmental Protection Agency, "Rotating Biological Contactors (RBCs), A Checklist for a Trouble-Free Facility," September 1983.
39. U.S. Environmental Protection Agency, "Review of Current RBC Performance and Design Procedures," Project Summary, June 1985.
40. Eckstein, D., "Steep Slope Sewer Construction Criteria," Public Works, May 1987.
41. U.S. EPA, "An Emerging Technology, Sequencing Batch Reactors," September 1983.
42. U.S. EPA, "Sequencing Batch Reactors," Summary Report, October 1986.
43. *Journal Water Pollution Control,* Vol. 51, No. 2, February 1979, pp. 235–305.
44. "Treatment of Wastewater by Batches Saves Money," *Chemical Engineering,* January 1985.
45. Levine, R., et al., "Analysis of Full-Scale SBR Operation at Grundy Center," Iowa, Journal Water Pollution Control Federation, Part 1 of 2, Vol. 59, No. 3, March 1987.
46. Levine, R., "Technology Assessment of Sequencing Batch Reactors, Project Summary," U.S. EPA, March 1985.
47. Morris, R. C., "Simulating Batch Processes," *Chemical Engineering,* May 16, 1983.
48. Culp, R., Wesner, G. M., and Culp, G., *Handbook of Public Water System,* Van Nostrand Reinhold, N.Y., 1986.
49. *Wastewater Treatment and Plant Design,* Water Pollution Control Federation, Second Edition, 1982.
50. Passavant Corporation, "Cross Counterflow LME/Inclined Plate Separator," Birmingham, Alabama.
51. U.S. EPA, "Handbook, Estimating Sludge Costs," October 1985.
52. American Society of Civil Engineers (ASCE) *Manual and Report on Engineering Practice.* No. 60.
53. Water Pollution Control Federation (WPCF) *Manual of Practice NOFD-5,* "Gravity Sewer Design and Construction," The ASCE and the WPCF, April 1982 Revision.
54. R. S. Means Company, Inc., Annual Edition, "Means Site Work Cost Data," 1987, 1988, 1989, 1990.
55. Kemmer, F., *The NALCO Water Handbook,* Second Edition, McGraw Hill, New York, 1988.
56. U.S. EPA, "Tertiary Granular Filtration, Problems and Remedies," August 1984.

57. Letterman, R. and Cullen, T., "Slow Sand Filter Maintenance: Costs and Effects on Water Quality, Project Summary," U.S. EPA, August 1985.
58. U.S. EPA, "Wastewater Treatment and Disposal, Total Containment Ponds," March 1985.
59. U.S. EPA, Office of Water, "Aquaculture Systems for Wastewater Treatment, An Engineering Assessment," June 1980.
60. Trabajos De Investigacion Realizados En Eris, Arturo Pazos, Escuela Regional De Ingenieria Sanitaria, Y Recursos Hidraulicos," Universidad De San Carlos De Guatemala, Noviembre, 1986.
61. U.S. EPA, "A Practical Technology, Land Application of Sludge," September 1983.
62. Loehr, R., et al., *Land Application of Wastes,* Vols. I and II, Van Nostrand Reinhold, New York, 1979.
63. Ranney Western Corporation, Kennewick, Washington, 1985.
64. "Brazilian Stabilization pond research suggests low-cost urban applications," World Water, July 1983.
65. Kroftan Co., Sandfloat, System, 1988.
66. Anderson, D., et al., "Technology Assessment of Intermittent Sand Filters, Municipal Environmental Research Laboratory, Office of Research and Development," U.S. Environmental Protection Agency, April 1985.
67. Martin, E. J. and Johnson, J. H., *Hazardous Waste Management Engineering,* Van Nostrand Reinhold, New York, 1987.
68. Clifford, D. and Lin, C. C., "Arsenic (III) and Arsenic (V) Removal from Drinking Water in San Ysidro," New Mexico, U.S. Environmental Protection Agency, September 1988.
69. American Society of Civil Engineers and the Water Pollution Control Federation, "Sewage Treatment Plant Design," United Engineering Center, New York, 1961.
70. Health Education Service, "Recommended Standards for Sewage Works," 1973.
71. Imhoff, K. and Fair, G., *Sewage Treatment,* John Wiley & Sons, Inc., New York.
72. Metcalf, L. and Eddy, H., *American Sewerage Practice, Vol. III, Disposal of Sewage,* McGraw-Hill, New York, 1916.
73. Eckenfelder, W., *Principles of Water Quality Management,* CBI Publishing Co., Inc., Boston, 1980.
74. New York State Department of Health, "Manual of Instruction for Sewage Treatment Plant Operators," Health Education Service, New York.
75. U.S. EPA, "Process Design Manual, Land Treatment of Municipal Wastewater," EPA625/1-81-013, Center for Environmental Research Information, Cincinnati, Ohio, October 1981.
76. U.S. EPA, "Process Design Manual, Land Treatment of Municipal Wastewater, Supplement on Rapid Infiltration and Overland Flow, Center for Environmental Research information, Cincinnati, Ohio, October 1984.
77. U.S. EPA, "Cost of Wastewater Treatment by Land Application," EPA-430/9-75-003, Office of Water Programs, Washington, D.C.
78. Engineering Improvement Recommendation System (EIRS), EIRS Bulletin 88-03, U.S. Army Corps of Engineers, March 31, 1988, and Newsletter update, January 10, 1990.
79. Logsdon, A. S., Sorg, T. J., and Clark, R. M., "Capability and Cost of Treatment Technologies for Small Systems, JAWWA, Vol. 82, No. 6, June 1990.
80. U.S. EPA, Process Design Manual, Wastewater Treatment Facilities for Several Small Communities, EPA-625/1-77-009, October 1977, Environmental Research Information Center.
81. Arora, M. L., Barth, E. F., and Umphres, M., "Technology Evaluation of Sequencing Batch Reactors," WPCF Journal, Vol. 57, No. 8, August 1985.
82. Updegraff, K. F., "Technology Assessment of Draft Tube Submerged Turbine Aeration, Presented Field Evaluation of Innovative Alternative," Technologies Seminar Series, U.S. EPA, 1985.

PART II INTRODUCTION

The technologies in this section are presented because they are relevant for application to small systems in the U.S. and for developing countries. Note that in some cases additional information is presented in Part I. The technologies are considered for application on the basis that the results from applications in practice thus far are promising. Each has been applied several times in the U.S. and/or other parts of the world.

Most of the technologies in this section apply to wastewater, but since a number of them relate to pretreatment for water reuse, the distinction between applications for wastewater and water supply becomes clouded. All of the technologies are applicable to treatment of contaminated water (water supply sources or wastewater). Many sludge treatment technologies are applicable to both water and wastewater treatment sludges. Figure II.1 (page 296) shows schematically how water reuse may be applied (II-9).

Many developing countries and areas in developed countries are water short. Application of water treatment technologies is a continuum, starting with highly contaminated water and continuing to less contaminated. Systems should be designed for meeting water quality criteria for various water uses and not on the basis of presumed water supply vs. wastewater distinctions. Water used for carrying waste is undoubtedly water destined for drinking at another time and place.

There is necessarily, some overlap with Part I. Part II presents specific examples taken from studies which are underway (refs: II-1, II-2).

INNOVATIVE TECHNOLOGIES

An innovative technology is defined as a water treatment process or technique which has not been fully proven for the proposed application and which offers a significant advancement over the state of the art. In order to qualify as innovative, a technology should meet two conditions. First, the technology or its application must include an inherent risk (performance and/or cost) which is outweighed by a corresponding benefit, thereby making the risk acceptable. If a technology or its application is fully proven, there would be no "risk" involved, and it could not qualify as innovative. The term innovative may also relate to application of an existing technology; if a specific application of a proven technology has not been made at sufficient scale, the application itself may qualify as innovative.

The second condition is that the technology should meet one or more of a series of criteria which measure its advancement over the state of the art. Six criteria may be considered: 1) cost reduction, perhaps in the order of 15% on a life cycle basis; 2) reduction in the use of energy, again on the order of 15%; 3) improved removal or destruction of a toxic substance; 4) improved operational reliability; 5) improved environmental benefits; 6) improved

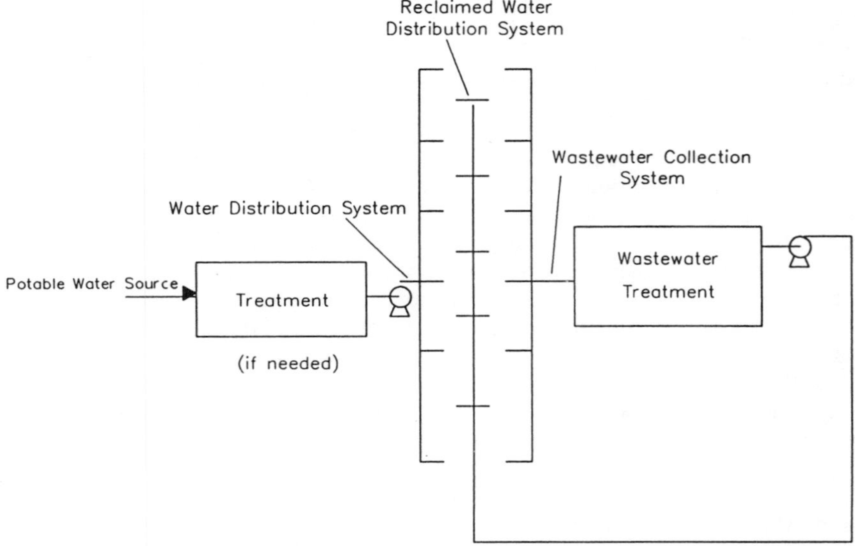

Figure II.1. Dual distribution system schematic.

potential for joint treatment of industrial and municipal wastes, or for water reuse.

The following data may lend perspective to the application of innovative technologies. The distribution of the full range of innovative technologies which have been applied within the last 5 to 10 years is (all technologies are not included):

aeration	11.4%
clarifiers	10.9%
disinfection	9.4%
energy conservation and recovery	5.0%
filtration	5.3%
lagoons	7.8%
land application of effluent	4.4%
nutrient removal	7.7%
oxidation ditches	7.0%
sludge technologies	10.7%

Table II.1 presents a summary of some innovative technology applications in the U.S., broken down by applications in the northern and southern portions. Some, notably countercurrent and draft tube aeration, oxidation ditches, and ultraviolet disinfection have probably been used often enough to classify them as alternative technologies.

Table II.1. Summary of Innovative Technologies Funded Five or More Times

TECHNOLOGY	SOUTHERN U.S., HAWAII AND TERRITORIES	NORTHERN U.S., ALASKA
Anoxic/Oxic Systems	1	4
Counter-Current Aeration	21	—
Draft Tube Aeration	6	15
Fine Bubble Diffusers	2	8
Flocculating Clarifiers	—	5
Hydrograph Controlled Release Lagoons	18	1
Integral Clarifiers	1	4
Land Treatment	11	6
Microscreens	3	3
Oxidation Ditches	21	15
Phostrip Activated Sludge	1	4
Sequencing Batch Reactors (SBR)	7	9
Single Cell Lagoons with Sand Filter	—	10
Small Diameter Gravity Sewers	1	9
Solar Heating	48	
Swirl Concentrators	1	7
Trickling Filter/Solids Contact	4	6
Ultraviolet Disinfection	5	37
Vacuum Assisted Sludge Drying Beds	5	6

ALTERNATIVE TECHNOLOGIES

An alternative technology is a fully proven method of water, wastewater, or sludge treatment that has not been used frequently. It is a technology that: 1) provides for the reclaiming and/or reuse of water, 2) productively recycles treated water constituents, 3) eliminates the discharge of pollutants with water or treatment sludges/residues, and 4) recovers energy. Alternative technology emphasizes land treatment of contaminated water and sludges, sludge/residue handling and disposal techniques that reuse or reclaim pollutants, on-site methods of disposal at the household, industrial park, or small community level and alternative methods of waste conveyance that are especially applicable to small communities.

Composting of sludge and land treatment of wastewater and sludge are perhaps the best known alternative technologies. Some other technologies, although proven, are less known because of infrequent use. Effluent treatment alternative technologies include aquifer recharge, aquaculture, revegetation of disturbed lands, horticulture, direct nonpotable reuse, and total containment ponds. Energy recovery alternative technologies include self-sustaining incineration and anaerobic digestion with greater than 90% methane recovery and use. For small community systems, alternative technologies include individual or cluster on-site treatment, septage treatment, small diameter collection and conveyance systems such as pressure sewers, and some centralized treatment systems.

Alternative technologies are distributed as follows:

land treatment	29.1%
onsite treatment (small systems)	11.7%
collection systems	19.5%
energy recovery form sludge	7.1%
sludge treatment	24.8%
other	7.8%

Applications

Table II.2 lists the alternative technologies which have been implemented in the U.S. within the past several years. These data illustrate the recent heavy use of land treatment for both water and sludge, and energy recovery.

Table II.2. Summary of Alternative Technology Projects

TECHNOLOGY	SOUTHERN U.S., HAWAII AND TERRITORIES	NORTHERN U.S., ALASKA
Land Treatment:		
Aquaculture/Wetlands/Marsh	6	14
Overland Flow	29	20
Rapid Infiltration	26	43
Slow Rate	172	116
Preapplication Treatment or Storage	69	61
Other Land Treatment	2	20
Collection Systems:		
Pressure Sewers/Effluent Pump	25	57
Pressure Sewers/Grinder Pump	31	107
Small Diameter Gravity Sewers	34	117
Vacuum Sewers	4	14
Energy Recovery/Sludge:		
90% Methane Recovery/Anaerobic Digestion	42	85
Self-Sustaining Incineration	5	8
Sludge Treatment:		
Land Spreading of POTW Sludge (Publicly Owned Treatment Works)	74	286
Preapplication Treatment	14	25
Composting	12	41
Other Sludge Treatment or Disposal	11	26
Other:		
Aquifer Recharge	1	1
Direct Wastewater Reuse	15	5
Total Containment Ponds	37	96

System design considerations must include water and energy recovery techniques. Even areas of high rainfall can make use of water reuse technologies because of the seasonal nature of high rainfall volumes. High intensity rainfall common to some areas does not help shortage problems significantly. Even capture of high intensity rainwater is probably not cost effective because of high evaporation rates in warm climates.

Any water or wastewater management program can make effective use of energy and materials recovery (even simple recovery of the soil conditioning benefits of sludge), regardless of the current economic and water availability status of the area of interest.

OTHER APPROPRIATE TECHNOLOGIES

In the following sections, technologies are briefly described which have applicability for small systems in developed countries and developing countries. In each case an attempt has been made to include as much information as possible to allow an evaluation. It may be seen that in general, less information is provided for each technology than in Part I. The information provided is based on actual applications and is provided in the categories of:

- Description—short written description including a diagram in many cases.
- Application—notes on the applicability of the technology and guidance on applicability elsewhere, where possible.
- Benefits—general performance information. Usually, the full results are not available as yet for a particular application. Also, benefits probably outweighed potential disadvantages or the technology would not have been selected for the project.
- Status—notes are sometimes provided for the status of the technology for other locations than the one being tested. This gives perspective on the general use patterns.
- Findings—If a project is in the early stages, data and information about progress is presented.

CASE STUDIES OF OTHER APPROPRIATE TECHNOLOGIES

CASE 1 BARDENPHO PROCESS

C-1.1 Description

The Bardenpho process was originally developed as a four stage system for BOD and nitrogen removal from wastewater where partial removal of phosphorus also occurs. In order to maximize phosphorus removal, an anaerobic stage is added to the front of the four stage process. Nitrogen and phosphorus removal is achieved by carefully controlling the concentration of oxygen in each of the five stages. Nitrogen removal is accomplished by biological denitrification, while phosphorus removal is by microbial uptake into the waste activated sludge. The process is shown schematically in Figure C-1.1.

C-1.2 Application

The Bardenpho process is applicable to wastewater systems where phosphorus and/or nitrogen discharge is of concern. The basic four stage system can be used when the limiting discharge is nitrogen and not phosphorus. By adding the fifth stage, the system can be used where the discharge is phosphorus limited.

C-1.3 Benefit

Chemical addition is not necessary to remove nitrogen and phosphorus but are probably required if low nutrient concentrations in the effluent are necessary. Capital and maintenance costs are lower for non-chemical systems since chemical handling facilities are not required. Operating costs may be reduced if additional digestion equipment is not included; thereby capital, operating, and maintenance costs of sludge treatment and disposal are lower. Operation of the process is similar to conventional activated sludge system operation but is more complicated because of the multiple phases of operation. The long solids retention time provides process stability.

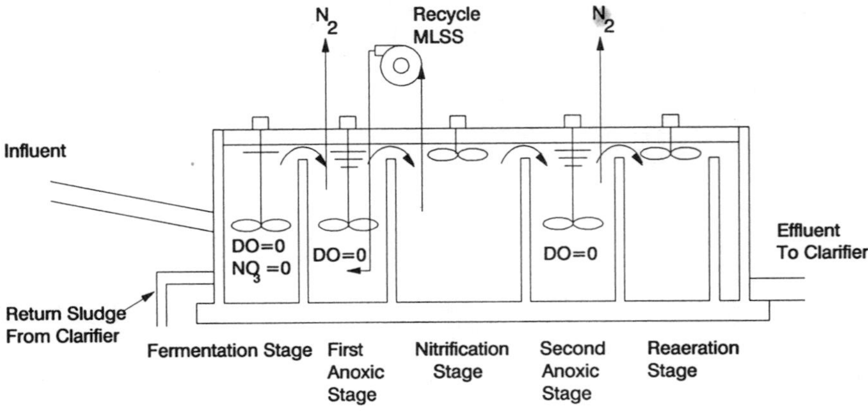

Figure C-1.1. Bordenpho process flow diagram.

C-1.4 Status

There are nine wastewater treatment facilities in the U.S. using the Bardenpho process. Worldwide, there are another 40 systems in operation. There are six facilities under construction currently and another eight being designed. The operating systems consistently report good removal of nitrogen and phosphorus. However, alum must be added to enhance phosphorus removal in cases where effluent standards require consistent phosphorus levels at or below 2.0 mg/L.

CASE 2 BIOLOGICAL AERATED FILTERS (BAF)[1],ONEONTA, ALABAMA

C-2.1 Description

Prior to the addition of the biological aerated filters (BAF), the Oneonta, AL treatment system was a single lagoon. The BAF units were added to achieve effluent limits that are BOD, ammonia nitrogen and suspended solids limited. The treatment system is 2.2 mgd (0.1 m^3/sec) consisting of two pond cells, an aerated channel, eight BAF units, and chlorination prior to discharge. It serves a population of approximately 4,500. The BAF units are high rate, attached growth, aerobic treatment units which use a patented catalyst bed to remove soluble and suspended organic material. (See Figure C-2.1.)

---

[1] The BAF system was developed and is patented by OTV of Paris, France. EMICO in Salt Lake City, Utah, has exclusive marketing rights for the patented BAF system in the United States.

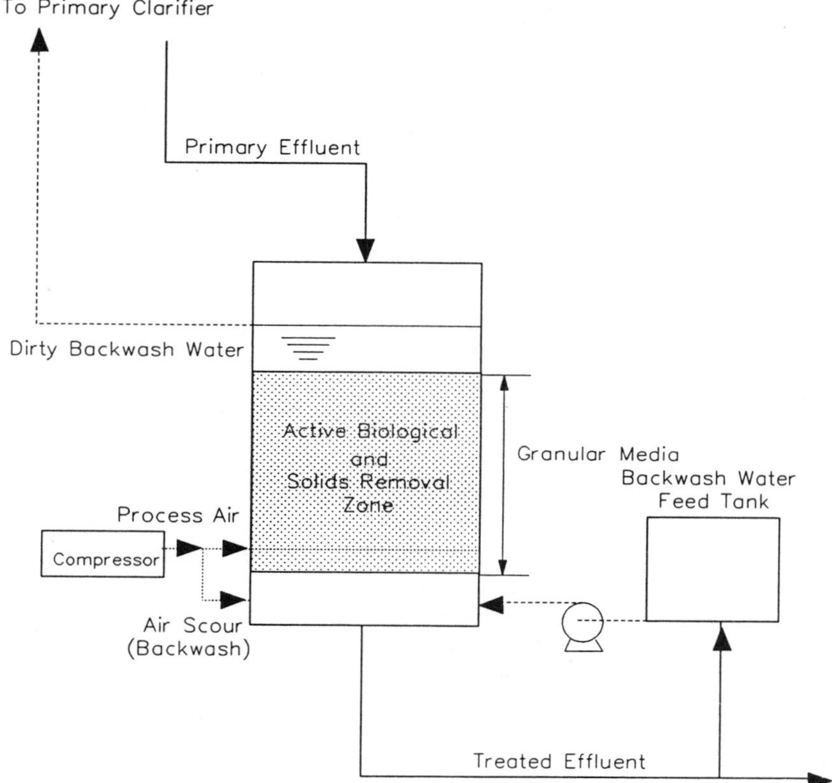

Figure C-2.1. Biological aerated filter.

C-2.2 Application

The BAF process could be used in many systems where improved BOD and SS removal are required, especially where low effluent limits exist. BAF units may also be applicable where nitrification is required. BAF units can be especially attractive if land is limited. Table C-2.1 presents some design parameter ranges.

C-2.3 Benefit

Capital costs are reduced because secondary clarifiers, effluent filters, and costly aeration basins are not required. BAF units require less land than conventional systems resulting in another potential saving. O&M cost savings are possible because energy requirements are potentially less than for

Table C-2.1. Typical BAF Design Ranges

Effluent Quality	
BOD_5	10–20 mg/L
Suspended Solids	10–20 mg/L
Dissolved Oxygen	0.5–2.0 mg/L
Removal Efficiency	
BOD_5	80–96%
Suspended Solids	85–96%
Bed Depth	5–6 ft (1.5–1.8 m)
Organic Loading Rate	140–350 lb BOD_5/day/1000 ft^3 of media volume (2.2–5.6 kg/day/L)
Detention Time	30–80 min
Sludge Production	0.4–0.7 lb/lb BOD_5 removed
Airflow	2–5 scfm/ft^2 (600–1500 L/min/m^2)

conventional systems; there are no chemical requirements; personnel requirements are reduced by automating process cycling. BAF units can reduce BOD to below 10 mg/L and effluent ammonia to 1 mg/L, when properly designed according to influent BOD loading. Effluent suspended solids are generally less than 10 mg/L. BAF systems are simple to operate and possess certain features which should reduce costs: land requirements one-fifth to one-tenth of conventional systems, no secondary clarifier or filter required, single source of sludge, simplicity of operation (II-3).

C-2.4 Status

The system has been achieving effluent concentrations which are better than the required limits since start-up. Studies are being conducted to optimize performance and also to reduce power costs. BAF systems have been successfully operating in France since 1978. In the United States, there are four BAF systems in operation, one under construction and one under design.

CASE 3 TEACUP GRIT REMOVAL SYSTEM, CALERA, ALABAMA

C-3.1 Description

The wastewater treatment system in Calera, AL consists of twin Teacup solids classifiers with stacked static screens, counter-current aeration, clarification, and chlorination. The system has a design flow of 750,000 gpd (33 L/s). The effluent is BOD, ammonia nitrogen, and ss limited. The Teacup solids classifier (see Figure C-3.1) removes grit by a combination of centrifugal and gravity forces. Flow enters tangentially near the top, creating

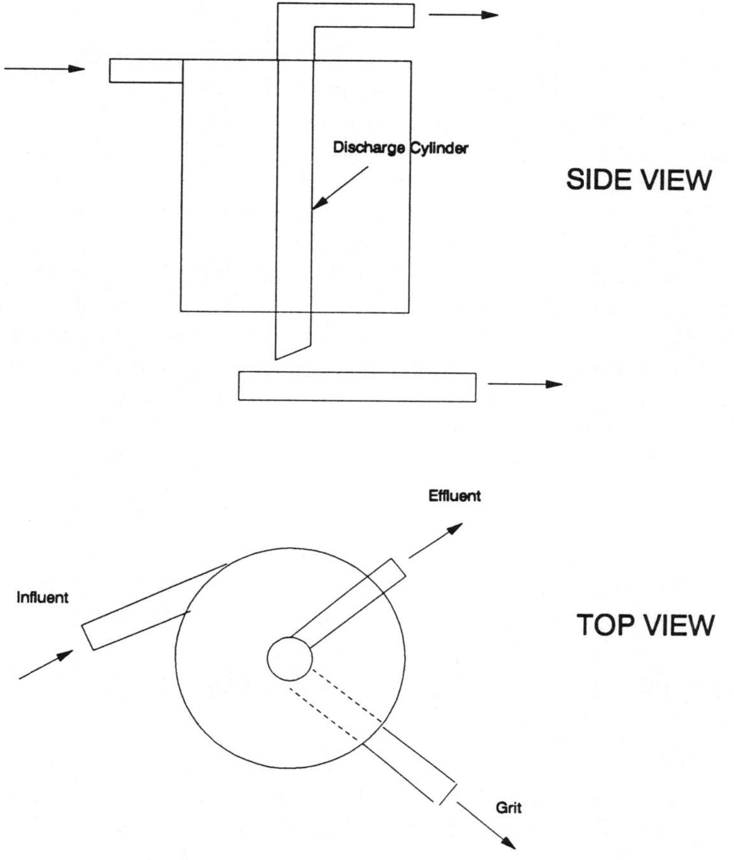

Figure C-3.1. Teacup grit removal/solids classif.

a free vortex and resultant centrifugal forces. Grit particles settle toward the bottom where the free vortex boundary layer sweeps them to a central well. Acceleration within the boundary layer separates the particles by density. The denser grit particles are separated and removed, while the less dense organics tend to remain with the wastewater.

C-3.2 Application

The Teacup solids classifier is applicable to a variety of wastewaters, including municipal treatment systems, food processing wastewater reclamation and industrial cooling waste reclamation. The system can be used in any application where grit accumulation and/or damage is a problem.

C-3.3 Benefit

If grit is not effectively removed from wastewater before it enters a treatment system it adds sludge volume, additional sludge solids and abrasives which can cause excessive wear on mechanical equipment. All of these situations increase operation and maintenance costs; therefore removal of grit decreases costs and maintenance time.

The Teacup solids classifier removes 95% of the grit under peak flow conditions. The grit removed is generally less than 15% organics, which results in less odor than typical biological sludge. Since grit is still about 15% organics however, careful disposal is necessary, and treatment may be required before disposal. The Teacup solids classifier is a hydraulic system which saves energy and reduces maintenance. The Teacup has no moving parts which reduces maintenance costs. The aerated discharge maintains dissolved oxygen levels.

C-3.4 Status

The Teacup solids classifier at Calera is performing as designed. The system is removing more than 95% of the grit in the influent, and there are no odor problems.

CASE 4 PRESSURE SEWER TECHNOLOGY

C-4.1 Description

There are two basic variations of on-site pressure sewer systems: septic tank effluent pump (STEP) units and grinder pump (GP) units. Both are shown in Figure C-4.1. STEP systems consist of a septic tank, a wet well with an effluent pump, and accessories such as valves and level control system. GP systems have a pumping chamber storage tank and a grinder pump with accessories similar to a STEP system. Both system variations pump wastewater into small diameter, sealed sewer lines. As illustrated, the STEP systems produce a wastewater with lower organic loading than conventional sewers due to pretreatment in the septic tank; GP systems produce a wastewater with higher than normal organic loading due to little or no dilution from infiltration/inflow (I/I).

C-4.2 Application

Pressure sewers allow small and/or widely dispersed communities to add collection areas as sporadic growth occurs. This type of system is well suited to applications where the treatment plant must be located upgradient of the collection system but can also be used effectively in areas of slight or widely

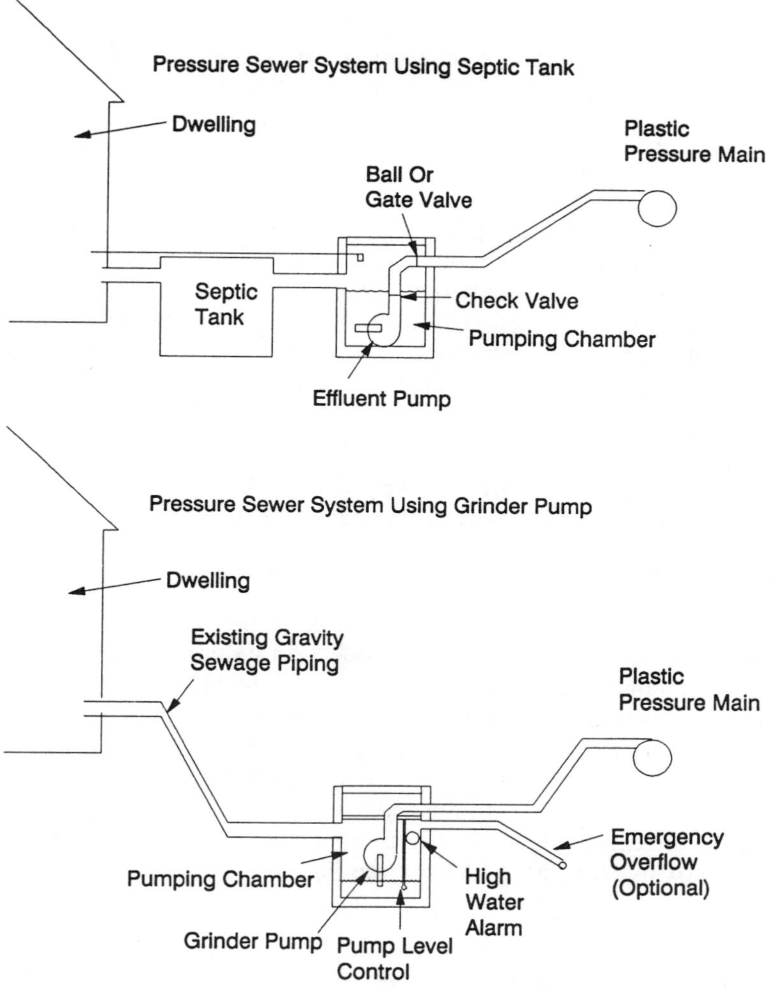

Figure C-4.1. Pressure sewer technology.

varying topography. Pressure sewers can also be advantageous in areas with shallow bedrock or high ground water tables.

C-4.3 Benefit

Initial costs are lower than gravity sewers due to easier installation (smaller diameter pipes are used); shallower, narrower trenches are possible; a non-critical variable grade may be designed which can be adjusted for specific site conditions. System expansion can be accomplished one house at a time without the need to install large collector lines based on future expansion projects. The sealed pipe system reduces infiltration and inflow, thereby relieving operational and treatment problems and saving capital costs for larger systems.

C-4.4 Status

Pressure sewer systems are applicable in numerous communities where construction conditions are not conducive to installation of conventional gravity systems and/or growth patterns of the area warrant this type of system. Capital costs must be low enough to offset higher operating costs in order to consider application. Construction with corrosion resistant valves, water level sensors, and switches should increase long term reliability and ultimately decrease O&M costs.

CASE 5 GRINDER PUMP WASTEWATER COLLECTION SYSTEM, GREENE COUNTY, VIRGINIA

C-5.1 Description

The Greene Mountain Lake subdivision is located in rough terrain area downhill from an existing wastewater collection system. To connect the two systems, a small low pressure system with individual grinder pumps at each residence was designed. Each residential station has a 60 gal (0.2 m^3) storage tank which is pumped at a predetermined tank capacity by a 2 hp (1.5 kw) packaged grinder pump. The collection system includes 1.5 to 4 in. (3.8 to 10 cm) low pressure mains connected to a central pump station which discharges to the existing gravity collection system. The system is designed to serve approximately 120 residences (see Figure C-5.1).

C-5.2 Application

Any area of wastewater generation that is topographically isolated from existing collection and treatment facilities can benefit from this technology provided the pumping required to overcome the topography is cost effective. Additional applications might include state parks, recreational and second home developments, or business parks. Low gradient areas (e.g., beach communities) might also benefit by using this system.

C-5.3 Benefit

The grinder pump, low pressure wastewater collection system may reduce the number and/or size of major pump stations required for an equivalent gravity collection system. The collection lines can also be located in existing

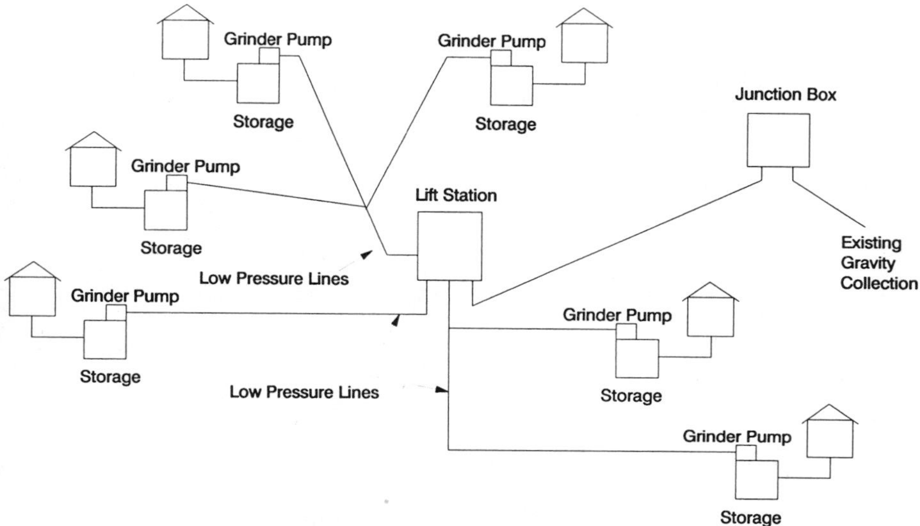

Figure C-5.1. Grinder pump flow schematic, Greene Co., VA.

road rights-of-way at shallow depths and avoiding stream channels. Small shallow lines following the mostly uphill topography provided a cost savings in this project. The closed nature of the system also reduces inflow and infiltration, providing an additional cost savings to the system operation.

C-5.4 Status

The Greene County wastewater collection and treatment facilities have recently been completed and are operational.

CASE 6 SMALL DIAMETER EFFLUENT SEWERS, MT. ANDREW, ALABAMA

C-6.1 Description

The Mt. Andrew, AL small diameter effluent sewer system was installed in 1975 and serves a subdivision community of 31 houses. The system consists of modified septic tanks, small diameter transport lines, and a lagoon for final treatment. The system uses 2 and 3 in. (5 and 7.6 cm) PVC gravity lines and a 3 in. (7.6 cm) pressure/gravity line grade; the effluent from theses houses is pumped to the collection line. Collection lines were installed along the existing grades independent of the elevation and without manholes or cleanouts. The collection lines are upgradient at several points (see Figure C-6.1).

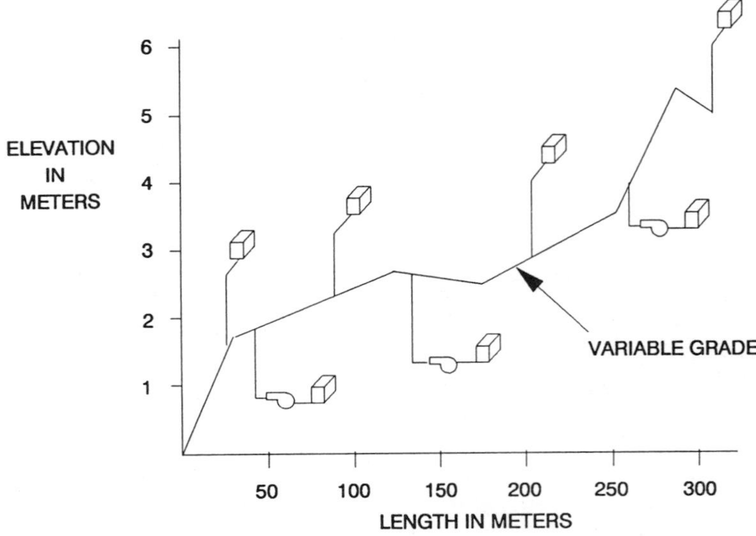

Figure C-6.1. Grinder pump flow schematic, Greene Co., VA.

C-6.2 Application

Small diameter effluent sewers are best suited to reasonably small user groups which will not be experiencing large amounts of growth.

C-6.3 Benefit

The benefits of the system to Mt. Andrew are: 1) lower installation costs due to the use of small diameter pipe as well as pipe installation that follows existing contours which eliminate costly deep cuts and lift stations; 2) a reduced number of manholes, cleanouts, and associated infiltration/inflow; 3) negligible maintenance costs due to smaller pipe sizes and an essentially closed system. Advantages and disadvantages are given in Table C-6.1 (II-8).

C-6.4 Status

The small diameter effluent sewer system at Mt. Andrew has been operating satisfactorily since 1975. The transport lines have proven to be very reliable with only minimal maintenance requirements. The modified septic tanks have functioned as designed, although rapid solids buildup in the primary section of the tanks occurred due to their initial undersizing resulting in more frequent pumping than anticipated.

Table C-6.1. Small Diameter Gravity Sewers

ADVANTAGES	DISADVANTAGES
Low construction cost	Each Service connection requires septic tank
Relatively low operation/maintenance requirements	Septic tank cleaning and septate disposal required
Reduced grit, grease, and solids present in waste flow	Additional pump stations may be required in hilly areas
Sewers can be installed at lesser gradients	Potential odor/corrosion problems
Cleanouts used instead of manholes	May require mainline flushing
Reduced infiltration/exfiltration	

CASE 7 COMMUNAL TREATMENT SYSTEM, MAYO PENINSULA, MARYLAND

C-7.1 Description

The decentralized wastewater treatment project developed for the 8 mile (14.8 km) Mayo Peninsula, Maryland, includes three treatment approaches. One approach is on-site septic systems in areas with suitable soils. The second approach is cluster soil absorption systems where septic tank effluent is collected from several homes and conveyed to an area with soils suitable for a communal infiltration field. The final approach (shown in Figure C-7.1) is a 0.9 MGD (0.04 m^3/sec) communal treatment system for the majority of the peninsula.

The communal system starts with collection and discharge of septic tank effluent into seven acres of recirculating sand filters. Following this sequentially are: a 7 acre (2.8 Ha) constructed bulrush and cattail wetland equipped with intermediate ultraviolet disinfection; an 8 acre (3.2 Ha) constructed

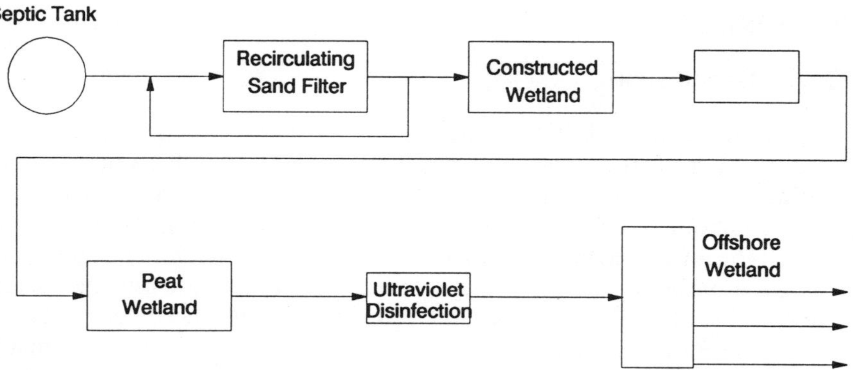

Figure C-7.1. Communal system flow diagram, Mayo Peninsula, MD.

peat wetland with final ultraviolet disinfection; final discharge into constructed, offshore, submerged wetland.

C-7.2 Application

Decentralized systems are feasible for rural areas with widespread clusters of population. Systems similar to the Mayo Peninsula project enhance the current lifestyle and do not contribute to unplanned growth. Areas striving to maintain a simplified infrastructure could benefit from a decentralized waste treatment plan.

C-7.3 Benefits

Following a history of failed septic tank systems with accompanying flooding and adverse effects on well water quality, local residents encouraged development of a system which would treat the residential wastes but would not contribute to rapid development of the area. The decentralized system will achieve the community goals while reducing initial costs significantly when compared to conventional systems.

C-7.4 Status

The system is completely installed and operational as of 1988.

CASE 8 CONSTRUCTED WETLANDS SYSTEMS TECHNOLOGY

C-8.1 Description

A constructed wetlands (CW) system is essentially a lateral, subsurface flow trickling filter. Primary or secondary treated wastewater flows into a long, shallow trough filled with a stone base and topped with a layer of pea gravel (shown as a dotted cross section on Figure C-8.1) supporting rooted aquatic plants. The illustrated system is treating septic tank discharge. The biological treatment of the wastewater is restricted to the aerobic root zone below the pea gravel surface. Open surface and root/rhizome produced aeration provide the necessary oxygen. Degradation of organic material by bacteria in the root zone produces substances (e.g., metabolites) which are assimilated by

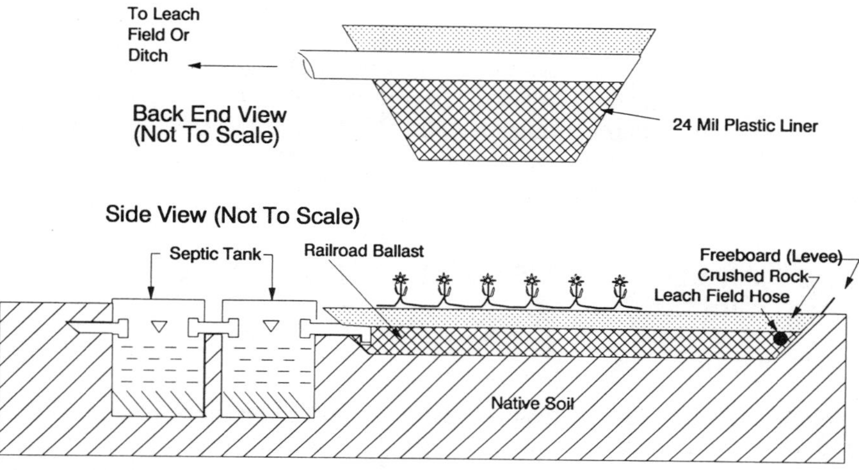

Figure C-8.1. Constructed wetlands.

plants. In turn, microorganisms utilize plant metabolites and dead plant material as a food source.

C-8.2 Application

CW systems have a wide range of applications for small to medium size residential, commercial, and industrial waste streams. Following primary treatment to prevent gravel clogging, the CW system can serve as a secondary or tertiary level of treatment. The most promising application may be the replacement of septic tank drain fields. CW systems are also being used to treat river water contaminated with organic pollutants, acid mine drainage, and agricultural runoff.

C-8.3 Benefit

The CW concept has the potential to lower capital and O&M costs compared to conventional mechanical treatment alternatives. The process is flexible and can be designed to meet specific treatment needs, including the removal of toxics and nutrients. Reed grasses used in CW systems should have a wide range of tolerance for temperature, salinity, and toxicity. This will expand the system's applicability. The CW system requires less land area compared to a floating marsh treatment system. The system also has a nice appearance, and the biomass produced may have some economic value.

C-8.4 Status

The National Space Technology Laboratories Station in Mississippi is operating three CW systems. Several systems are currently being designed or constructed in Alabama, Louisiana, and Mississippi. The U.S. Public Health Service is designing a system at their hospital facility in Corvallis, Mississippi.

CASE 9 PULSED BED FILTRATION (PBF), CLEAR LAKE, WISCONSIN

C-9.1 Description

The primary purpose of this field test was to evaluate the ability of PBF to reduce organic loading to secondary biological treatment systems thereby increasing the performance. The filter selected uses a shallow bed of fine sand with an air diffuser just above the bed's surface to keep solids in suspension (see Figure C-9.1). An air pulse is generated periodically through the backwash and underdrain system that resuspends trapped solids and/or distributes them throughout the bed. The filter is backwashed through the underdrain system after a set number of pulses. A semiautomatic grease cleaning system restores the sand to its original condition. The BPF was tested in the

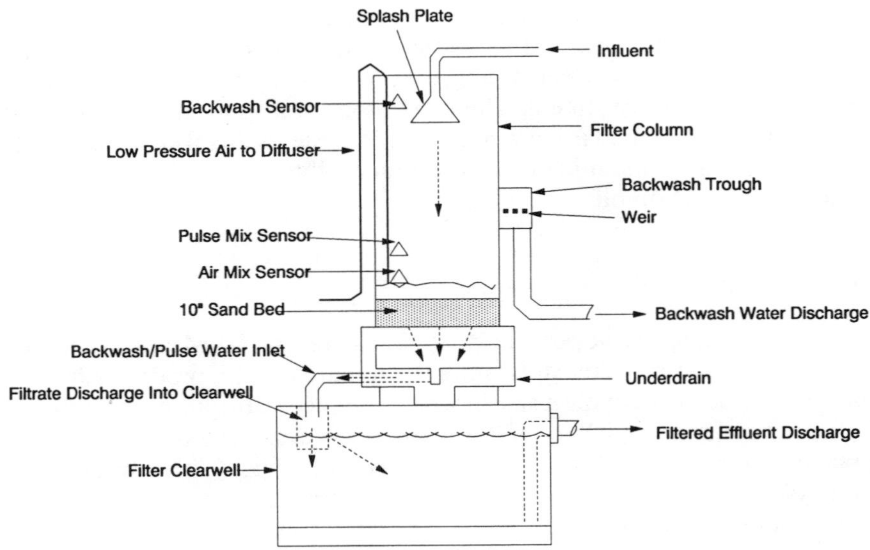

Figure C-9.1. ZIMPRO pulse bed filtration, Clear Lake, WI.

primary filtration mode utilizing primary clarifier and/or roughing filter effluent.

C-9.2 Application

In addition to primary effluent filtration, PBF has proven effective in the filtration of raw water supplies, process waters, wastewater streams, cooling tower water, and boiler feed water.

C-9.3 Benefit

The benefits of primary filtration include the removal of large quantities of solids, increased capacity of existing secondary biological treatment facilities, and reduction of biological treatment sludge.

C-9.4 Findings

Throughout the two-month field test, the PBF reduced SS by an average of approximately 52%, with a corresponding average reduction of approximately 24% in total BOD_5 at the trickling filter effluent. The best results were achieved during the third of five test periods when the discharge to the PBF was changed from the combined primary/roughing filter effluent to roughing filter effluent only. The additional biological activity in the roughing filter produced a higher proportion of larger particle sizes which were more amenable to filtration.

CASE 10 ANOXIC/OXIC (A/O) BIOLOGICAL NUTRIENT REMOVAL, FAYETTEVILLE, ARKANSAS

C-10.1 Description

A pilot scale test study was operated at Fayetteville, AR to determine if the A/O process could achieve desired operational performance under design conditions. The pilot plant was 1 gpm (0.063 L/sec) capacity, sized to provide the same retention time as the full scale plant. Microorganisms in the A/O process solubilize phosphorus in the absence of oxygen in the anaerobic cells. Soluble phosphorus uptake occurs in the oxic cells; organic matter is converted to cell matter, carbon dioxide, and water; ammonia is oxidized to nitrite and nitrate. The process is shown in Figure C-10.1.

C-10.2 Application

The A/O process is applicable to wastewater systems that have phosphorus and/or nitrogen discharge limit which are not stringent.

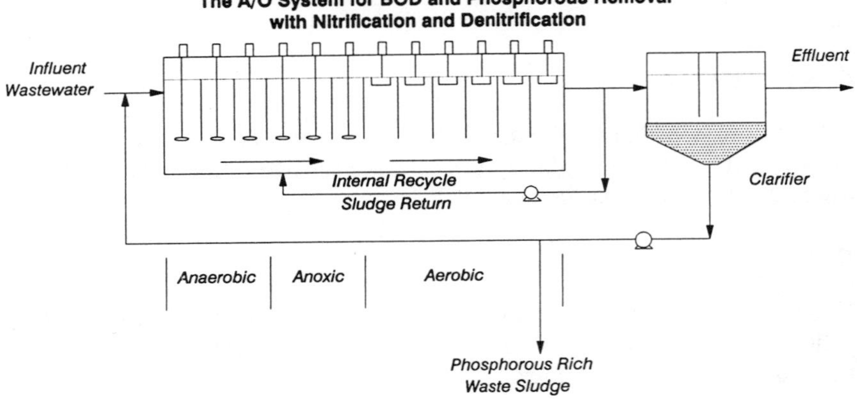

The A/O System For BOD and Phosphorous Removal
Without Nitrification

Influent
Wastewater

Effluent

Clarifier

Sludge Return

Anaerobic

Aerobic

Figure C-10.1. Anaerobic biological nutrient removal, Fayetteville, AR.

C-10.3 Benefit

The A/O process can save costs because oversized clarifiers are not required for phosphorus removal, separate nitrification and denitrification basins are not required for ammonia removal, and chemical storage/handling facilities are not required. Since the only chemicals required are relatively small amounts of alum, operating and maintenance costs are reduced. Stringent effluent limits for BOD, suspended solids, ammonia, and phosphorus reduction can be met with relatively simple operating controls. The A/O process substantially reduces sludge volumes when compared to conventional systems.

C-10.4 Findings

The pilot plant generally achieved excellent BOD, suspended solids, ammonia, and phosphorus removal. Effluent concentrations of BOD, ammonia, and suspended solids were consistently at or below permit limits. Alum had to be added to the oxic basin effluent during low flows to reach the 1 mg/L phosphorus effluent limit. Effluent phosphorus ranged from 0.5 to 3.1 mg/L without alum addition. The field test demonstrated that the full scale facility will be capable of meeting effluent limits.

CASE 11 SEQUENCING BATCH REACTORS (SBRs)

C-11.1 Process Description

All of the treatment steps occur in one tank in the SBR process, as depicted in Figure C-11.1 (also see Chapter 42 for further discussion). The tank is first filled with raw primary wastewater and then aerated to convert the organics into microbial mass thereby treating the wastewater. The aerators are turned off after treatment allowing the solids to settle. During this IDLE period clarifier effluent is withdrawn, and solids are wasted. The SBR process is then ready to begin again. The dynamics of this process follow batch rather than continuous flow kinetics. The process is consequently easier to operate, and degradation of organics is probably more complete following each cycle than is possible in a flow-through system.

C-11.2 Application

SBRs are well suited for small communities which require wastewater treatment systems that are economical to build, simple to operate and maintain, and reliable in meeting secondary effluent quality limitations.

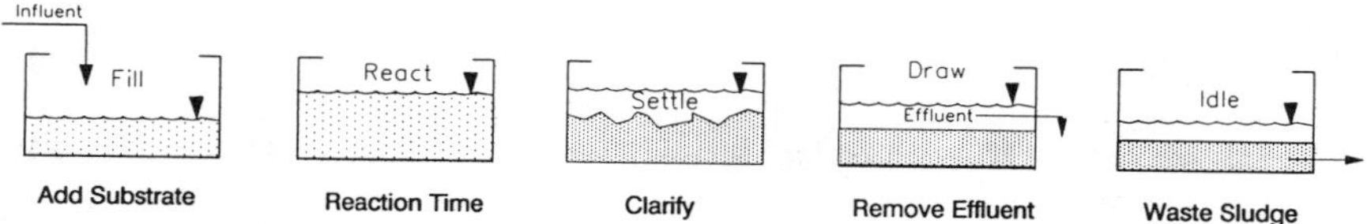

Figure C-11.1. Typical sequencing batch reactor sequence (one cycle).

C-11.3 Benefit

SBR systems require less land area but may require more operator attention than conventional activated sludge treatment systems. Biological treatment and clarification are conducted in one basin thereby eliminating secondary clarifiers and the associated piping and mechanical systems.

C-11.4 Status

Full scale SBR systems are operational in Culver, Indiana and Poolesville, Maryland. Recent data suggest that SBRs can produce excellent BOD and SS removal with minimal energy input. SBRs can also be operated in a mode which will remove substantial nitrogen and phosphorus.

CASE 12 INTRACHANNEL CLARIFICATION (ICC)

C-12.1 Process Description

The ICC concept combines a secondary clarifier with an oxidation ditch. The unique feature of ICC is that as wastewater enters the clarifier, effluent is withdrawn from the clarifier, and sludge is returned to the ditch without pumping. Figure C-12.1 shows one type of ICC within an oxidation ditch.

C-12.2 Application

ICC is applicable for use by communities of all sizes seeking to reduce the costs associated with a conventional oxidation ditch process. It is important to design adequate mixing and aeration capacity, scum removal systems where flow barriers occur, adequate sludge handling capacity, and adequate structural support for the mixing and aeration systems. One manufacturer recommends not using an intrachannel clarifier if the peak-to-average flow ratio exceeds 2.5. Table C-12.1 shows typical operating conditions for the pilot plant (II-5).

C-12.3 Benefit

The advantages of intrachannel clarifiers can include reduced construction and O&M costs and a reduction in land area requirements. Common wall construction reduces concrete requirements. Hydraulic head differences and gravity are used to force wastewater into the clarifier and return sludge back into the ditch. Pumping requirements are thereby reduced. The need

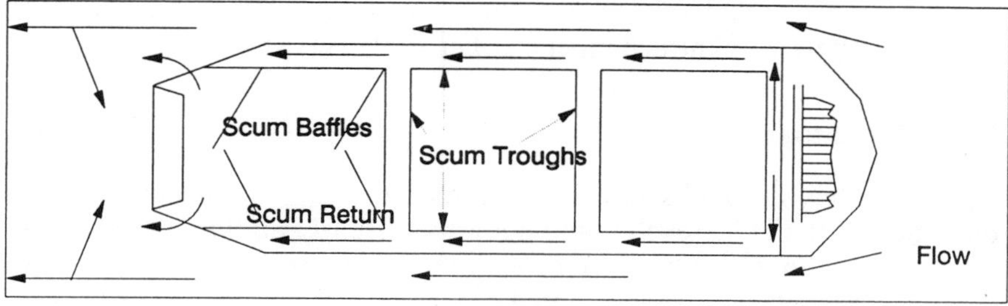

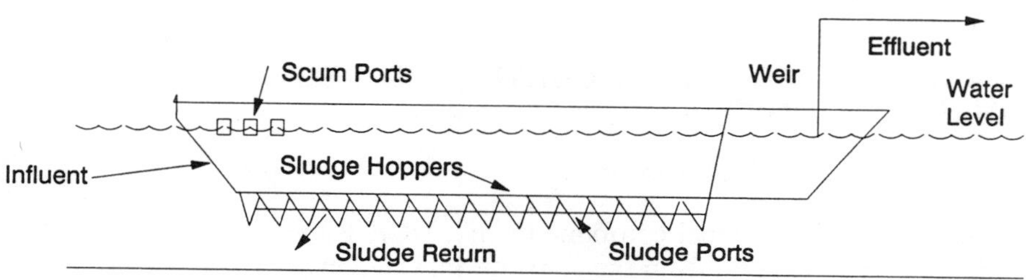

Figure C-12.1. Typical boat clarifier.

for control of sludge return is eliminated, and sludge age is easily controlled by wasting mixed liquor from the ditch or from the intrachannel clarifier. The operational problems reported may thus be more representative of start-up problems rather than long term design deficiencies. Problems with several systems have been encountered with obtaining adequate flow velocity in the oxidation ditch. Proper operation of the clarifier is dependent upon adequate wastewater flow velocity around the ditch. Several facilities have reported that inadequate velocity has caused solids to settle in the ditch resulting in sludge bulking and excess scum accumulation. Changes in mixer design or mixing systems have since corrected velocity problems. Insufficient aeration has occurred in several systems. In general, aeration systems

Table C-12.1. Typical Pilot Plant Operating Conditions

Sludge Age	10 Days
Overflow Rate	600–800 gpd/ft^2
Mixed Liquor Concentration	3000–3500 mg/L
Aeration	10 hr
Effluent Quality	<20 mg/L BOD_5 <20 mg/L SS

which have performed well in conventional oxidation ditch systems provide adequate aeration in intrachannel clarifier systems (II-4).

C-12.4 Status

Approximately 80 ICC systems are currently in design, construction, or operation in the U.S.; seven manufacturers currently market ICC systems. Twelve operational systems are in existence including Morgan City, LA; Sedalia, MO; Owensboro, KY; and Thompson, NY. The current performance data for these systems shows that effluent BOD and SS concentrations of 20 mg/L can be achieved where adequate mixing is provided.

CASE 13 HYDROGRAPH CONTROLLED RELEASE (HCR) LAGOONS

C-13.1 Process Description

There are three principal components of a HCR lagoon: a storage lagoon which receives effluent from the conventional lagoon system, a stream flow monitoring system and an effluent discharge structure. The effluent discharge structure releases the treated wastewater from the storage lagoon in proportion to the streamflow as measured by the monitoring system. The size of the storage lagoon is determined by the streamflow characteristics. A schematic diagram is presented in Figure C-13.1. The system is basically pollution control by dilution.

C-13.2 Application

The HCR concept is applicable to systems where the receiving stream's assimilative capacity does not permit continuous discharge from a conventional lagoon system. In such cases, the HCR lagoon is used in combination with the conventional lagoon system. HCR lagoons will not be a cost effective alternative to other treatment systems in all cases. Design considerations which must be evaluated include: site availability, receiving stream effluent requirements, receiving streamflow pattern. Due to the relatively large area required for construction of an HCR lagoon, lack of a suitable site near the treatment plant may not permit cost effective construction of the HCR lagoon. In the case of receiving streams which have stringent year-round effluent requirements or low flow patterns in comparison to the wastewater treatment plant flow, the variable discharge characteristic of HCR lagoons may not be effective.

In general, capital costs for HCR lagoons are dependent upon the following factors: storage volume required, pond liner requirements, and land

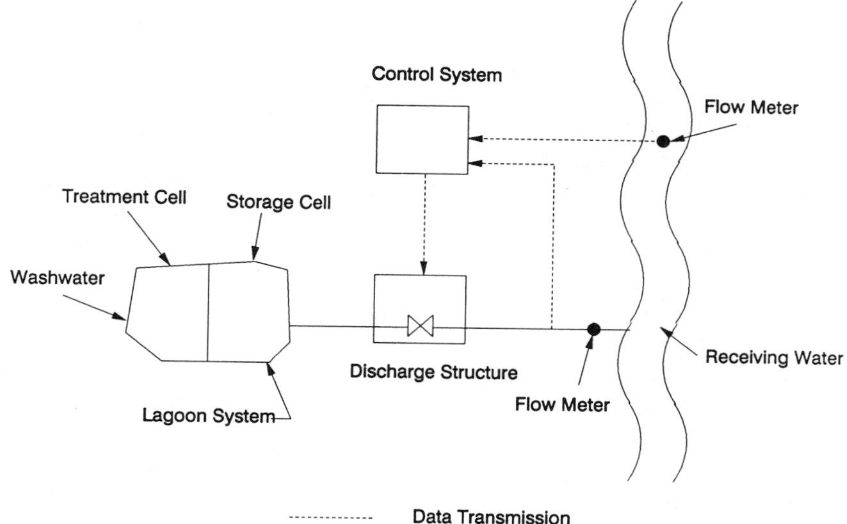

Figure C-13.1. Hydrograph controlled release lagoon schematic.

costs. The total storage volume required is related to both the treatment plant flow and the receiving streamflow pattern. If the receiving stream has a relatively high flow in comparison to the plant flow, a storage volume equal to 30 days of the average plant flow may be adequate. Conversely, a relatively low stream to plant flow ratio may require a storage capacity in excess of 120 days.

The pond liner requirements are site specific and depend upon the nature of the existing soils, proximity of ground water, and local requirements regarding permissible pond leakage. Typical pond liner materials include clay, plastic liners, and soil additives such as bentonite. Asphalt or concrete liners may also be used. Clay seals or soil additives are generally less costly than other liners if clay is readily available or the soils are suitable for use with an additive (II-6).

C-13.3 Benefit

A HCR lagoon system can be used to make the maximum use of a stream's assimilative capacity since effluent discharge is proportioned to streamflow thereby allowing the use of low cost, easy-to-operate lagoon systems where higher levels of treatment might otherwise be required. Planning and manual override may be required to manage forced discharges of the system when it is full. Extended periods of low streamflow will interfere with the effectiveness of the HCR system.

One of the key design aspects is the proper sizing of the storage lagoon relative to the flow and water quality characteristics of the receiving stream. Design of the storage cell is based upon performing a water balance for the storage cell, with the first step being the determination of the amount of wastewater which may be discharged as a function of the stream flow. An analysis of water quality changes should be performed considering flow characteristics of the receiving stream. A time span between discharge events is then determined and a storage volume is calculated (II-7).

Four different discharge systems are currently in use: motor operated valves, motor driven sluice gates, floating weirs, and a series of variable sized pumps.

C-13.4 Status

Over 18 HCR systems are currently in design, construction, or operation, primarily in the Southeastern U.S. There have been no major operational problems related to the HCR components. Examples of operational systems are Linden, AL; Heidelberg and Canton, MS; and West Monroe, LA.

CASE 14 VACUUM ASSISTED SLUDGE DEWATERING BEDS (VASDB)

C-14.1 Process Description

In a VASDB system, the sludge is first chemically conditioned and then distributed onto porous media plates. After an initial gravity drying phase, a vacuum is created beneath the beds thereby drawing off additional water. After the sludge begins to crack it is allowed to air dry before being removed. A cross section of a typical VASDB is shown in Figure C-14.1.

C-14.2 Application

VASDB systems can be used to dewater municipal and industrial sludges unless they are highly viscous or contain high concentrations of grease or fine solids. The process may be effective on more granular chemical sludges, such as from water treatment plants. No examples of the latter were found. See Table C-14.1 for design criteria (II-10, II-11).

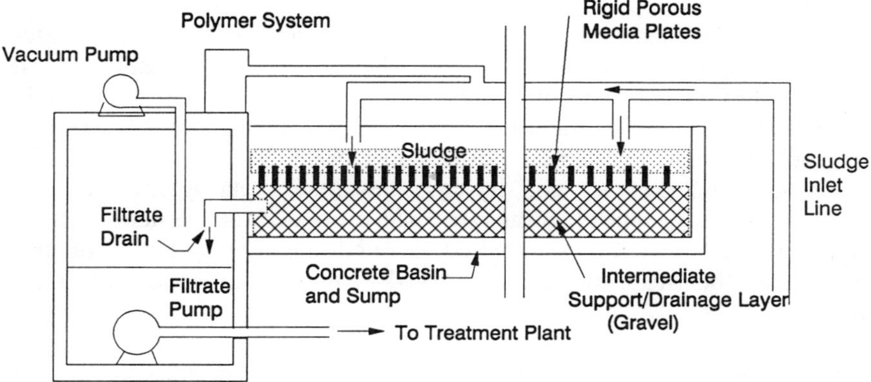

Figure C-14.1. Vacuum assisted sludge dewatering bed cross section.

C-14.3 Benefit

VASDBs may reduce the area required for drying beds by as much as 90% compared with conventional drying beds. Cycle times for dewatering are also less thereby reducing the effects of weather or sludge drying and increasing treatment capacity in available spaces.

Table C-14.1. Design Considerations for VASDB

Bed Geometry and Size	Standard sizes are 20 ft × 20 ft (6.1 × 6.1 m) or 20 ft × 40 ft (6.1 × 12.2 m); media plates are generally 2 ft × 2 ft (0.61 × 0.61 m) or 2 ft × 4 ft (0.61 × 1.2 m). The number of VASDBs per facility range from one to four depending on operation flexibility and schedule desired.
Bed Loading Factors	Manufacturers suggest 1–2 lb dry solids/ft^2 (4.9–9.8 kg/m^2) per cycle for digester sludge; less for waste activated sludge, and higher for Imhoff tank sludge or lime stabilized sludge. Facilities have been able to increase loading by decanting clear supernatant from sludge bed, then adding more sludge.
Feed Sludge Tank	Since optimum polymer dosage is a critical parameter to sludge dewatering, it is suggested that a feed sludge tank be included in designs where polymer mixing and addition are provided.
Polymer Makeup and Feed System	A polymer feed pump which can be adjusted during operation is necessary since sludge characteristics may change during bed loading.
Polymer-Sludge Mixing	Mixing is provided in some systems by air injection, residence time, or a series of 90 degree or 180° elbows.
Sludge Pumping and Distribution	A uniform sludge loading on the plates must be maintained, otherwise premature loss of vacuum may result. It was suggested that bed flooding at the start or use of dual discharge headers may improve uniformity.

Costs were developed to compare covered and uncovered VASDB systems to uncovered, roofed, and totally enclosed sand drying beds, for a wastewater treatment plant generating 2000 lb/day (910 kg/day) of aerobically digested sludge solids. It is important to note that a VASDB system is normally designed to yield only a liftable dewatered sludge cake in contrast to the sand drying bed which yields a much drier sludge cake.

Figure C-14.2 presents sand drying bed estimated total system costs as a function of sludge solids loading rates. The estimated total system costs for equivalent capacity uncovered and covered VASDB systems are also presented. The intersections in Figure C-14.2 indicate the following (II-10, II-11):

- An uncovered VASDB system would be more cost effective than an uncovered sand drying bed at loading rates of less than 16 to 17 lb/ft²/yr (78.4 to 83.3 kg/m²/yr).
- A covered VASDB system would be more cost effective than a covered sand drying bed at loading rates of less than 38 to 39 lb/ft²/yr (186 to 191 kg/m²/yr).
- A covered VASDB system would be more cost effective than a totally enclosed sand drying bed at loading rates of less than 62 to 63 lb/ft²/yr (304 to 309 kg/m²/yr).

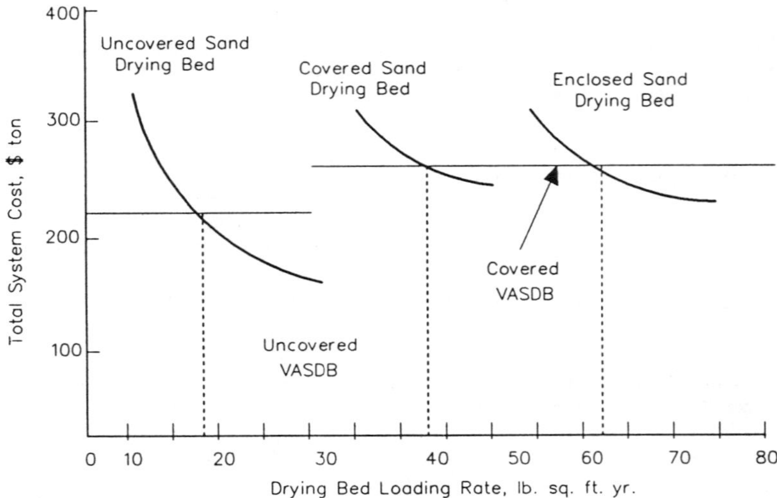

Figure C-14.2. Estimated sand drying bed and VASDB system total costs as a function of solids loading rate for systems processing 365 tons of sludge solids per year.

C-14.4 Status

Treatment systems utilizing VASDBs include Portage, IN; Sunrise City, FL; Lumberton, NC; and Grand Junction, CO. Data from operational systems indicate that solids concentrations of 8 to 23% can be produced with cycle times ranging from 8 to 48 hours.

CASE 15 COUNTER-CURRENT AERATION (CCA) SYSTEMS

C-15.1 Process Description

In CCA, the aeration system moves with respect to the solids unlike conventional systems where the aeration system is stationary. In one of the six configurations of a CCA system, shown in Figure C-15.1, the aeration system rotates around a circular tank about once per minute. The rotation creates a longer bubble flow path which may result in a greater oxygen transfer.

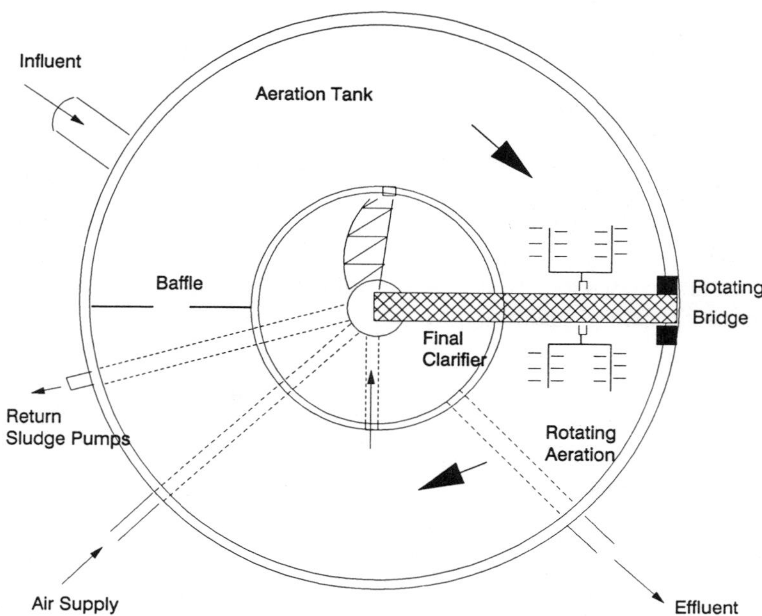

Figure C-15.1. Counter-current aeration system.

C-15.2 Application

CCA systems can be cost competitive for plant sizes over 0.15 MGD (2630 L/sec).

C-15.3 Benefit

CCA may reduce the land area and energy requirements for extended aeration systems. Oxygen transfer efficiency may also be higher with CCA systems than with other aeration systems.

Capital costs are similar to activated sludge system aeration. Operation and maintenance costs will be similar to other extended aeration processes with the exception of power. Table C-15.1 shows the dramatic energy savings that can be achieved by using the countercurrent aeration process (which also uses the fine bubble diffusers) (II-12).

C-15.4 Limitations and Advantages

- Generally not cost competitive for plant sizes under 150,000 gpd (0.007 m^3/sec).
- Significant power savings over other activated sludge processes.
- Requires less land than other extended aeration processes (e.g., oxidation ditch).

C-15.5 Process Considerations

- Concentric clarifier/aeration tanks possible up to 1.25 MGD (0.055 m^3/sec)
- Careful tank construction and rotation equipment placement is required.
- Standard design may require some additions or modifications for maximum operation flexibility and safety.

Table C-15.1. Power Use Comparison

POWER	OXYGEN TRANSFER[1] lb O_2/HP/HR	USE HP/MGD
Extended Aeration with Coarse Bubble Diffusers	3.0	83
Oxidation Ditch	3.4	73
Mechanical Aerators	3.6	69
Extended Aeration with Fine Bubble Diffusers	5.0	49
Countercurrent Aeration	6.0	41

[1] With clean water

C-15.6 Status

CCA systems are currently in design, construction, or operation at over 20 locations in the United States. Over 500 systems are operational worldwide. Operational systems in the U.S. include Grand Island, NY; Loudon, TN; Rome and Clayton County, GA; and Tuskegee, AL. Operational data from these and other operating facilities demonstrate the energy savings in operating these systems.

CASE 16 VACUUM COLLECTION SYSTEM, CEDAR ROCKS, WEST VIRGINIA

C-16.1 Description

A vacuum collection system consists of a special vacuum valve which allows a mixture of air and wastewater to enter the vacuum system from each residence. The vacuum valve opens automatically when wastewater accumulates in the storage reservoir below the valve and remains open for a preset interval to allow the wastewater and air to enter the vacuum system. The air/wastewater mixture is drawn towards the collection station by pressure differentials between the vacuum valves and a vacuum pump station which maintains the vacuum throughout the system. Figure C-16.1 shows a schematic diagram of a vacuum sewer system.

C-16.2 Application

The system consists of three main trunks which are controlled separately from the vacuum station to allow isolation of problems or installation of a new service without disruption of the other branches. Two hundred vacuum valves were installed in the Cedar Rocks system with one valve serving two homes in some cases. The collection station operates an average of 4.5 hr/

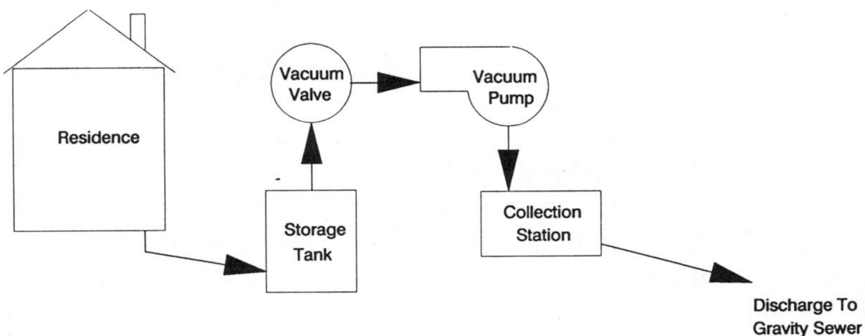

Figure C-16.1. Vacuum sewer system schematic diagram.

day. A vacuum is applied to the collection system by a vacuum pump through a fiberglass collection tank. An 800 gal (3 m^3) vacuum reserve is also used for moisture collection. A collection tank receives the wastewater from the three mains. Sewage collected from the Cedar Rocks area is then discharged to the Wheeling, West Virginia, wastewater collection system.

The advantages and disadvantages of vacuum sewers are:

ADVANTAGES	DISADVANTAGES
Low construction cost	Vacuum must be maintained
Reduced infiltration/exfiltration	High operation/maintenance requirements
Shallow sewer depths	Malfunctions difficult to locate
Cleanouts used instead of manholes	Not adaptable to hilly terrains
Reduced water use when vacuum toilets utilized	

C-16.3 Status

A gravity collection system was proposed for Cedar Rocks, West Virginia, in the original wastewater facilities plan for the area. The Cedar Rocks vacuum collection system began serving 250 users in December 1984. Although some problems were encountered during the construction phase, they were readily solved, and the system has been operating satisfactorily since start-up.

CASE 17 WETLANDS/MARSH SYSTEM, CANNON BEACH, OREGON

C-17.1 Description

The three lagoons and chlorination facilities were modified to include the addition of an aeration basin and a new chlorine contact chamber. A portion of the adjoining forested wetlands is used to polish the secondary effluent before discharge. The wetlands, marsh system was designed to serve approximately 7,000 people. The system operates from June 1 to October 31, with all of the treatment plant effluent going into the marsh. The wetland, marsh system is not used during the other months because increased flows during the winter rainy season provide sufficient dilution in the nearby Ecola Creek. The marsh system covers 16 acres (6.5 ha) and consists of two 8 acre (3.2 ha) cells used in series. The average depth is 2 ft (0.67 m). Winter flooding structures allow periodic flushing of the marsh. The site plan is shown in Figure C-17.1.

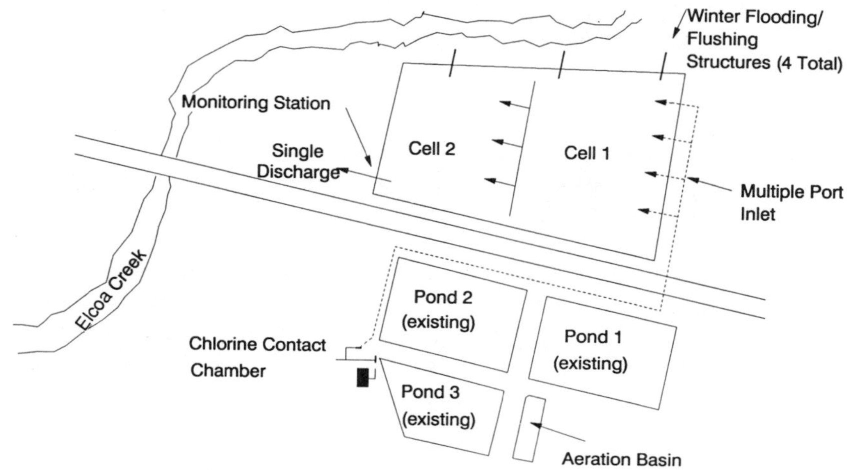

Figure C-17.1. Cannon Beach wetlands marsh treatment system.

C-17.2 Application

The Cannon Beach, Oregon, stabilization pond treatment system could not meet the stringent effluent discharge requirements of 10 mg/L SS and BOD. Higher flows in the summer resulting from a tripling of the summer population caused the noncompliance. To solve the problem, the city selected an artificial marsh and aquaculture system to expand the existing wastewater treatment system. However, because the selected site was a wooded wetland, the plan was altered to employ a natural wetlands, marsh in the treatment system.

The primary objective of the project was to meet the discharge requirements. Secondary objectives were to minimize disturbance to existing wetland habitat and allow continuing usage of the site by wildlife.

C-17.3 Findings

Operating data available for 1985 proved that effluent discharge limits can consistently be met. Average BOD in the influent to the marsh was 12.5 mg/L, while the average BOD in the effluent from the marsh was 4.1 mg/L. This represents an average BOD removal efficiency of approximately 70%. The average SS concentration in the influent to the marsh was 41 mg/L, while the average in the influent from the marsh was 9 mg/L. This represents a SS removal of approximately 80%.

CASE 18 SPRAY IRRIGATION AND WASTEWATER RECYCLING SYSTEM

C-18.1 Description

In 1974, the county began a planning process that evolved into a unique system for recycling the county's wastewater into its water supply system. Figure C-18.1 presents the flow diagram for the system. The major component of the system is a 19.5 MGD (0.86 m³/sec) spray irrigation system. The irrigation system is located in the headwaters of Pates Creek, which is the backbone of the county's water supply system. Effluent from the Flint River and the R. L. Jackson activated sludge treatment facilities are pumped to a 12-day storage pond at the spray irrigation site. Three 15,000 gpm (950 L/s) pumps then distribute the wastewater through 18,300 sprinklers onto the 2,400 acre (971 ha) site. The irrigation site, which is planted in pine trees, is divided into seven cells. Each cell is irrigated 1 day/wk for 12 hr at a hydraulic loading rate of 2.5 in./wk (6.35 cm/wk). The site is located approximately 7.5 miles (4.65 km) upstream of the Clayton County water reservoir. The wastewater applied to the site percolates into the ground water and reappears as streamflow in Pates Creek. At design flows, the wastewater represents approximately 84% of the water flowing into the water supply reservoir

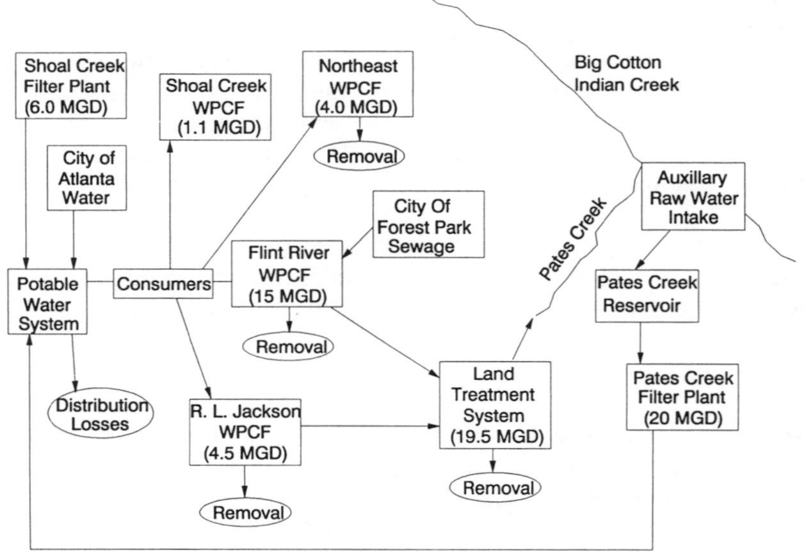

Figure C-18.1. Clayton County, Georgia, wastewater recycling system flow schematic.

during low flow conditions and approximately 33% during normal flow conditions.

The second segment of the recycling system is the discharge of 4.0 MGD (0.175 m^3/sec) of advanced treated effluent into Big Cotton Indian Creek. Clayton County operates an auxiliary water intake on Big Cotton Indian Creek that pumps water back into the reservoir. At design flows during low flow conditions, wastewater represents approximately 62% of the flow in Big Cotton Indian Creek at the auxiliary intake.

C-18.2 Application

Clayton County, Georgia, is a metropolitan Atlanta county. The topography and geology of the county create unique water supply and wastewater treatment problems. Two ridges divide the County into three drainage basins. Because of this, all streams within the borders of the County are headwaters and are too small to serve as a water supply. Consequently, Clayton County's water supply is located in an adjacent county. In addition, each stream has a limited capacity to assimilate wastewater.

C-18.3 Findings

An extensive monitoring program has provided substantial data on the system. With the exception of chlorides, no change from background levels of all constituents monitored has been detected during five years of operation of the system. Chlorides in the groundwater at the site have increased from 6 mg/L to 15 mg/L, which is far below the threshold limit of 250 mg/L for drinking water.

CASE 19 OVERLAND FLOW (OLF) SYSTEM, KENBRIDGE, VIRGINIA

C-19.1 Description

In the OLF process, wastewater is applied at the top of uniformly graded terraces. Renovation of the wastewater occurs as it flows in a thin film over the vegetated soil surface. Typically, 40 to 80% of the applied wastewater runs off and is collected in ditches at the bottom of the slope. A schematic diagram of the OLF process is presented in Figure C-19.1. The existing wastewater treatment facilities were incorporated into the design as preapplication treatment. A 15 million gallon (57,000 m^3) pond was added for storage during inclement weather. Effluent from the preapplication treatment system flows to the storage pond and is then pumped to the overland flow terraces. The final design required 22 acres (8.9 ha) of overland flow

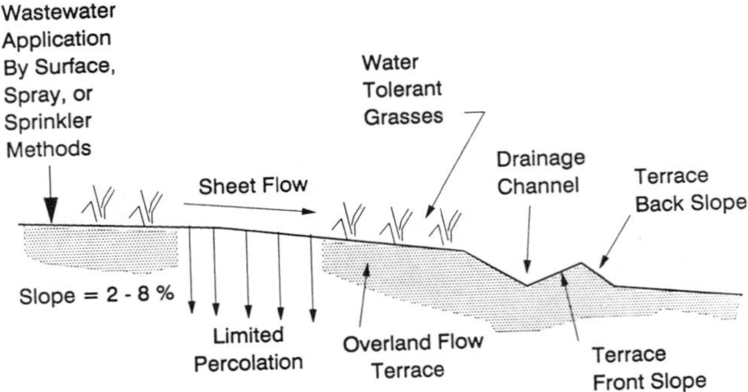

Figure C-19.1. Schematic diagram of overland flow process.

terraces, with an application rate of 3.5 in./week (8.9 cm/wk). Fourteen independently controlled overland flow terraces were designed. The wastewater is applied to the terraces by an 8 inch (20 cm) diameter slotted pipe. Figure C-19.2 shows the layout of the overland flow system. The cover crop is a mixture of water tolerant grasses. From January 1986 to June 1986, the system produced an average effluent BOD of approximately 8.5 mg/L and an average SS of approximately 6.1 mg/L. Grass is cut and removed from the terraces, thereby removing solids and nutrients from the system discharge.

C-19.2 Application

Kenbridge, Virginia, upgraded its existing trickling filter wastewater treatment system in an economic and effective manner. The effluent from the existing treatment facility was discharged into Seay Creek, which is a tributary to the water supply reservoir for several communities. The trickling filter system was not capable of meeting the discharge limitations of 28 mg/L BOD and 30 mg/L SS at the design flow of 0.3 MGD (0.0132 m^3/sec).

OLF can produce advanced treatment quality effluent by treating screened, primary, or secondary wastewater. Operation and maintenance costs are low and land and storage volume requirements are less than those for slow rate land treatment. OLF can be used in areas with low permeability soils where land area is somewhat limited and is not prohibitively expensive. A site evaluation of nearby property revealed that an available 100 acre (40.5 ha) tract was well suited for land treatment by overland flow.

This form of land treatment can be used in areas with low permeability soils where land area is somewhat limited but not prohibitively expensive. The site was located adjacent to the existing treatment plant in a rural area

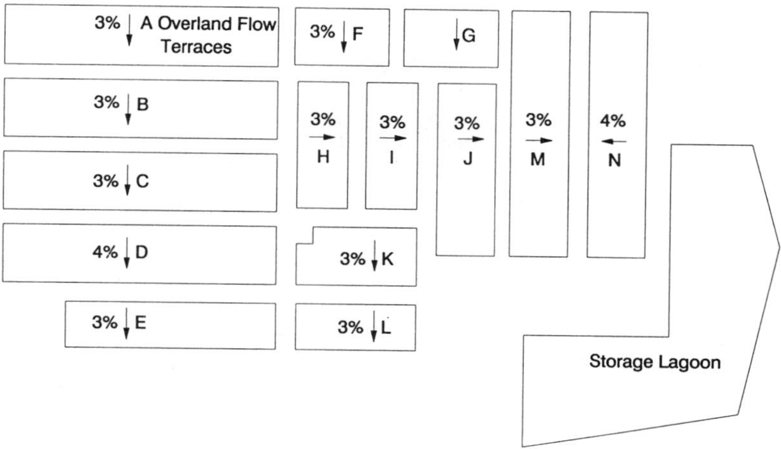

Figure C-19.2. Kenbridge, Virginia, overland flow system.

with little potential for future development. The shallow subsoils at this site had a permeability of less than 1.3 in./hr (3.3 cm/hr).

C-19.3 Findings

An analysis of the overland flow concept compared to an aerated lagoon system showed that the overland flow system would be more cost effective. The total construction cost for the facility was approximately $1.1 million.

C-19.4 Status

Numerous OLF systems are in operation, including systems in Cleveland, MS; Davis, CA; Kenbridge, VA; and Raiford, FL. Effluent biochemical oxygen demand and suspend solids concentrations of less than 10 mg/L can be achieved. Significant reductions in nitrogen and phosphorus can also be achieved.

CASE 20 IN-VESSEL SLUDGE COMPOSTING SYSTEM, (IVC), EAST RICHLAND COUNTY, SOUTH CAROLINA

C-20.1 Description

East Richland County's variation of the IVC shown in Figure C-20.1, is to cure the sludge in piles on the ground instead of in a closed vessel. The system has been operational since March 1986. Five tons per day (4.5 mt/day) of

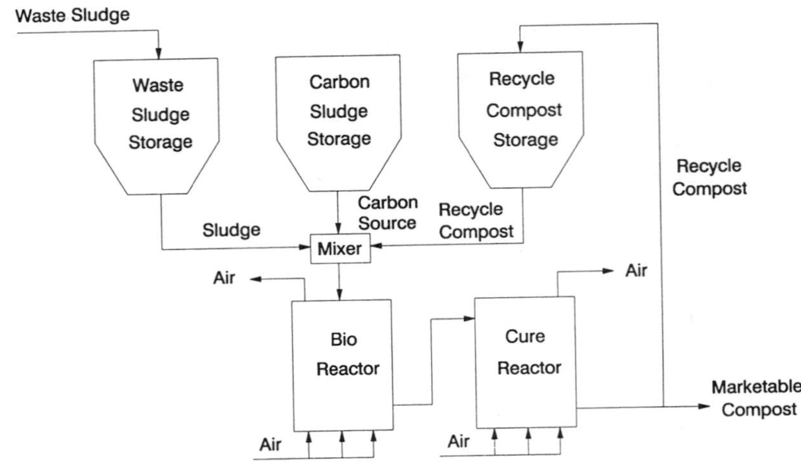

Figure C-20.1. In-vessel sludge composting schematic.

sludge is produced by the extended aeration wastewater treatment process. The sludge is dewatered to approximately 17% solids by belt filter presses before entering the compost system. The compost system produces approximately 14 tons (12.6 mt) of compost per day. The county currently has a renewable one year contract to sell the compost for $12.50 per ton.

As illustrated, waste sludge is discharged to a storage bin. A mixture of the sludge, a carbon source such as wood chips and recycled compost is fed to the bioreactor. The mixture is held in the bioreactor for approximately 14 days to allow complete decomposition of the sludge and to destroy disease causing organisms. The compost system is fed to the reactor to obtain further solids stabilization and conversion of organic materials to humus. Air is fed into the reactors to maintain an aerobic process. Composting is a thermophilic, aerobic decomposition process whereby complex organic constituents of sewage sludge are broken down microbiologically into simpler compounds. The composting reaction can be illustrated as follows (II-14):

$$\begin{array}{l}\text{Fresh}\\ \text{organic}\\ \text{waste}\end{array} + O_2 \xrightarrow[\text{metabolism}]{\text{Microbial}} \begin{array}{l}\text{Stabilized}\\ \text{organic}\\ \text{residue}\end{array} + CO_2 + H_2O + \text{Heat}$$

Heat generated during the process reduces the number of pathogenic microorganisms in the sludge. The stabilized organic residue (the end product of the composting process) possesses physical and chemical properties which make it useful as a soil amendment in landscaping, reforestation, land reclamation and land development projects.

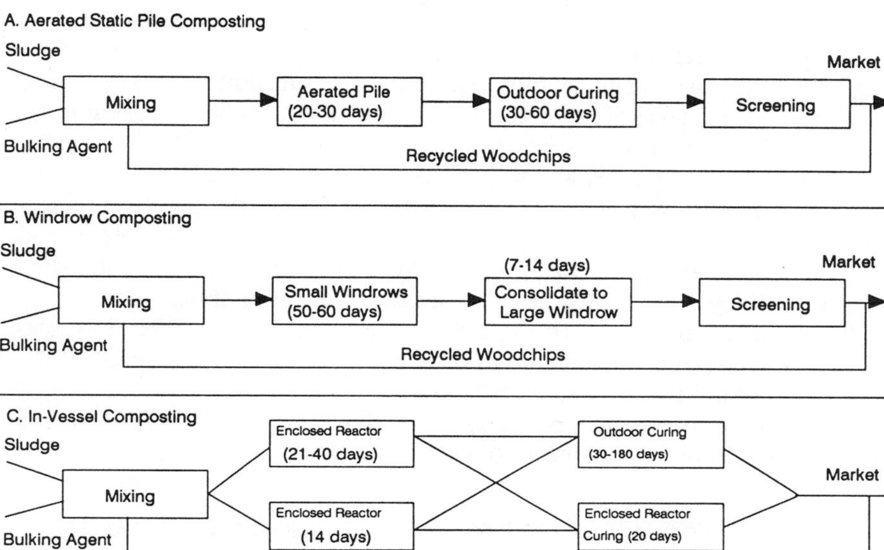

Figure C-20.2. Schematic representation of static pile windrow and in-vessel composting.

Composting in the U.S. until recently, has been carried out only using nonenclosed windrow or aerated static pile methods. Problems related to odor control and land area requirements, and the reduced efficiency of these methods during adverse weather conditions have prompted the use of in-vessel, or enclosed, composting (IVC) methods.

A schematic comparison of IVC and non-enclosed composting methods is presented in Figure C-20.2. IVC provides enhanced odor control, reduced land area requirements, and better operations control during adverse weather relative to non-enclosed methods. IVC also offers greater potential for recovery and subsequent reuse of the heat generated during the composting process.

C-20.2 Process Controls

Efficient composting requires sludge with a solids content of 18 to 30%, a volatile solids content exceeding 50%, a pH of 6 to 9, and a carbon to nitrogen ratio (C:N) of 25 to 35:1. The moisture content and pH are often dependent on sludge processing prior to composting.

The C:N ratio and moisture content of sewage sludge are commonly altered through the addition of bulking agents or amendments prior to composting. Bulking agents provide a carbon source to increase the C:N ratio and also increase the solids content and subsequently the porosity of the composting mass. A sludge: bulking agent ratio sufficient to provide a feed mixture in the range of 35 to 40 tons (31.5 to 36 mt) solids is desirable for optimal IVC efficiency. Numerous materials are used as bulking agents, including: wood chips, sawdust, peanut or rice hulls, paper, and shredded automobile tires. Sawdust and wood chips, however, are the most widely used bulking agents. The choice of bulking agents is commonly determined by availability and cost. Most IVC facilities have turned to reuse of finished compost (recycle) as a carbon source and the use of a bulking agent strictly for enhancing compost porosity. This has resulted in lower operating costs but has added the need for additional equipment to mix or convey the recycle stream into the system.

There are numerous types and sizes of IVC systems on the market. These systems incorporate derivations of either agitated (dynamic) or nonagitated

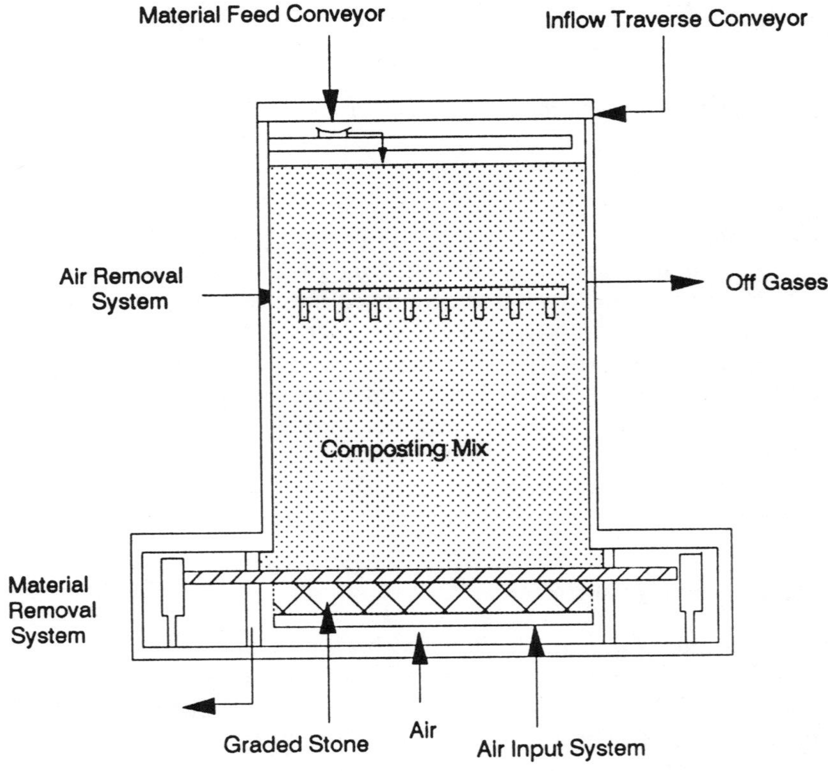

Figure C-20.3. Typical nonagitated IVC system.

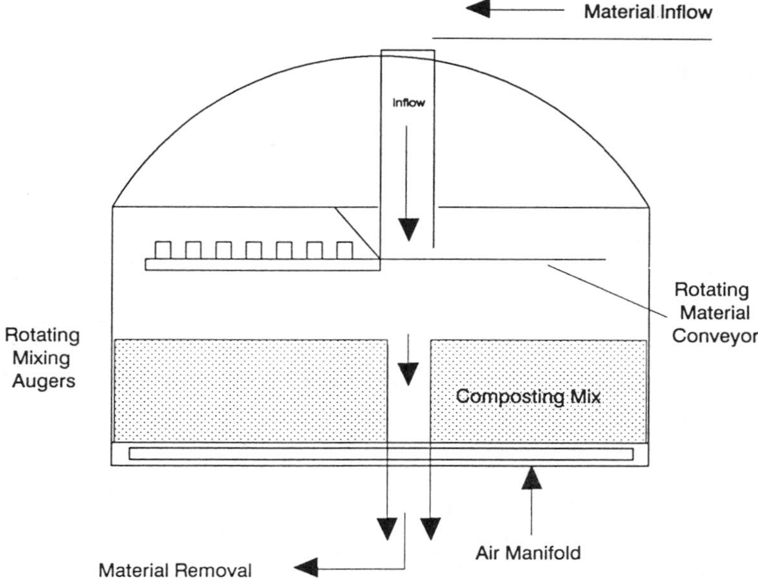

Figure C-20.4. Typical agitated IVC system.

(static) reactor types (Figures C-20.3, C-20.4). Reactor sizes for static systems typically have a capacity range from 5 to 60 dry tons/day (4.5 to 54 mt/day); agitated systems are commonly designed for 30 to 90 dry tons/day (27 to 81 mt/day). Static systems, often referred to as plug flow reactor systems, can either be circular or rectangular. These load from the top of the reactor and discharge from the base relying on gravity to move the composting mass through the system. Dynamic systems also include both circular and rectangular digester reactors. These systems periodically mix the sludge/amendment mixture in order to maintain uniform heat and air distribution within the composting mass. Movement of the compost through the agitated systems occurs by means of mechanical manipulation. Dynamic IVC systems use loading, unloading and aeration devices similar to those of plug flow reactors.

The IVC process can be divided into factors affecting: 1) aeration and moisture removal, 2) odor control, and 3) retention, curing and discharge.

Air serves three purposes in IVC. These include: 1) temperature control, 2) moisture removal, and 3) as a source of oxygen for microbial degradation (compost stabilization). Temperature control and moisture removal are achieved through regulation of air pressure and speed. Considerations to minimize or collect moisture resulting from condensation within the reactor vessel are considered during IVC design. Inadequate removal of condensate from the reactor vessel may result in the return of moisture to the composting mass, extending the time needed for composting or increasing the amount of energy needed for adequate air flow.

Five to fifteen percent oxygen is required in all zones of the reactor for complete composting. Oxygen levels below or above this range can result in significant pathogen reduction and/or odor control problems (II-13).

C-20.3 Application

Sludge composting is the decomposition of organic constituents to a stable humus-like material. In-vessel composting encases this age old process in confined vessels. The result is a marketable compost product without the odor and storage problems sometimes associated with other composting systems. Compost is most acceptable for resource recovery when quality is satisfactory. A typical chemical analysis is given in Table C-20.1.

Compost is often marketed in order to produce income to help offset production costs. Examples of successful compost marketing operations are listed in Table C-20.2. However, many composting facilities rely on *give-away* programs to distribute their finished product.

Table C-20.1. Typical Chemical Analysis for Municipal Sludge Compost from Montgomery Co., Md.

Cation Exchange Capacity (CEC) (meq/100 g)	40
% Moisture	60
pH	6.0–7.0
Essential Plant Nutrients[1]	
Calcium	2.7%
Copper	68 ppm
Iron	0.9%
Magnesium	0.2%
Manganese	383 ppm
Total Nitrogen	1.0%
Total Phosphorus	0.3%
Total Potassium	0.02%
Sodium	0.05%
Zinc	203 ppm
Nonessential Metals[1]	
Arsenic	0 ppm
Cadmium	2.7 ppm
Lead	180 ppm
Mercury	0.5 ppm
Nickel	0.2 ppm

[1] Average concentration, dry weight basis

Table C-20.2. Market Values for Compost

LOCATION	BRAND NAME	WHOLESALE PRICE ($/yd^3)
Bangor, ME	Bangor Compost	3.75
Montgomery Co., MD	Corn Pro	4.00
Hampton Roads, VA	—	6.00
Durham, NH	—	7.00
Philadelphia, PA	Earth Life (Philorganic)	8.25
Missoula, MT	Eco-Kompost	35.00
Columbus, OH	Corn-Til	9.00
Windsor, Ontario Canada	Growth-Rich	17.00

C-20.4 Status

Initial planning studies to select a sludge treatment alternative for the East Richland County Public Service District wastewater treatment facilities recommended sand drying beds followed by landfilling. However, county officials wanted to evaluate a system that would provide resource recovery and revenue generation. A subsequent cost effectiveness analysis determined an in-vessel composting system similar to the one shown in Figure C-20.1 to be the lowest cost alternative.

C-20.5 Summary

The conversion of present sludge management techniques to IVC has been relatively recent and on a limited basis. As a result, early IVC operations have suffered from problems relating to all aspects of composting. The strengths

Table C-20.3. Advantages and Disadvantages of IVC Systems Over Conventional Sludge Management Options

ADVANTAGES	DISADVANTAGES
• Elimination of detrimental effects of adverse weather conditions on the composting process • Improved potential for odor control • Reduced land area requirements • Reduced health risks to employees and residents near composting facilities • Reduced energy consumption and elimination of air quality problems associated with sludge incineration • Production of an ecologically safe and economically beneficial end product	• Substantial capital expenditure necessary • Materials handling and order control systems not specifically designed for sludge applications are often inadequate • Demand for compost is seasonal, this may create the need for compost storage areas and additional handling equipment • IVC operation may require upgrading of existing dewatering facilities • Computerized nature of aeration and temperature control systems require technical expertise.

and weaknesses associated with IVC as compared to other sludge management options are presented in Table C-20.3. These problems, as with other emerging technologies, are expected to be corrected with time (II-13).

CASE 21 METHANE RECOVERY SYSTEM, CHARLOTTE, MICHIGAN

C-21.1 Description

Figure C-21.1 shows a typical methane gas recovery system. In this example, methane gas generated by the anaerobic sludge digestion process is captured and pumped to a gas storage tank. The gas is then used to fuel engines which generate electricity, and to fuel boilers which heat water and produce steam. The electricity is used to operate other plant equipment. The hot water and steam are used to heat raw sludge entering the digester and to heat work areas in the treatment plant. Boilers and engines are dual-fuel equipment since a supplemental fuel is necessary. Methane has a net heating value of 970 Btu/ft^3 (36×10^3 kJ/m^3) at standard temperature and pressure. Digester gas has a net heating value of approximately 600 Btu/ft^3 (22×10^3 kJ/m^3) since it is only 65% methane.

Utilization options depend on the quantity and quality of digester gas. The characteristics of digester gas from a typical anaerobic digester are shown in Table C-21.1. Gas characteristics from a specific digester depend on the

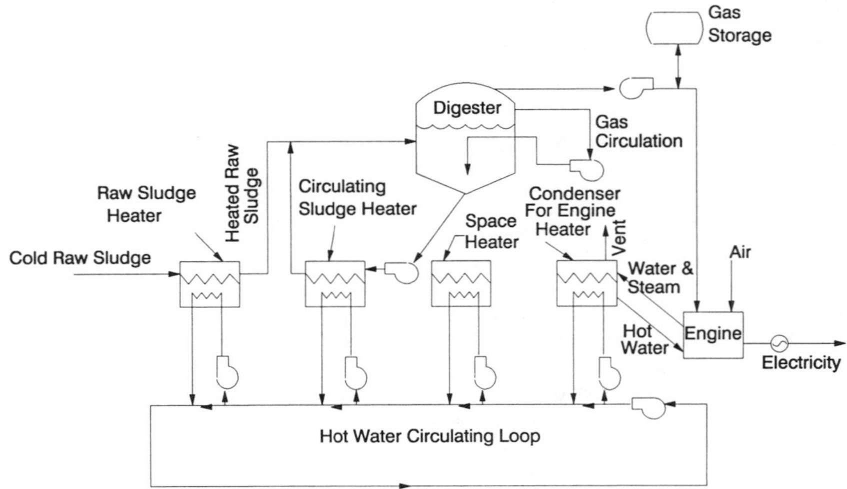

Figure C-21.1. Methane gas recovery schematic.

Table C-21.1. Characteristics of Digester Gas

Digester Gas Quantity	
8–12 ft^3/lb volatile solids added	
12–18 ft^3/lb of volatile solids destroyed	
0.6–1.25 ft^3/capita/day	
Digester Gas Quality	
Methane (CH_4)	65–70%
Carbon dioxide (CO_2)	25–30%
Nitrogen (N_2)	Trace
Hydrogen (H_2)	Trace
Hydrogen sulfide (H_2S)	Trace
Water vapor	Trace
Heat value, BTU/ft^3	550–600

nature of the sludge and the rate at which the sludge is fed to the microorganisms (II-15).

The energy value of the methane generated from the anaerobic digestion process often exceeds the energy requirements of the digestion process for mixing and heating the sludge. It is this excess energy in the form of methane gas which can be used to supply other energy needs including:

- production of steam or hot water,
- fuel for internal combustion engines or gas turbines,
- domestic or industrial gas supply.

Digester gas may require treatment depending on ultimate use of the gas. It is generally preferable to use recovery systems that can operate on untreated digester gas. Minimal treatment is required for combustion in boilers or internal combustion engines. However, gas sold to local gas utilities requires treatment to upgrade the gas to pipeline quality.

C-21.2 Application

Charlotte, Michigan, city officials selected anaerobic digestion followed by land application to farm land for treatment of the sludge produced by the city's wastewater treatment plant.

Methane gas is a natural byproduct of the anaerobic sludge digestion process. In order to properly operate the sludge digestion system, raw sludge must be heated which takes energy. It was decided that use of the methane as an energy source to heat the sludge would increase the efficiency of the treatment system and save operating costs. A recovery system was designed to use the methane for heating of the raw sludge and for fueling an engine to generate electricity.

Possible limitations of methane recovery include:

- Applicability—a variety of sludges from municipal wastewater treatment can be stabilized by anaerobic digestion; however, decreased plant performance may result from additions of some chemical sludges and activated sludges because the additional solids do not readily settle after digestion.
- System Reliability—The microorganisms that generate methane are sensitive and do not function well under fluctuating operating conditions. The process must be carefully evaluated for use at treatment plants where wide variations in sludge quantity and quality are common. A backup source of fuel is necessary to assure continuity of operation.
- Gas Characteristics—Impurities in digester gas can cause operational problems, increased maintenance costs, and give rise to air emission problems. Hydrogen sulfide and its combustion byproducts can cause corrosion in energy recovery systems if effective treatment of the digester gas is not achieved.

C-21.3 Status

Construction of the Charlotte, MI wastewater treatment plant was completed in September 1980. The plant is designed for an average daily flow of 1.2 MGD (0.05 m^3/sec). A total of approximately 2,500 dry tons (2268 mt) per day of sludge is digested. This results in an average methane production of approximately 12,000 ft^3 (340,000 L) per day. A total of approximately 8,700 ft^3 (250,000 L) per day of methane is used, resulting in an average equivalent cost savings (natural gas) of approximately $18,000 per year.

CASE 22 ASSESSMENT OF DUAL DIGESTION SYSTEM (DDS)

C-22.1 Introduction

The dual digestion system (DDS) is a sludge stabilization process utilizing both aerobic (with pure oxygen) digestion and anaerobic digestion. The major advantages of using DDS over many of the conventional sludge stabilization processes are the increased generation of biogas and the potentially higher degree of pathogen destruction. A study of three facilities utilizing DDS conducted in 1984 showed DDS to be a promising sludge stabilization alternative for those plants using the pure oxygen activated sludge process (II-16).

C-22.2 Description

The DDS technology includes the use of a pure oxygen aerobic digester with a one day detention time (the Step I reactor) followed by one or more anaerobic digesters with approximately 8 days detention time (the Step II reactor). Figure C-22.1 shows a schematic representation of the process. PSA refers to oxygen generation and/or supply. The technology relies on the conservation of the heat generated in the Step I reactor by the biological oxidation of the sludge, allowing the reactor to reach mesophilic or thermophilic temperatures. The higher temperatures result in significant increases in the rate of volatile suspended solids reduction. The Step II anaerobic digestion process provides further digestion and the generation of methane gas.

The three wastewater treatment facilities studied include Hagerstown, MD; Nutbush Creek, NC; Lackawanna, NY. All of the plants previously used conventional two stage anaerobic digestion for sludge stabilization and currently use pure oxygen activated sludge as their secondary wastewater treatment process. The oxygen is supplied by a pressure swing adsorption (PSA) oxygen generation system with a liquid oxygen backup. The additional oxygen requirements of the DDS are met by incrementally increasing the capacity of the PSA system.

The process modifications to incorporate DDS include the construction of a Step I reactor and modification of what was formerly the first stage digester to be the Step II reactor. The former second stage digester was retained and is used to provide additional anaerobic digestion and solids separation. This digester is referred to informally as the Step III reactor but would not gener-

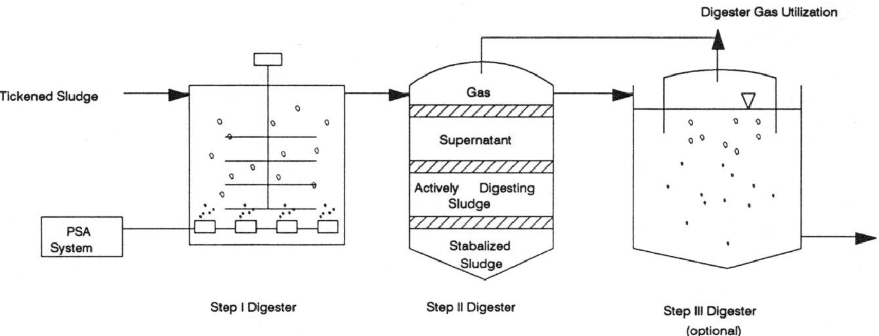

Figure C-22.1. DDS schematic at Lackawanna WWTP.

Table C-22.1. DDS Summaries

	TREATMENT PLANT		
	LACKAWANNA	HAGERSTOWN	NUTBUSH CREEK
Startup Date	Feb. 1983	May 1984	Dec. 1982
Average Daily Flow (MGD)	4.5	8.0	4.12
System Capacity ton/day	3.0	17.2	6.0
DDS Feed Mode	Continuous	Intermittent	Intermittent
DDS Detention Time, days			
Step 1	1.17	0.9	1.25
Step 2	7.4	12.0	10.4

ally be constructed as part of a new system. A new floating cover was added to the Step III digester to provide storage for the biogas generated.

The sizes and operating conditions of the three facilities are presented in Table C-22.1.

C-22.3 Performance

Table C-22.2 summarizes the DDS performance at the Lackawanna and Hagerstown wastewater treatment facilities.

Manufacturers' claims suggest that a 10% reduction in solids can be expected in the Step I reactor. Both Lackawanna and Hagerstown experienced even higher reductions. The overall volatile solids reductions at Lackawanna were between 26.9 and 44.4% during the study. Overall volatile solids reduction observed at Hagerstown ranged from 51.1 to 56.6%.

Using the solids reduction data and oxygen feed rate data, the specific oxygen feed rate to remove volatile solids was calculated at Lackawanna. Feed rates of 2.88 to 3.82 kg O_2 were necessary to remove one kilogram of volatile solids. The oxygen feed rate was not measured at Hagerstown.

Biogas production was measured at Hagerstown and Lackawanna. Compared to the manufacturer's estimates of 0.74 to 1.12 m^3 biogas per kg volatile solids destroyed, the values at Lackawanna were higher than expected, 1.18 to 1.51 m^3/kg volatile solids destroyed; at Hagerstown they were lower than expected, 0.22 to 0.79 m^3/kg volatile solids destroyed. No determination could be made as to why the gas production at Hagerstown was lower than projected.

Dewaterability of the sludge was not quantitatively evaluated. However, the minimum 20% total solids required by the local landfill were met with the processed sludge at both the facilities.

Table C-22.2. DDS Performance

PARAMETER	LACKAWANNA	HAGERSTOWN
Total Solids (%)	5.3	6.8
Volatile Solids Fraction (%)	62	79
Total Solids Loading (kg/d)	1,941	3,154
Temperature: (°C)		
Step I Effluent	50.2	49.4
Step II Effluent	42.4	35.9
pH:		
Step I Effluent	7.3	6.5
Step II Effluent	7.4	7.6
Alkalinity: (mg/L as $CaCO_3$)		
Step I Effluent	3,434	NM
Step II Effluent	4,830	7,518
Step I Oxygen Feed (kg/d)	619	NM
Step II Biogas Production (m^3/d)	296	484
Total Inert Solids Reduction (%)		
Step I	0.9–6.6	2.8–6.3
Step II	14.2–16.2	11.5–28.1
Overall	7.1–17.0	14.0–30.6
Total Volatile Solids Reduction (%)		
Step I	13.7–19.9	10.8–16.4
Step II	14.3–30.6	45.2–48.0
Overall	26.9–44.4	51.1–56.6

Note: NM: Not Measured.

C-22.4 Operation

DDS equipment requirements are similar to a conventional anaerobic treatment process. The major additional item is an oxygen generation system (which was already included in the plant design at both of the plants studied) to supply the requirements of the pure oxygen activated sludge process although incremental increases in the system capacity were required for the addition of DDS.

The second major additional piece of equipment required is a mass transfer device to achieve enhanced oxygen transfer in the Step I digester. A motor driven, slowly rotating, custom designed vertical shaft extending the depth of the digester is used. The shaft is designed with five pairs of tapered arms to sheer the rising oxygen bubbles. Baffles are mounted on the side of the digester. The equipment is fabricated of stainless steel and was developed based on the experience at the Hagerstown demonstration project.

Mechanical reliability is a problem associated with the mass transfer device. Because of the counteracting forces of the rising gas bubbles and rotating blades, uneven forces are exerted on the shaft bearings. Failure of the shaft has occurred at Lackawanna. However, there are several options available to mitigate the potential mechanical problems. For example, at the Nutbush Creek facility, the total design volume was split between two digesters to provide some degree of backup and reliability.

C-22.5 Process Operation and Control

Operation of the DDS has been proven to be relatively simple. Compared to conventional anaerobic digestion, only one additional sample point is required for process monitoring. The oxygen feed rate is the main process control variable that must be monitored to control the Step I operating temperature. The design of a consistent sludge thickening process is also important to the performance and ease of operation of DDS. The raw sludge pumping rate depends on the solids production as well as the thickening process so that a total solids concentration of approximately 5% is maintained.

Based on actual operating experience, several modifications have been made to improve DDS operations. These include:

- step I odor scrubbing system,
- telescoping valve air purge to prevent clogging,
- PSA oxygen generation totalizer,
- DDS oxygen feed totalizer,
- DDS automatic feed rate control.

C-22.6 Costs

The net present worth values of the estimated capital costs for the DDS system are summarized in Table C-22.3. Note that the capital cost for the oxygen generation system represents only the incremental cost for increasing the PSA system capacity. If the cost for a stand-alone oxygen generation system had been used, there would be no advantage in using DDS.

Comparisons of costs are site specific. Table C-22.3 shows that for Lackawanna, total costs for DDS are comparable to conventional systems.

Operating costs for DDS were difficult to determine since budgets for electrical power, labor, and chemicals were not broken down by process at these facilities. Measurements of electrical power were recorded during the study to calculate power consumption and yearly electrical cost. The annual power requirement can be expected to be nearly 20% greater during operation in the thermophilic mode.

C-22.7 Summary

The two facilities that were studied in detail demonstrated the successful utilization of DDS technology. Although solids reduction exceeded the goal,

Table C-22.3. DDS Cost Estimate at Lackawanna

	DDS		CONVENTIONAL 2-STAGE ANAEROBIC	
	YEARLY	PRESENT WORTH[1]	YEARLY	PRESENT WORTH[1]
Capital Cost[1]	—	$1,000,000	—	$ 818,000
Labor	$76,000	836,000	$67,000	741,000
Power	2,000	22,000	2,300	25,000
Chemical	—	—	1,000	12,000
Maintenance and Spare Parts	—	130,000	—	151,000
Digester Heating	—	—	4,000	45,000
Digester Gas for Space Heating	10,000	115,000		
Total Present Worth		$2,100,000		$1,792,000

[1] 20 years at 7%.

long term consistent performance was not shown at one facility. This may be attributable to sludge quality, process operation, or control.

One important advantage to using DDS is the higher degree of pathogen destruction achievable at the higher operating temperatures.

CASE 23 ABURRA WATER TREATMENT PLANT (II-17)

C-23.1 Location and Address

It is located on the right hand slopes of the Medellin River, south of Machado and near the Medellin-Bogota highway, in a place called Croacia, Columbia. The area of the lot is 15.3 ha and has elevations ranging from 1730 to 1800 m above mean sea level. The plant proper has an average elevation of 1752 m above mean sea level. The Empresas Publicas de Medellin, owners of the project, can be contacted for more information.

C-23.2 DESCRIPTION OF THE TECHNOLOGY

C-23.2a Source of Raw Water Supply

The source for the raw water supply will be the water originating from the new Rio Grande reservoir, with a useful capacity of 110 million m^3. From the influent works the water will flow through a tunnel 15.8-km long to the Hydroelectric Central of Niquia, from which the pumped water will flow by gravity through a pipeline to the Aburra Plant.

C-23.2b Treatment System Adopted

The system adopted for the treatment of the waters from the Rio Grande Reservoir, includes the processes of rapid mixing, flocculation, sedimentation, filtration and disinfection. If the quality of the waters requires it in the future, it is envisioned that the plant will convert to the direct filtration system, flocculation and sedimentation will not be used.

Metering. The metering for the volume of flow to the plant will be done through three Parshall flumes, each one with a capacity of 3.0 m^3/s.

Rapid Mixing and Coagulation. The coagulants will be added to the water in the Parshall flumes, which were designed to produce a hydraulic jump. The flume energy will be taken advantage of in the rapid mixing of the coagulants with the raw water. The primary coagulant will be alum, which will be applied along or in combination with some type of polyelectrolyte. Also, lime could be added before the Parshall flumes, to control the pH for optimum coagulation.

Flocculation. In the first stage, flocculation will take place in six rectangular concrete tanks. Each flocculator is made up of three zones, in each one of which there will be different velocity gradients. These could vary from between 80 and 20 sec^{-1}. In the first and second zone the velocity gradients will be obtained through mechanical oscillating systems, such as wooden paddles. In the third zone, with a velocity gradient of 20 sec^{-1}, flocculation is developed by means of concrete baffles that produce a horizontal flow component.

Sedimentation. In the first stage, the sedimentation process will take place in six rectangular concrete tanks. Each unit will be equipped with flat asbestos-cement plates, inclined 55° to the horizontal. The entire surface area of the settlers will be used. The rate of sedimentation compared to conventional systems is much higher. The clarified water from each settler will be collected by means of six fiberglass flumes equipped with submerged orifices spaced over the length of the unit.

The sludge that is deposited over the flat plates will move by gravity as a result of the inclination of the plates, and then deposited at the bottom of the settlers. The deposited sludge will then be extracted hydraulically by means of a net work of perforated PVC pipes, interconnected with a system of two siphons. The siphons are located at the edge of the settler and adjacent to the flocculators and use water from the settlers.

Filtration. Twelve double-celled filters are used. Each cell measures 18 × 4 m, for a total of 1728 m^2. The filtering bed is composed of three layers: an upper layer of anthracite, 55-cm thick, an intermediate layer of sand, 25-cm thick, and a lower layer of gravel, 45-cm thick. The filtered water recovery system is composed of concrete beams, shaped like inverted V's with orifices conveniently spaced, located near the apex. The beams are located above the title bottom of each filter. During filtration, discharge will occur into a central conduit located between the two cells of each filter, and then into a spillway to storage.

Flow is controlled by means of two weirs at the entrance to each filter, one for controlling the flow of water to each unit, and the other, a control weir common to all filters, that will allow the interconnecting canal of the filters to maintain a minimum level.

Backwashing uses filtered water. Backwashing is controlled by means of operating two gates: shutting off the intake of settled water, and opening the outlet for wash water. This operation could be done manually, or by means of electronic devices (operation based on water levels or predetermined schedules).

The filters could function in two modes of operation: on a constant volume of flow, or with a declining filtration rate. In both cases, the water level in the filters will be variable. Switching modes will be accomplished by means of a gate specially designed for this purpose.

Disinfection. The disinfection of the filtered water will come about by means of chlorine application at the entrance of the storage tank. The system has been designed to allow application of chlorine at the tank outlet.

Chemical Stabilization. In the event that it becomes necessary to stabilize the filtered water, a bed of lime will be applied at the entrance of the storage tank.

Chemical Products for Treatment. Alum and/or polyelectrolyte will be used for coagulation; lime will be used for control of pH for coagulation and chemical stabilization; potassium permanganate will be used to control odors and taste of water before treatment; activated carbon will be used to control odor and taste after filtration; and chlorine will be used for disinfection (postchlorination).

C-23.3 Design Criteria

Plant Capacity	
Design (1st stage)	6.0 m^3/s
Maximum (2nd stage)	9.0 m^3/s

Rapid Mixing

Parshall Flumes	3 Units
Width of throat (gorge)	2433.6 mm

Flocculators

Tanks	6 Units
Dimensions (L × W × H) each unit	16.2 × 20.7 × 5.2 m
Water Volume per unit	1744 m^3
Detention time	29 min
Oscillating agitators	6 units
Range of "G" value	20 to 80 sec^{-1}

Plate Settlers

Tanks	6 units
Dimensions (L × W × H)	33.7 × 20.7 × 5.8 m
Water Volume	4046 m^3
Total retention time	67 min
Retention time in plates	10 min
Surface loading rate	159 $m^3/d/m^2$
Weir loading	183 $m^3/d/m$
Length of weirs	472 m
Plate Angle	55°

Filters

Number of units (double-celled)		12
Area (2 cells)		144 m^2
Filtration rate		300 $m^3/d/m^2$
Backwashing rate		0.90 m/min
Volumes of Flow of backwash/filter		130 m^3/min
Surface washing rate		0.15 m/min
Volume of Flow of surface wash		22 m^3/min
Filtering Medium		
Anthracite:	Layer thickness	0.55 m
	Effective size	0.90 to 1.10 mm
	Uniformity coef.	1.45 to 1.55
Sand:	Layer thickness	0.25 m
	Effective size	0.50 to 0.55 mm
	Uniformity coef.	1.45 to 1.55
Gravel:	Layer thickness	0.50 m

Surface Wash Pumps

Number of units	2
Capacity per unit	21.5 m^3/min
Power per unit	112.5 kw

Wash Water Recirculation Pumps

Number of units	2
Capacity per unit	21.2 m^3/min
Power per unit	94 kw

Chemical Product Doses

Aluminum	30.0 mg/L
Coagulating polymer (optional)	0.5 mg/L
Polymer aids coag. (optional)	0.5 mg/L
Polymer aids filt. (optional)	0.5 mg/L
Chlorine (postchlorination)	4.0 mg/L
Lime	30.0 mg/L
Activated Carbon (optional)	10.0 mg/L
Potassium Permanganate (optional)	0.25 mg/L

C-23.4 Limitations

Mechanical agitation systems are used in the hydraulic flocculators to help manage high raw water solids at start up.

C-23.4.1 Advantages and Disadvantages. The design of the high rate settlers and filters allows them to be smaller, which in turn represents a savings in labor costs. Similarly, the "autowashing" system for the filters represents a great savings by making conventional systems unnecessary.

The only identifiable disadvantage is the imbalance that the treatment process may experience as a result of the mechanical failure of the flocculating or dosing systems, although these have backup systems.

C-23.5 Costs

The approximate cost of the labor and equipment is as follows:

• Preparation of access road and lot	$ 750,000
• Construction	4,500,000
• Supply of equipment and materials	5,000,000
TOTAL	$10,250,000

C-23.6 History of Operation

As of March 1987, the plant was still under construction.

C-23.7 Other Criteria

For this particular plant, high rate settlers employing inclined plates, as well as high rate filters and autowashers were adopted as appropriate technology, due to the fact that this technology has extensively proven itself in Latin America and especially Columbia. In addition to the savings, simple operation of the plant is possible without the use of imported systems.

CASE 24 PROCESS NAME "UNIPACK" CENTRAL FILTRATION WATER TREATMENT PLANT (II-17)

C-24.1 Locations: Los Garzones, Cordoba and Croachi, Cundinamarca, Columbia

C-24.2 Description of the Technology

The system consists of a tank in which the mixing, flocculating, settling, and filtering processes all take place simultaneously and on a continuous basis.

The water to be treated is mixed with chemicals in the mixing chamber. The products of this reaction are generally insoluble solids of colloidal size, that take long to grow and settle. In *UNIPACK* this process of growing and settling is accelerated by the mixing of untreated, with previously treated water which contains suspended solids. This acceleration is much like the minimizing effect that solids have on oversaturated liquids. In *UNIPACK* this effect is obtained from the raw water inflow that keeps the previously formed sludge recirculating and makes the solid phase disperse finely into the liquid phase. The solid-liquid contact enhances precipitation of the suspended particles in the water. The special hydraulic design creates a flow regimen that results in solid-liquid mixing in a circular pattern towards the clarification zone.

The task that the balanced floating sludge blanket performs is very important, in that in addition to facilitating an excellent solid-liquid contact that provokes excellent flocculation, it acts like primary filtration. This offers greater efficiency and yield from the chemical addition. The sludge is easily controlled by a periodic manual purge.

Pretreated water is passed through a submerged high rate filter which eliminates the trace quantities of turbidity and results in a clear effluent. The filter has hydraulic collectors and diffusors balanced to avoid canalization and sludge clogs. It can be backwashed automatically.

Backwashing is accomplished automatically by means of a siphon which activates when head or the filter bed reaches a predetermined level (usually 1.20 to 1.50 above the filtered water outlet). The design calls for the backwash pipe to transport 21 gpm/ft^2 (5100 m^3/m^2/min) of filtering area. When the water level in the storage tank reaches the level of the siphon breaker, the siphon lets air into the backwash pipe; breaking the siphon, and ending the wash operation.

C-24.3 Application

The plants are designed for flow rates that vary between 1 L/sec and 60 L/sec. Typical applications are: water treatment for municipalities, that include removal of turbidity, softening as well as filtration and disinfection; water treatment for industries as diverse as textiles, nutrition, paper milling, mining, etc.

C-24.4 Design Criteria

Tolerances: The plant can operate at 10% of its design flow and handle overloads of up to 15% of its design flow. The design makes it an auto-operated system requiring a minimum of attention from the operator and a minimum of mechanical parts.

C-24.5 Limitations

The system design requires that the raw water arrives with a pressure equivalent to a head of 6.5-m static head.

C-24.6 Advantages of the System

The "UNIPACK" design and its operation principles are novel concepts in the field of water treatment:

1. It does not require electrical energy, a characteristic that makes it especially useful in many locations.
2. The automatic operation of the filter makes it a useful application where trained personnel are scarce, or where the budget cannot accommodate a highly trained operator.
3. The overflow (loading) rates are very low, a characteristic that makes it useful with all water sources, including rivers with changing turbidity.
4. The construction cost is lower than other treatment plants due to its compact size.
5. Maintenance cost is much lower than for conventional systems.
6. There are few mechanical parts subjected to wear and frequent replacement.

C-24.7 Costs

Capital: The estimated cost for LPS is currently from $2,700 to $4,400 per L/sec, including laying the foundation and start-up. Plants with a greater volume of flow may have a lower unit L/sec cost.

Cost of Operation: The unit water treatment cost at the Choachi location (everything operates by gravity and there is no pumping), is estimated at 2.4 cents/m^3 (only one operator). In Garzones (Cordoba) double pumping is required; from the river to the plant and from the plant to the elevated tanks. The average cost is 7 cents/m^3.

C-24.8 History of Operation

The first plants of this type were constructed by ACUATECNIA in Puerto Lleras in 1968 with a capacity of 15 L/sec. Today the plant continues to function adequately. Some problems which have been corrected include short circuiting and air binding in the overflow.

C-24.9 Other Criteria

The system is more appropriate in small communities since it can be operated by untrained personnel. The operator can be native to the area and can be trained in a few days. As mentioned before, the system has no parts subjected to wear and does not require electrical energy. Also, a single operator can be in charge of the entire system.

CASE 25 TECHNOLOGY OXIDATION POND (II-17)

C-25.1 Location and Address: Municipio de Tabio, Sabana de Bogota, CAR. Carrera 10 #61-82, Bogota, Columbia, South America

C-25.2 Description of the Technology

In the Cota municipality in 1971 a pond was constructed to serve 2322 inhabitants (estimated for 1980). In other words the design period was for 10 years. Today, Tabio has 3700 habitants. Tabio is located 46 km from Bogota.

The total area of the pond is 7.9 acres (3 ha). A loading rate of 54 grams of BOD per day per capita was assumed. Two ponds were constructed in series in an attempt to achieve plug flow conditions.

The volume of flow of the combined wastewater reaches a distribution chamber of 1 m^2 area, through a 24 in. diameter pipe to the settler. An overflow effluent structure discharges into the Chicu river at about 20 L/sec.

The entrance volume of flow is conducted to two settlers constructed in parallel, that are operated alternately. Their dimensions are 21.5 × 10.4 × 0.42 m. The retention time is 6 hours.

An inclined bar screen, at 60 degrees with rods separated by 1 in. spaces is used. The settled flow is distributed through 3 to 6 in. PVC pipes to the first stabilization pond. The ponds are connected with 12 in. diameter pipe, 13.4 m long.

C-25.3 Design Criteria

- Design Load: 129 kg/Ha/day
- Retention time: 28 days
- The following two ponds were designed:

PARAMETER	1ST POND	2ND POND
Length	138 m	90 m
Width	47 m	47.5 m
Area	1693 m^2	2100 m^2
Depth	2 m	1 m
Ret. time	23 days	5 days
Pond type	Facultative	Facultative

C-25.4 Limitations (Available land area was limited.)

C-25.5 Advantages and Disadvantages

The first pond with a depth of 2 m was not as efficient as anticipated, due to the fact that at this depth, the zones near the bottom of the pond had temperatures near 12°C. This temperature inhibits anaerobic activity.

Another disadvantage is that, since the sewage is combined, great variations in the load concentrations are possible.

C-25.6 Cost

The annual operating and maintenance costs are as follows:

ITEM	
Caretaking	$1160
Maintenance	717
Repairs	134
Others	89
Total	$2100

Maintenance includes sludge removal from the settlers. The pond should be cleaned every ten years.

If the pond were constructed today, its cost would be about $120,000.

C-25.7 History of Operation

Between January and June of 1985, the pond achieved 83% BOD removal. The system produced effluents less than or equal to 48 mg/L, 95% of the time.

Significant treatment (50% BOD removal) occurs in the primary settler. The ponds continue removing BOD, but efficiency is low because of solids accumulation. The average removal of total and fecal coliforms was in the order of 92 to 98% respectively; values that are considered high. The concentration of coliforms in the effluent is still very high, thus not appropriate for agricultural use. The removal of suspended solids was 73%.

CASE 26 TECHNOLOGY: OXIDATION DITCH (II-17)

C-26.1 Location and Address: Municipio de Cota, Sabana de Bogota, CAR, CRRA 10 #16 82, Bogota, Columbia, South America

C-26.2 Description

A modified biological process using activated sludge and described as completely mixed, extended aeration. The population is 3100, in a region described as high plaines at 2659 m with an average temperature of 14°C.

Experimental laboratory studies were done at three pilot plants, where a 0.16 lb BOD/lb MLVSS/day, food to microorganism ratio (F/M) was obtained. This allowed for satisfactory removals of BOD with minimal oxygen utilization. The efficiency of the system did not decrease when the organic loading was increased so as to obtain a 0.2 to 0.25 lb BOD/lb MLVSS/day F/M. Larger quantities of oxygen and greater sludge production were needed at the increased loadings.

Adopting a concentration of solids in the aeration tank of 3000 mg/L and projecting an effluent of 219 mg/L, a hydraulic retention time of 11 hours was calculated. Afterwards, this value was revised to 18 hours. A design volume of 312 m^2 was selected. The required oxygen was calculated at 152 Kg O_2/day. Settling tank dimensions are 40 m^2 area, and 3 m in depth. The drying beds are rectangular in shape; 18 m long and 6 m wide.

Sludge drying is by infiltration and evaporation. The BOD removal reached 90% and the total suspended solids 94%. Removal of total and fecal coliforms was 98% and 99%, respectively.

C-26.3 Design Criteria

- Elliptical ditch
- MLVSS: 3000 to 5000 mg/L
- Aeration time: 12 to 35 hours
- Cell residence time: 15 to 35 days (sludge retention time)
- Optimum F/M ratio: 0.16 day^{-1}

C-26.4 Limitations

Pretreatment facilities had to be constructed to reduce incoming solids. Better recirculation of the sludge was necessary in order to reach the LMVSS value in the design.

C-26.5 Cost

The annual operation and maintenance costs, are as follows:

Caretaking and Operation	$2550
Energy	3800
Maintenance	230
TOTAL	$6580

Total construction cost of the plant: $100,000

C-26.6 History and Operation

During the first four years of operation (starting in 1981) it was necessary to construct and modify the treatment unit in order to achieve optimization for the Cota area. Good removal of organic load in terms of BOD and suspended solids only began in 1985. An extensive shake down period was required.

REFERENCES FOR PART II

II-1. Innovative and Alternative Technology Projects, 1986 Progress Report, USEPA, Office of Municipal Pollution Control, Washington, D.C., September 1986.

II-2. Innovative and Alternative Technology Projects, 1987 Progress Report, USEPA, Office of Municipal Pollution Control, Washington, D.C., September 1987.

II-3. The Biological Aerated Filter: A Promising Biological Process, USEPA, Energy Technology Publication, July 1983.

II-4. Intrachannel Clarification: A Project Assessment, USEPA, Office of Municipal Pollution Control, Emerging Technology Publication, June 1983.

II-5. Intrachannel Classification: An Update, USEPA, Office of Municipal Pollution Control, September 1986.

II-6. Hydrograph Controlled Release Lagoons: A Promising Modification, USEPA, Office of Municipal Pollution Control, July 1984.

II-7. Hydrograph Controlled Release Lagoons: An Update for a Promising Technology, USEPA, Office of Municipal Pollution Control, May 1987.

II-8. Alternative Wastewater Collection Systems: Practical Approaches, Energy Technologies Series, USEPA, Office of Municipal Pollution Control, December 1983.

II-9. Water Reuse Via Dual Distribution Systems, USEPA, Office of Municipal Pollution Control, May 1985.

II-10. Vacuum Assisted Sludge Dewatering Beds: An Alternative Technology, USEPA, Office of Municipal Pollution Control, Emerging Technology Series, November 1984.

II-11. Vacuum Assisted Sludge Dewatering Beds (VASDB): An Update, USEPA, Office of Municipal Pollution Control, September 1986.

II-12. Counter-Current Aeration: A Promising Process Modification, USEPA, Office of Municipal Pollution Control, July 1983.

II-13. Composting: A Viable Method of Resource Recovery, USEPA, Office of Municipal Pollution Control, June 1984.

II-14. In-Vessel Composting: A Technology Assessment, USEPA, Office of Municipal Pollution Control, July 1987.

II-15. Methane Recovery: An Energy Resource, USEPA, Office of Municipal Pollution Control, September 1983.

II-16. Dual Digestion: Assessment of an Emerging Technology, USEPA, Office of Municipal Pollution Control, September 1986.

II-17. Personal Communication: from Julio Burbano, with input from German Rarvirez, Bogota, Columbia, April 1988.

INDEX